Fundamentals of Digital Communication Systems

Fundamentals of Digital Communication Systems

Prof. Dr. R.K. Singh

ARCLER
PRESS

www.arclerpress.com

Fundamentals of Digital Communication Systems

Prof. Dr. R.K. Singh

Arcler Press
4164 Lakeshore Road
Burlington, ON L7L 1A4
Canada
www.arclerpress.com
Email: orders@arclereducation.com

ISBN: 978-1-77956-979-0 (Hardcover)

ABOUT THE AUTHOR

Dr.R. K. Singh, born on 11th June, 1969, is a Professor and worked as Vice Chancellor, Patliputra University, Patna, Bihar, a state university of Bihar. Having throughout first division career, he obtained his Ph.D. from Allahabad University and M. E. from BITS, Pilani. He is recipient of numerous awards and honours. He is engaged in teaching, research, administration and consultancy for the last 28 years. Apart from this, he has worked as Director, B.T. Kumaon Institute of Technology, Dwarahat, Almora, Uttarakhand an autonomous Institution of Government of Uttarakhand and appointed as Director since 01/10/2015 at B.T. Kumaon Institute of Technology, Dwarahat, Almora,

Dr. Singh started his esteemed career as teaching assistant at Birla Institute of Technology and Science (BITS), Pilani (one and half years, Oct. 4, 1990 - July 1993) and as RA & Scientist(P) at CEERI, Pilani (more than one and half years, July 1993 – March 20, 1995), as an Assistant Professor March 21, 1995 – July 30, 1998) at BT Kumaon Institute of Technology (BTKIT), Dwarahat, formerly known as Kumaon Engineering College (KEC), Dwarahat, Almora (more than 3 years) and Associate Professor July 31, 1998 – July 30, 2006) at BT Kumaon Institute of Technology, Dwarahat (8 years). He became professor (July 31, 2006 – till date) at the age of 37 years, at BT Kumaon Institute of Technology, Dwarahat. He held the positions of HOD ECE, Principal, BT Kumaon Institute of Technology, Dwarahat, Officer on Special Duty, Registrar and Controller of Examination, at Uttarakhand Technical University, Dehradun.

Prof. Singh has received felicitations and awards by professional bodies such as Rastriya Samman Purskar- presented by Indian Society for Intellectual Development in presence of His Excellency, ShriSatya Narayan Reddy, Ex. Governor, U.P. and His Excellency, Shri Bhism Narayan Singh, Ex. Governor, Bihar, Jewel of India award- presented by International Institute of Education and Management in presence of His Excellency, Shri Bhism Narayan Singh, Ex. Governor, Bihar and Hon. Minister, Health, Govt. of Delhi, Shri Yoga Nand, Rastriya Ekta Award- presented by Global Economic Council, New Delhi, in presence of His Excellency, Shri Madan Lal Khurana, Governor, Rajasthan on 28th April, 2004 and Arch of Excellence Award-2004- presented in National Seminar-All India Achievers Conference in presence of His Excellency, Shri Satya Narayan

Reddy, Ex. Governor, and U.P. He is member of institution of Engineers (India). He has worked as Programme Officer of National Service Scheme at BTKIT, Dwarahat.

Prof. Singh has held positions of responsibility such as - Member, various committees of UTU University and Govt. of Uttarakhand, First Appellate Authority, BTKIT, Dwarahat (for Right 2 to Information Act, 2005); Dealing World Bank assisted Technical Education Quality Improvement Program for BTKIT; Expert Member, of various committees of several Institutions and organizations of national and international repute. Technical Expert for AICTE, NBA, and UGC of Government of India

Prof. Singh has guided 22 Ph.Ds. and about 15 M.Tech.thesis. He has participated in various National and International conferences, and published more than 120 research papers in national and international journals of repute. He has also authored eight books. Besides these, he has undertaken numerous consultancies for state government and various PSUs. He has provided technical assistance as member to groups of expert set up for various tasks.

Prof. Singh is member of the various prestigious bodies like Institution of Engineers (India), Kolkata, IEEE, etc. Prof. Singh has visited various International countries as key note speaker and session chair-man. Some of visited countries are Peris France, Hong Kong, Zurich, Switzerland, Kuala lumpur, Malaysia and Bangkok, Thailand where he Delivered Key note speech and chaired the session of International Conference.

TABLE OF CONTENTS

LIST OF FIGURES

PREFACE

Communication is at the heart of human advancement. All forms of interaction, from voice conversation to text messaging, video calls, and streams of data between machines, rely on effective systems of communication. As technology has advanced, communication has transitioned from older analog systems to current digital systems capable of carrying vast amounts of information rapidly, reliably, and safely. Digital communication is the basis of the internet, mobile networks, satellite systems, and multimedia services and is thus one of the most powerful technologies in shaping contemporary life.

The shift to digital systems was spurred by demands for accuracy, toughness, and flexibility. Digital communication makes it possible to send information with fewer errors, compress it for efficiency, and encrypt it for security. It also allows for smooth convergence of voice, video, and data in ways previously impossible with analog systems. Global positioning systems and cloud computing, smart cities and 5G wireless networks, among many others, today depend on digital communication concepts. Familiarity with the basics is critical for engineers, researchers, and students who seek to contribute to the development of future technology.

Fundamentals of Digital Communication Systems has been authored with the aim of delivering a comprehensive and systematic grasp of the topic. It starts with the conceptual underpinnings and subsequently moves on to technical details, such as coding, modulation, transmission, and information theory. Every chapter logically develops from what was discussed in the earlier chapter, providing a learning path from fundamentals to sophisticated ideas. The book is not merely designed to explain theoretical principles but also to demonstrate their practical relevance through applications and contemporary trends. The organized sequence of steps guarantees that students can reinforce their conceptual understanding in conjunction with gaining practical knowledge in contemporary communication systems.

Chapter 1 deals with the Introduction to Digital Communication. It discusses why digital systems are preferred compared to analog systems, highlighting advantages such as immunity to noise, data compression, and integration with computing technologies. It also presents the fundamental elements of a communication system—transmitter, channel, and receiver—and how information is translated into digital form.

Chapter 2 is on Signals and Spectra. It introduces readers to signal representation, time-domain and frequency-domain analysis, and the significance of spectral properties in communication systems. Fourier analysis, bandwidth, and the use of spectral efficiency in developing dependable systems are discussed in the chapter.

Chapter 3 deals with Source Coding and Data Compression. The chapter emphasizes methods of minimizing redundancy within transmitted data. The chapter discusses coding schemes like Huffman coding, run-length encoding, and contemporary multimedia compression algorithms. Through minimizing excess data, these methods maximize storage and transmission efficiency.

Chapter 4 discusses Pulse Modulation Techniques. The topics covered are pulse amplitude modulation, pulse code modulation, and delta modulation. These techniques are the basis for digital representation of analog signals and are also major building blocks for contemporary telecommunication systems.

Chapter 5 discusses Digital Modulation Techniques. ASK, FSK, PSK, and QAM modulation techniques are discussed in detail. The chapter compares their performance with respect to bandwidth efficiency, noise immunity, and complexity of implementation and gives an insight into how these are used in wireless, optical, and satellite communication.

Chapter 6 discusses Transmission Through Communication Channels. It explains how signals travel through channels and the issues created by attenuation, distortion, and noise. Chapter 6 also discusses channel models and introduces techniques to overcome channel impairments like equalization and diversity techniques.

Chapter 7 deals with Error Control Coding. Accuracy in digital transmission is very important, and the chapter discusses coding schemes such as block codes, convolutional codes, and cyclic redundancy checks. It also discusses contemporary techniques such as turbo codes and LDPC codes, which find extensive use in modern communication systems.

Chapter 8 covers Information Theory and Channel Capacity. The chapter discusses the intrinsic limits of communication, defined by Shannon's theory. Theoretical concepts such as entropy, mutual information, and channel capacity are defined, giving a theoretical basis on which to measure system performance.

Chapter 9 discusses Detection and Estimation Theory of Digital Communication. In this chapter, students become aware of optimum detection methods, decision mechanisms, and techniques for estimation. Maximum likelihood detection, Bayesian estimation, and matched filtering are the highlights of the chapter that form the basis of credible signal recovery in noise.

Chapter 10 deals with Applications and Future Trends in Digital Communication. This final chapter relates theory to practice by analyzing the role of digital communication in areas like mobile networks, satellite systems, data centers, and the Internet of Things. It also emphasizes future directions such as quantum communication, 6G systems, and communication networks based on artificial intelligence.

This book has been designed for undergraduate and graduate-level students of electrical engineering, electronics, computer science, and allied fields. It will also prove beneficial to researchers, industry experts, and tech enthusiasts who want to reinforce their knowledge of digital communication systems. For students, it provides systematic advancement from basic principles to advanced subjects. For professionals, it is a guide to actual applications in wireless networks, data transmission, and new technologies. The book has been presented in simple and clear language to make it accessible while offering depth for those looking for a rigorous academic grounding.

CHAPTER 1

INTRODUCTION TO DIGITAL COMMUNICATION

Contents

INTRODUCTION

In defining digital communication, it should be understood as a multifaceted phenomenon that is comprised of various means of interacting via digital platforms. Essentially, digital communication can be defined as sending information electronically via technologies such as the internet, smartphones, and social media. This dynamic communication allows for real-time exchanges, which in themselves transform the traditional paradigms of communication based predominantly on print and co-presence. As society becomes increasingly engulfed in the digital world, the extent of the changes transcends convenience; it also brings about new ways of thinking and learning. This transformation requires a mature understanding of the socio-economic shift resulting from the onset of digital communication. A single, universal definition is elusive, as the applications and functions of digital communication are diverse in cultural and geographical contexts (Sasvári et al., 2012). Also, the shift from print to electronic media is a watershed in the way information is presented, consumed, and interacted with.

Figure 1.1. *Digital communication.*

Source: https://external-content.duckduckgo.com/iu/?u=https%3A%2F%2Fwww.labcompr.com%2Fwp-content%2Fuploads%2F2021%2F01%2Fdigital-communication-1.png&f=1&nofb=1&ipt=d24faaec66af1266639be0dda9655bc79dcb52ff1f355cb889f129cc2c252b15.

The infrastructure of digital communication combines elements borrowed from outside its boundaries, notably graphical user interfaces and type. These factors have evolved significantly in screen-based communication, which introduces complexities and dynamics not encountered in traditional media. With digital typographic practices on the ascendancy, it is most important to delineate frameworks that effectively capture the inherent differences of print and screen communication. This intricate interplay of typography and digital media underscores the importance of interdisciplinarity to fully take into consideration the evolving landscape of communication since extant theories are generally short of capturing the distinctive characteristics of screen media (Yee et al., 2004). The implications of thinking about digital communication extend far beyond the technological; they involve cultural and social dimensions as well. The emergence of new communication technologies has transformed social interaction, giving rise to new culture and norms defining contemporary society. This is expressed in the way individuals negotiate their online identity and engage with their diverse publics. Secondly, interactivity as a component of communication, where the users are involved in the process actively, is a departure from passive reception typical of traditional media. Such interactivity is central to digital communication, as it facilitates collaboration and gives people a sense of community. As digital communication landscapes continue to evolve, there is a need to critically reflect on its definitions and meanings so that they reflect the richness and complexity of contemporary communication conduct in our fast-paced interconnected world (Sasvári et al., 2012) (Figure 1.1).

Explanation of Digital Communication

The evolution of electronic communication has revolutionized the way individuals and communities interact in modern times. Unlike traditional forms of communication, which were usually limited to a specific geographic location and time-sensitive, electronic communication eradicates these obstacles, allowing for instant intercommunications over vast distances. The emergence of the internet has facilitated a plethora of platforms, such as social networking sites, email, and instant messaging, which not only unite individuals but also redefine cultural interaction norms. This shift represents a paradigm shift in human connection and depicts the manner in which digital technologies have not just maximized communication speed but also facilitated new forms of relationships and cultural expression and facilitated the creation of a globally connected world. In digital communication research, one cannot help but consider the impact of digital communication on community building and interpersonal relationships. According to research, digital platforms offer new modes of social interaction that are likely to lead to the build-up of communities centered on interest rather than geographical location. This is a follow-up to Walter J. Ong's treatise on communication as a theme in Western society, where the intersection of people and technology leads to dense weavings of interconnectivity that redefine community formation and sustenance (Soukup et al., 2004). Furthermore, as human beings communicate asynchronously through various media, the nature of communication shifts to become more inclusive so that several voices may form part of debates otherwise relegated to solitude. Inclusivity has the potential to strengthen connections within communities so that more stable networks of cooperation and mutual support emerge. Perhaps most significant in digital communication is also how it contributes to the shaping of creativity and cultural production. The use of computational modeling has been recognized as a way to examine the dynamics of creativity in virtual communities, and that is through the field of art. By simulating people's behavior towards seeking novelty, scientists have begun to learn how these interactions can potentially spur movements and styles of art (Bown et al., 2006). This kind of research points out the group character of digital communication, where the pursuit of novelty is not localized in the individual artist but rather facilitated by the clustering of similar artists. The shared pursuit of novelty forges communities of connection, which only serve to confirm that digital communication not only facilitates discourse but also plays a crucial role in the creation of cultural forms in contemporary society.

Comparison with Traditional Communication

The evolution of internet communication has significantly altered the picture of how the general public receives news and information from traditional forms of communication. Traditional media, such as printed newspapers and broadcast television, were previously under a linear model where information traveled in the direction of the producer to the consumer. On the other hand, digital communications have introduced a dynamic, interactive platform to the equation, allowing users to not only consume but also create and distribute content. This has been dramatized with the development of mobile news consumption, which has merged so seamlessly into the everyday user experience. The integration of mobile news into broader news media repertoires is indicative of a complex relationship in which the public will remain reliant on traditional sources but supplement their diet of information with digital sources, expressing a twin dependency on both media (Courtois et al., 2015). Also, the interactivity of online communication gives rise to new opportunities for user engagement that do not necessarily follow from traditional media forms. With the advent of platforms such as interactive television (ITV), consumers are now permitted to participate in creating their own media experience. The distinction between low and high interactivity shown in ITV consumption characterizes how traditional television maintains its linear form yet accommodates for a minimum level of user input. Advanced television applications, which extend existing programs through choices and tools for customization, are a relatively passive experience. On the other hand, more interactive and immersive experiences depict how user input may frame programming content, with more active viewership. The interactive elements invite users into novel and engaging ways that challenge the consumer-producer paradigm in traditional media (Dijk et al., 2003). The juxtaposition of old and digital communication therefore indicates a paradigm shift that encompasses not technology alone, but also user attitudes and expectations. Online sites allow participants to be empowered through tools for engagement, as well as through tools for customization of media experience. This content democratization is a stark contrast to the gatekeeping of conventional media, in which editorial prerogative dictated much of what would be heard and seen by audiences. As audiences move toward digital formats, reliance on brand content that has been trusted by their audiences persists—beyond spotlights the paradox of familiarity in a fast-evolving digital world. Consequently, understanding such trends is imperative because communicators will position themselves accordingly, knowing about the double effect that traditional and digital media exercise on contemporary patterns of communication (Courtois et al., 2015; Dijk et al., 2003) (Figure 1.2).

Figure 1.2. *Digital communication comparison with traditional communication.*

Source: https://www.melinakmiller.com/wp-content/uploads/2020/10/Traditional-Vs-Digital-2.jpg.

Importance in Modern Society

In contemporary times, electronic communication is a decisive factor in societal relations and information sharing. The pervasiveness of technology has enabled the accessibility of enormous amounts of information to be a common phenomenon. Individuals today navigate a complicated digital world, where technology's commitment plays a vital role in structuring communication processes. Digital competence is required, as it allows society to communicate with knowledge in quick, easily available forms. Further, the development of online resources such as virtual libraries and databases illustrates how institutions respond to the dynamic needs of users in this rapidly evolving info space. Through the development of such environments, society becomes capable of expanding its collective knowledge base, thus highlighting the importance of digital communication as a pillar of modern life (Kulczycka et al., 2018). In addition, the evolution of media and communication channels indicates the extent to which digital technology has become entrenched in the consumption of information. Individuals now receive an overwhelming amount of messages through numerous channels, a process which has changed individuals' consumption and enjoyment of information. With the maturity of the Internet age, multidimensional and redundant information flows freely, and individuals tend to privilege the most salient messages. Social media platforms have evened the ground in information distribution, positioning personal sites alongside mainstream media sources in terms of audience reach. An average Facebook user, for instance, can distribute material to thousands, a near equivalent of circulation figures in mainstream newspapers. This innovation has not only revolutionized communication dynamics within society but has also put to question the quality and moral worth of information distributed (Rexhepi et al., 2014). The consequences of digital communication go beyond simple information sharing; they redefine the role of media in society. As traditional medias authority wanes, there is an imperative to reflect on who is informing the public and what is relevant information in this digital age. The culture of user-generated content threatens entrenched norms of truth and ethics in media. As individuals continue to rely more on new media sources for their information, the likelihood of receiving false information rises, which requires consumers to exercise critical discrimination. Appreciation of the dynamics within this new media environment is thus crucial if problems associated with information overload are to be addressed and credibility of communications processes maintained within modern society. Lastly, this is an indication of how important it is to create media literacy so that individuals can make informed decisions about what information they absorb and spread (Rexhepi et al., 2014).

1.1. HISTORICAL DEVELOPMENT

The development of digital communication over time has paralleled significant technological progress as well as other changes in society. From its inception, digital communication has relied on an advanced interplay of networking technologies and protocols decades in the making. The transition from analog to digital means was a revolutionary period, and it led to the creation of the internet during the latter half of the 20th century. This transforming shift permitted unprecedented interconnection and sharing of information globally. With the introduction of various means of communication, the implication on personal interactions and business communications was realized, setting cultural practices and norms. This evolution marked the services sector beginning to recognize digital platforms as main facilitators of communications, propelling innovation and economic growth, setting the stage ultimately for contemporary forms of digital interaction. The role of electronic communication among diaspora populations, and particularly that of newer arrivals, is characteristic of the richness and versatility of the technology. For example, research on the Karen refugee population in Britain illustrates how individuals engage in various internet-based mechanisms to remain connected with their homeland as well as become engaged in political and social activism. While coping with their displacement, the ability to engage in online communication allows them to maintain their cultural identity and generate solidarity among dispersed individuals. In addition, the use of traditional and hybrid modes of communication betrays an eclectical approach to maintaining ties to their global diaspora, where digital technology is not only a facilitator but also an element of cultural resilience in spite of barriers such as cost, education, and age-dependent access challenges (Castells et al., 2012). Besides, as we look at the broader effects of digital communication on social institutions, we can observe that the services sector has been revolutionized entirely with the use of digital technology. The shift towards a postindustrial economy reflects the significance of the services sector as a primary driver of economic activity. This transformation necessitates an exploration of how practices in strategic management are changing towards a more digital landscape. Although many conventional strategy techniques are dated, the emergence of digital platforms requires updated frameworks that consider the hyper-sector's pervasive coverage of all human activities. As scholars embark to explore this evolving field, they will be required to create specific goals for developing the sectors to ensure strategy is aligned with contemporary requirements and aid in promoting economic development (Britchenko et al., 2018) (Figure 1.3).

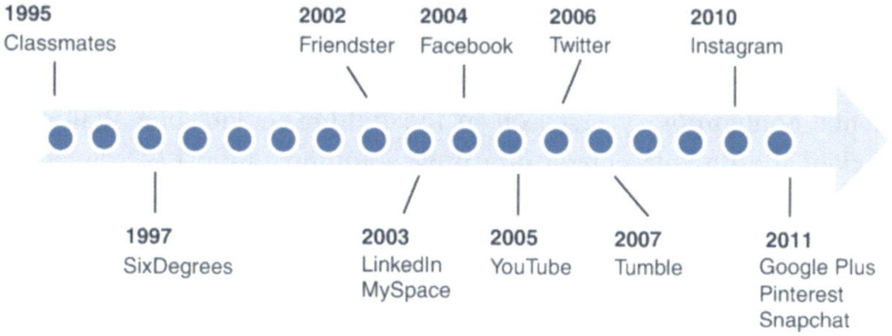

Figure 1.3. *A brief history of digital communications.*

Source: https://media.springernature.com/lw685/springer-static/image/chp%3A10.1007%2F978-3-030-14439-5_2/MediaObjects/468655_1_En_2_Fig3_HTML.png.

1.1.1. Early Forms of Digital Communication

The development of digital communication can be traced back to its inception, which predetermined the pace of the contemporary patterns of information exchange. The digital communication started as text-based, with electronic mail, invented in the early 1970s, revolutionizing the means of messages being communicated across distances. The users were able to send written messages via a computer network, which was a deviation from the traditional letter writing. Furthermore, the development of bulletin board systems (BBS) during the late 1970s and early 1980s provided platforms for users to share information, communicate, and establish communities through simple digital interfaces. Not only did these platforms provide new ways of interacting, but they also brought forth the growing need for digital literacy as society increasingly relied upon electronic means of communication. With the progressing technology, the initial digital communications began to involve multimedia elements, enhancing user experience and interaction.

The introduction of audio and video streaming in the early 1990s was a revolutionary advancement that introduced even more enriched content and new opportunities for communication. Instant messaging services such as AOL Instant Messenger and ICQ also became popular in the late 1990s, offering real-time interaction emulating face-to-face live conversations. These transformations were pivotal since they began blurring public and private communication spaces. Consequently, platforms began becoming hybrid spaces where socialization, information sharing, and entertainment coexisted. Such transformations not only diversified forms of digital communication but also emphasized studies on how such new forms affect social action and digital culture. The initial digital communication practices observed continue to influence contemporary research in the field. Scholars like Walter J. Ong have tracked the threads concerning how communication has shaped human experience throughout history, stating that understanding such threads is necessary for grasping interactions in contemporary times. As described in various studies, transition from text to multimedia communication is a part of a greater cultural shift towards interactive experiences. Contemporary Cultural Heritage programs utilize Next Digital Technologies to enable linking between digital content and physical objects, demonstrating how digital communication complexities transcend mere information exchange. Thus, early digital communication media not only gave rise to foundational technologies but also laid the ground for the immersive storytelling experiences we now see at institutions such as museums, aimed at maximizing cultural accessibility and engagement in new forms (Falco et al., 2017).

1.1.2. Evolution of Technology and Its Impact

The development in technology has greatly revolutionized the communication world to bring forth new paradigms of interaction across various fields. Technological advances in communication have primarily been brought about through the development of the internet and cellular phones, permitting immediate exchange of information, and facilitating world connectivity. This transformation can be best represented by the move from traditional methods of communication, including postal mail and landline telephones, to modern instruments such as instant messaging and social media. By this, individuals and companies are not only able to disseminate information more efficiently but can also cooperate beyond physical locations. The influence on interpersonal relationships, business culture, and even educational arenas is significant, with individuals adapting to these newer forms of interaction, in essence remapping the dynamics of communication in the era of technology. Within education, the use of technology in the classrooms has necessitated a re-examination of teaching methods, essentially altering the role of instructors and students. The development of electronic resources such as learning sites and virtual classrooms indicates that technology makes it more convenient and flexible for students to take advantage of learning opportunities, allowing them to complete materials at their own pace. The change is not without danger, however, since faculty members must typically learn new skill sets in order to function within them. The change in working practices within higher and further education calls for understanding how technology can impact staff relationships and roles, as discovered by research from the Work-with-IT project at the University of Strathclyde (Comrie et al., 2010). This new learning context emphasizes the dynamic relationship between technological progression and the need for responsive action within universities. Technological innovation has also been the central driver of the expansion of the services sector, highlighting it as an economic change agent. With economies transitioning toward a postindustrial stage, the services sector has come to dominate more and more the creation of gross domestic product (GDP) and overall economic activity. The presence of communication technology has enabled service businesses to operate efficiently and reach a broader geographic scope, which is crucial from the viewpoint of competitiveness in the fast-expanding internationalizing market. However, there remains a strong need for integrative strategic management models that can capture the complexities and interdependencies inherent in the services sector. Attention has been attracted by researchers to understand this sector not only as a cluster of firms but as a hyper-sector penetrating all spheres of economic activity, which requires greater study into the manner in which it must be developed strategically (Britchenko et al., 2018). The evolving landscape needs greater understanding into how technology continues to make the economy evolve (Figure 1.4).

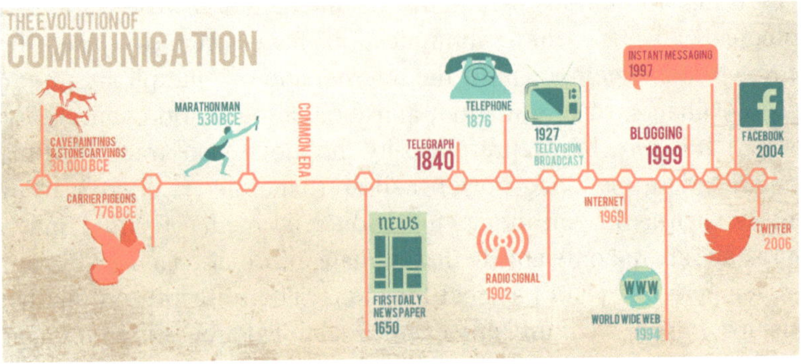

Figure 1.4. *Evolution of technology and its impact in digital communication.*

Source: https://sci10sectionm.wordpress.com/wp-content/uploads/2013/12/time-line.jpg.

1.1.3. Key Milestones in Digital Communication History

The history of digital communication is marked by a series of milestones that have fundamentally transformed the way human beings interact with technology and each other. The expansion of the Internet during the latter part of the 20th century was the most significant breakthrough. The Internet was originally developed for the military. It gradually evolved into an international network of millions of devices that facilitated the free exchange of information. This connectivity paved the way for the expansion of email, which emerged as one of the first major applications of the Internet, offering instantaneous communication that was a radical departure from the traditional postal services. These developments have not only augmented personal communication but have also revolutionized business practices, altering the face of global commerce and interaction. Another landmark development was the emergence of social media websites in the early part of the 21st century. These sites transformed online communication so that users could share content, engage in discussions, and form web communities. Sites like Facebook, Twitter, and YouTube democratized content creation and dissemination, allowing individuals to be more than consumers but active participants in generating public discourse. This transformation had a powerful impact on the dynamics of society so that information could go viral, influencing public opinion and mobilizing social movements globally. Thus, it becomes imperative to realize these transformations to comprehend the current scenario of digital communication and the social implications that are the result of greater connectivity. Additionally, mobile communication technology represents a turning point in the trajectory of digital communication. With the popularity of smartphones, it was very easy to browse and remain connected, further reinforcing the concept of a global village. The capacity of mobile apps and messaging platforms to enable real-time communication has transformed human-to-human relationships and provided new avenues for business and marketing engagement. The process of technological change in communication can be framed within the broad process of Information and Communication Technologies (ICTs) development, in which the process of technological innovation is viewed as transforming human relationships (Canela et al., 2017). In addition, the development of geospatial information infrastructures enhanced sharing and utilization of spatial information, which is at the heart of the Digital Earth vision and reflects the advanced nature of today's digital systems of communication (Aksulu et al., 2019).

1.2. TYPES OF DIGITAL COMMUNICATION

Digital communication has numerous modalities, each being created to facilitate distinct interpersonal interactions and the sharing of information on different platforms. Most notably, the prominent modes of digital communication include text messaging, email, social media, and video conferencing. These two provide asynchronous communication environments where individuals can forward messages that could be read and responded to at leisure. On the other hand, social media sites have revolutionized the world of communication by enabling individuals to publish content, opinions, and experiences in real time, thereby building community and interaction. These various types are greatly dependent on the infrastructure and support of the internet and illustrate how digital communication has evolved to meet modern connectivity demands, as also evident in the extensive overview of user studies in digital library development (Dobreva et al., 2012). In addition to traditional text-based communication, there have been multimedia approaches that have proliferated, particularly through the integration of video and audio elements. Video conferencing software, for example, Zoom and Microsoft Teams, has become increasingly vital in professional and personal life, particularly with remote working and cross-border collaborations. Such platforms enable individuals to have face-to-face meetings remotely, adding more depth to the communication through non-verbal cues and networked talks. These modalities also foster not only organizational communication but social gatherings because individuals can interface regardless of their geographical positions. The evolution of these platforms is a sign of heightened needs for multi-faceted networking tools that provide secure, flexible, and efficient communication solutions for different collaborative environments to provide an easy user experience (Barolli et al., 2011). Moreover, the dissemination of mobile communication has irreversibly shifted the way individuals interface with one another. Smartphones and tablets have rendered traditional forms of communication obsolete, providing users with the authority to access various media in an instant. Tools such as WhatsApp, Snapchat, and Instagram have undergone a shift in preference from text-based to visual, offering new ways for users to communicate through images, videos, and live streams. This shift reflects a larger trend by which digital communication moves beyond the exchange of information, becoming a site of collective creativity and individual expression. With continued diversification of communication methods, recognizing the specificities of each variety is critical to effective participation in a progressively digital era, setting the stage for future development in communication methods (Figure 1.5).

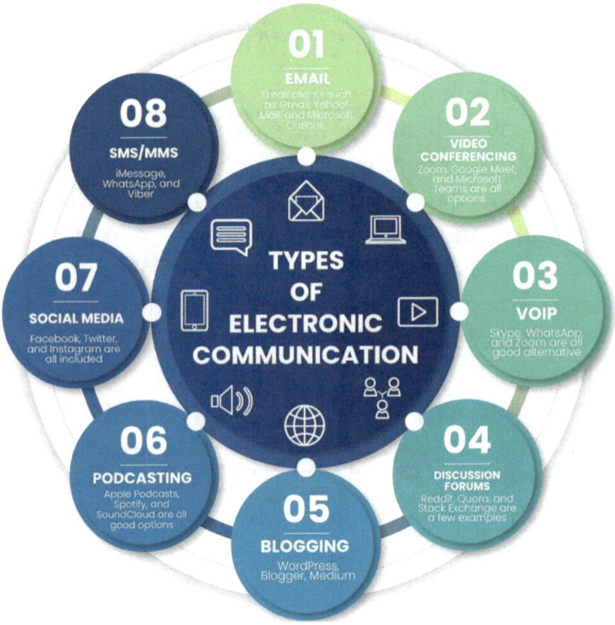

Figure 1.5. *Types of digital communication.*

Source: https://external-content.duckduckgo.com/iu/?u=https%3A%2F%2Fi.pinimg.com%2Foriginals%2Fb0%2F43%2Fc1%2Fb043c1637679facca94c63f85f25b7bb.jpg&f=1&nofb=1&ipt=8d8c7adc7e820f13b00d08fd6e44bdf008c1e57ae96c2bfeb3687171c042947c.

1.2.1. Text-Based Communication

Text-based communication has remapped the landscape of interperson communication in the age of electronics. Written communication was, until now, limited to letters and physical media, but electronic technologies have enabled immediate exchanges over distant channels. Social media platforms, emails, and messaging apps not only enable the rapid diffusion of ideas, but also the generation of unique language styles appropriate for electronic media. These advancements have introduced new norms of communication, such as brevity and the use of emojis, to prominence. These new norms are applied to convey tone and feeling when voice cues are missing. It is thus critical to learn these new means of text-based communication to negotiate communication environments in contemporary times. Secondly, the open access character of text-based communication facilitates information dissemination and learning democratization. Online platforms enable people from diverse backgrounds to share their ideas and thoughts, usually bridging geographical and socioeconomic divides. For example, digital learning tools such as talking books have been created to serve diverse learning requirements. These learning aids utilize text-based communication to disseminate information in an engaging platform, with synthesized speech for visually impaired students and overall improvement of the learning experience. This integration points to the inherent virtues of text-based communication for fostering universal learning cultures and information democratization, as pointed out in applications based on the OU OpenLearn schema (Brasher et al., 2009). Also evident is the evolution of digital libraries and repositories that is reflecting increasing pertinence in structured text-based communication for information dissemination. As the report section of the Discovery Workpackage has observed, digital libraries' digitization landscape is evolving with enhanced publications and interoperability standards. Such developments highlight the significance of text-based communication to maintain the integrity and accessibility of information resources and enable smoother exchanges between academic and public spheres. By connecting isolated communities and technologies, such as GRID, CRIS, and LTP, these initiatives highlight the necessity of the disciplined text communication of the Information Age. Lastly, the innovations within the field not only optimize access to information but also guarantee that valuable knowledge is still communicated effectively and inclusively (Hochstenbach et al., 2008).

1.2.2. Audio and Video Communication

The evolution of audio and video communication has greatly changed the face of digital communication. With technological advancements, audio communication has evolved from traditional telephone systems to sophisticated

VoIP services that are capable of carrying high-definition voice calls over the internet. This has been revolutionary in personal and business communication as it introduces greater flexibility and availability. To this effect, technologies such as Skype, Zoom, and Microsoft Teams bear witness to how communication through voice has integrated into our daily lives, allowing real-time speech irrespective of distances.

Figure 1.6. *Audio and video communication.*

Source: https://external-content.duckduckgo.com/iu/?u=https%3A%2F%2Fproavdc.com%2Fwp-content%2Fuploads%2F2023%2F12%2FAudiovisual-Communication.jpg&f=1&nofb=1&ipt=d3281a761e7bb8b05a526e72918731cfe3b4bb4af995b2eef3bb7c1b1d0af91a.

The transition has not been seamless, however, since bandwidth, latency, and cleanliness problems still persist; nonetheless, ongoing innovations continue to top them up to enhance the experience to be faster and easier. Parallel to this, video communication has emerged as a groundbreaking phenomenon in the digital era of interaction. Being able to see a conversation partner in real-time lends a vital sense of personal touch that is not possible to attain through audio. Video conferencing technologies have proliferated exponentially in terms of popularity, particularly documented during global events such as the COVID-19 pandemic when working remotely and virtual meetings became the new normal. Its use extends to social interactions, classrooms, telemedicine, and workplaces. Furthermore, the use of gamified aspects in video communication has facilitated greater interaction and engagement. According to the argument of recent reviews, sports media infotainment strategy has employed such aspects to attract and sustain audience attention, uncovering the potential of video communication beyond interaction (Rojas-Torrijos et al., 2020). Besides, the integration of video and audio technologies has led to the creation of a multidimensional communication space, whereby multimedia content dissemination has become crucial. Convergence enables smooth integration of all forms of media, hence yielding greater interactive richness on different terminal devices. Service providers are increasingly motivated towards developing platforms capable of delivering interactive multimedia services for various user requirements and devices. The technologies that drive this convergence are all about optimizing media and metadata flows, in addition to offering platform compatibility (Allamandri et al., 2007). As organizations continue to expand the limits of such interactive services, the audio and video communication future holds answers that will further erode the lines between traditional and digital communication, basically reconfiguring the manner in which we interact with one another (Figure 1.6).

1.2.3. Social Media Platforms

The role of social networking websites in revolutionizing communication dynamics cannot be overemphasized. Social networking websites have evolved from simple networking sites to powerful tools that redefine interpersonal communication and information sharing. Users, through their power to create and upload content, comment, and establish communities, are indeed revolutionizing the face of communication. This evolution has been accompanied by a culture of spontaneity and immediacy in how individuals communicate with each other, leaving far behind the one-to-many approach of traditional medias. Social medias democratization of information flow brings multiple voices into public discourse, though it causes problems about the credibility and reliability of the content shared. As such, it is essential to possess a clear picture of the social media implications in digital communication, where there is a critical examination of both its potential and issues in defining social interactions. Social media sites have, for example, emerged as significant strategic weapons in the business environment that enable organizations to effectively address various stakeholders. Organizations globally leverage these sites for various activities, including marketing, public relations, and customer care programs. The capability to engage in real time enables organizations to build meaningful relationships with their audience, inducing engagement and cooperation. As quoted, while most businesses assume social media to be an active platform for stakeholder engagement, the medium's actual usefulness is not always realized by them. Specifically, it has been observed that some organizations fail to leverage the entire potential of social media to build communication, and they fail to explore the potential of stakeholder engagement that can enhance their corporate reputation and customer relationship (Gómez et al., 2011). Therefore, an important scrutiny of social medias strategic role in corporate communication is required in understanding its performance. Along with this, the effect of social media use extends beyond mere engagement numbers to seep into cultural and societal trends overall. The widespread use of these platforms has caused consumers to evolve away from their role as passive receptacles for information to that of active participants in the digital conversation. This trend heralds a fundamental change in societal behavior as individuals demand immediate feedback and interaction from brands and peers alike. The environment that ensues makes organizations adapt and react to consumers' needs, employing social media for both marketing and as a means of genuine conversation. As outlined by Iosifidis, the outcome of adopting new communication technologies, like social media, is mostly influenced by how they are employed by their users. Hence, the ability of social media sites for building genuine relationships and political diversity relies on the active participation and responsible usage by individuals and groups within the digital communication landscape (Iosifidis et al., 2012).

1.3. TECHNOLOGIES ENABLING DIGITAL COMMUNICATION

The evolution of communication digitally has been made possible by a number of technologies, which have revolutionized the manner in which people communicate in their personal, working, and learning environments. One of the most important developments is the emergence of the Internet, which is the basis of communication digitally through the capability to link instantly globally. The global penetration of mobile phones further reinforces this possibility, allowing one to communicate anywhere and thus establishing a culture of availability and immediacy. Thus, the common barriers to communication are eliminated, and the user is made to cooperate and share information easily. For businesses, this technology context enhances working relationships since effective communication environments allow coordination among members, leading to productivity and ultimately quality results. This shows the increased recognition of teams as the organizational building block of modern businesses, emphasizing the importance of integrated communication technology within the workplace (Li et al., 2020). In the educational setting, technologies based on Web 2.0 tools have been particularly groundbreaking, changing the manner in which one learns and teaches. These tools provide collaborative learning environments in which students and instructors can engage in collective knowledge creation and sharing. Discussion forums, accumulation of resources, and peer review illustrate the numerous benefits of such tools, which not only enhance user interaction but also promote critical thinking and creativity among students. The review of such tools reveals some drawbacks as well; teachers may find it difficult to incorporate technology in an effective manner into conventional teaching techniques. Yet, as with initiatives like 'Peals in the Cloud,' there is significant potential for Web 2.0 technologies to support evidence-based teaching practice in helping stimulate scholarly thought and debate regarding teaching approaches (Alevizou et al., 2010). Increased reliance on communication technologies means that, in addition to learning what they can do, there is also a need to be aware of the constraints which they introduce. Although such technologies provide broad opportunities for connectivity and co-operation, they also have potential to increase problems of information overload and digital divide. The users may be overwhelmed by the never-ending flow of information across platforms, and hence would find it hard to concentrate and be productive. Also, variations in access to digital technologies can increase differences, particularly within schools where not all would have means of internet or devices. Thus, with the embracement of the innovations of digital communication technologies, stakeholders must also face these key challenges to ensure that the dividends of digital communication are equitably distributed and effectively used for personal and collective advancement (Figure 1.7).

Figure 1.7. *Technologies enabling digital communication.*

Source: https://cdn.airalo.com/images/60b34e87-90ae-4b0f-af2b-030408d009bc.jpg.

1.3.1. Internet and Its Infrastructure

The Internet is a giant communications network that connects millions of devices across the globe and enables the rapid sharing of information and interaction between users. It consists of many components that form its infrastructure, including fiber-optic cables, data centers, routers, servers, and others that all contribute to transferring data in a seamless manner. The evolution of telecommunications over time has revolutionized the structure of the Internet and accelerated the pace and connectivity to higher levels. As digital communication becomes more integral to day-to-day life and business, there is a need to understand the support infrastructure. For instance, the growing use of mobiles and apps helps to highlight the importance of robust Internet infrastructure in providing diverse services and affecting not just individual consumers but businesses and organizations seeking digital transformation as well. The interplay of technology advancement and end user remains key to defining the future of digital communication. The challenges faced in the building out of Internet infrastructure, particularly in regions like Africa, highlight the inequalities of access and the need for targeted interventions. Global technical system marginality, coupled with a history of substandard infrastructure, has long limited the ability to carry on productive digital communication in such places.

But new advances in information and communication technologies provide a unique opportunity to correct these shortcomings. Strategies to leverage accessible local resources in developing such necessary skills underscore the requirement of localized solutions that are suitable for the needs of the regions. As Little et al. (2006) has underscored, such technologies have the capacity to enhance new capacities that empower citizens and, in the process, generate inclusive digital access that can transform social dynamics. Sustained viable Internet infrastructure, therefore, requires a synergy between external partnerships and local engagement to ensure success and active participation.

As urban centers become more welcoming of clever technologies, the infrastructure of the Internet's becomes a fulcrum in Ubiquitous Eco City management. The concept promotes the integration of economic, socio-cultural, and urban growth with innovative communication technologies. Integration is important to generating environments that not only add to the well-being of users but also enhance urban management. Telecommunications technologies enable real-time data collection and monitoring, which can enable responsive city management. Moreover, the convergence of information and communication technologies usher in a new era with access to a vast array of services offered conveniently using multipurpose devices, as noted in Yigitcanlar et al. (2009). Such technological shift significantly alters urban residents' experiences with their surroundings, thus enabling smarter, connected urban spaces. Lastly, the ability of Internet infrastructure to facilitate these schemes will be a determining factor in shaping the urban destiny.

1.3.2. Mobile Communication Devices

The increase in mobile communication gadgets has revolutionized the digital world of communication as irreplaceable tools that open up doors to instant access despite distances. The gadgets, primarily smartphones, have converged a series of functions so that one can obtain information, communicate with one another, and make transactions at one's own will. The convergence of the technologies has been central to this so that cellular gadgets have been able to enable applications to encompass both telecommunication as well as internet functions. As highlighted in the book on reviewing urban management systems, the speed-up in the evolution of information and telecommunication technologies has significantly enhanced urban life by increasing connectivity and access to essential services. These advances underscore the positioning of mobile phones not only as communication tools but also as gateways to other online services that enhance the wellbeing of the consumers, thus their pivotal role in the current era of digital communication (Yigitcanlar et al., 2009). Moreover, mobile communication devices usage worldwide has achieved unparalleled connection, revolutionizing social, political, and economic interactions. With more than two billion wireless subscribers worldwide, mobile technology has become an essential part of life and no longer a luxury, bridging communication gaps which previously used to occur due to geographical or socio-economic grounds. User behavior has seen a massive shift; users prefer to utilize mobile devices not just for personal communication but even for conducting business and accessing learning material. As detailed in Digital Revolution discourses, mobile communication has democratized and improved the efficiency of information access. Such reforms empower the people and marginalized groups because mobile phones provide access to critical information about markets, opportunities, and resources, hence improving economic inclusiveness and engagement (Firestone et al., 2007). Also, the impact of cellular phones extends beyond a mere access to information; they significantly influence personal identity and social structure. As the world keeps embracing this Mobile Generation, various stakeholders in society including businesses, governments, and social organizations adjust their strategies to accommodate this move towards mobility. The interactivity and access of mobile communications enable new solutions, such as SMS-based public health services and e-commerce systems tailored to rural farmers. These serve as illustrations of the revolutionary potential of mobile communication devices in their role as support for economic development and social welfare. Their ongoing evolution is not just changing the communication practices but also profoundly changing how communities and individuals organize, collaborate, and thrive within an increasingly digital environment (Yigitcanlar et al., 2009; Firestone et al., 2007) (Figure 1.8).

Figure 1.8. *Mobile communication devices.*

Source: https://www.ipinternational.net/wp-content/uploads/2021/11/iStock-1274394138-1.jpg.

1.3.3. Software Applications and Tools

In the digital communications context, software applications and tools are fundamental equipment that is utilized to enhance communications and productivity. Communication applications such as email clients and instant messaging applications enable real-time communications that span geographical distances. The tools are able to support various formats, including text, voice, and video, and offer a variety of communication modes. As more users and organizations rely on them, they have evolved to include features such as shared workspaces, file sharing, and task management. These are critical in making communication and collaboration, which are the most vital aspects in the contemporary interconnected digital economy, a success. The rapidity with which software application developments are taking place has not only simplified day-to-day operations but also transformed the mode of information exchange, enabling users to engage and respond rapidly to varying communication needs. Furthermore, the digital world of communication has witnessed emerging trends in the development of specialized software tools specifically tailored for specific areas, as articulated in the learning technology society. These facilities address the individual needs of students and teachers, enabling a better and interactive learning process. One such instance is the application of learning management systems (LMS), which have turned out to be the cornerstone for providing online lessons, enabling teachers to upload content, track performance, and assess students' progress. Additionally, the availability of multimedia capabilities, such as simulation and interactive content, enhances learning, making it more captivating to students. By embracing such technologies, schools are better positioned to support different learning styles and active participation. According to Rius-Riu et al. (2001), continuous innovation in this area has the potential to further enrich the learning environment, with increasingly sophisticated mechanisms for learning and teaching. While the variety of packages designed for electronic communication is extensive, success hinges on user knowledge and proficiency. Stakeholders must understand tools available and how they may be utilized for specific communication goals. The creation of planning support systems, exemplified in the case of urban planning, is one such instance of this intricacy where software is not just collecting data but also acting as shared ground for collaborative decision-making among heterogeneous stakeholders. Effective use of these planning tools relies on the capacity of users to leverage visualization abilities and share knowledge across platforms in accordance with responsive social needs. Integration of software with web-based services is another indication of the trend toward collaborative planning and development practices. As noted by Batty et al. (2007), the future of online communication depends on developing these tools to a higher level and ultimately changing the face of participatory engagement in all areas.

1.4. BENEFITS OF DIGITAL COMMUNICATION

The introduction of electronic communication has transformed the interaction environment, bringing huge advantages for professional and personal communication. The most telling advantage is the rapidity with which information can be disseminated. Digital media contrast with traditional channels of interaction, such as postal communication, which is time-consuming, by allowing messages and information to be communicated in real-time. This acceleration increases productivity since it speeds up decision-making and encourages collaboration over geographic distances. With the addition of different methods of communication, such as emails, video conferencing, and instant messaging, businesses can streamline their operations, which ultimately increases general efficiency. Electronic communication also increases access to information, where people are able to engage with multiple sources and perspectives, thus increasing their understanding of complex issues that define the modern world. In addition, electronic communication facilitates the ease of building long networks beyond the limitations of conventional boundaries. Social networking sites, for instance, facilitate sharing with others in remote locations of the world, building relationships that might not otherwise have been possible. This sheer level of connectivity not only facilitates exchange of thoughts and cultural exchange but also strengthens support networks that can prove highly valuable in the event of adversity. In business, networking via digital routes has become vital, as it can provide access to collaboration, partnerships, and opportunities that were otherwise out of reach. In addition, organizations are able to utilize digital communication to develop a powerful brand presence, directly interacting with customers and gaining immediate feedback. This two-way communication ensures customer loyalty and satisfaction, which is particularly important in today's competitive market. The integration of big data analytics in digital communication campaigns is a revolutionary trend with extensive implications for business effectiveness. Using sophisticated data mining tools, competitors can have better knowledge about the behavior and preferences of consumers so they can provide personalized interaction. Increased awareness through these means not only enables firms to personalize their advertising campaigns but also increases customer satisfaction. For example, targeted campaigns can be created from the analysis of data gathered, making sure that the correct messages are received by the correct people at the best times. All this accuracy in communication not only guarantees maximum rates of engagement but also ensures a sense of bonding between customers and brands. Effectively, as noted in recent literature, this method is central to advancing long-term relationship marketing strategies to ensure profitability and sustainability amidst a continuously changing digital economy (Yigitcanlar et al., 2009; Baraulya et al., 2019) (Figure 1.9).

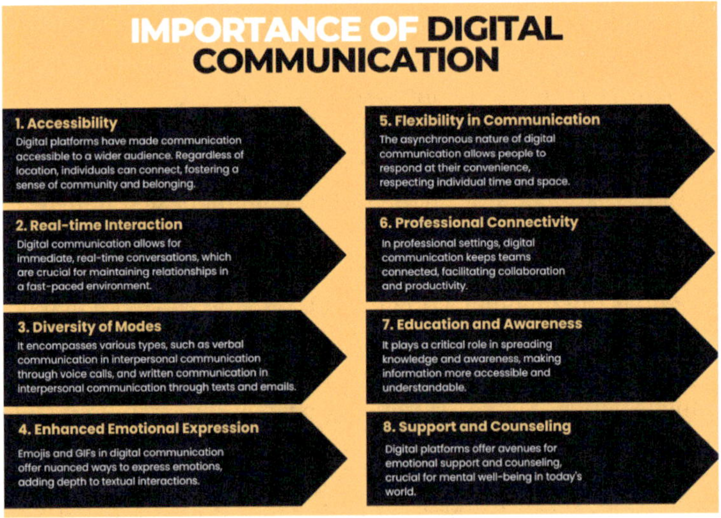

Figure 1.9. *Benefits of digital communication.*

Source: https://images.examples.com/wp-content/uploads/2023/12/Importance-of-Digital-Communication-in-Interpersonal-Communication.png.

1.4.1. Speed and Efficiency

The advent of digital communication has transformed the concepts of speed and efficiency in urban management systems. With cities turning into Ubiquitous Eco Cities, application of advanced information and telecommunication technologies is central to enhancing communication, as well as operation, efficiency. The technologies enable real-time data collection and processing, thereby ensuring more transparent and inclusive decision-making processes involving a broad array of stakeholders. The integration of these components allows for rapid response to urban problems, thus maximizing the quality of life in the city. Furthermore, convergence of telecommunication technologies facilitates seamless information exchange between networks, which is pivotal for optimum management of urban life. This revolutionary capability emphasizes the importance of speed and efficiency as underlying pillars of modern digital infrastructures, which play a key role in the sustainable urban living of the 21st century (Yigitcanlar et al., 2009). Speed, in the context of digital communication, transcends the external layer as data transfer rates. It encapsulates the ability to process and analyze information rapidly, a requisite for both traditional communication networks and emerging technologies such as quantum key distribution (QKD). The utilization of innovative technologies such as digital micro-mirror devices (DMD) is a strong example of such a need for high-speed processing. Recent research indicates that such devices are capable of switching between various modes at a speed of 4 kHz, far beyond the traditional spatial light modulators (SLMs) (Boyd et al., 2013). This significant improvement in the switching speed, apart from enriching user experience with bolder and more powerful means of communication, also works to secure data transmission. The delicate balance between efficiency and speed in such technologies is a precursor to robust communication systems that are capable of taking on increasingly rising demands. In addition, digital communication efficiency is always accompanied by speed since the two concepts infuse one another when applied. Effective systems are defined by timely delivery of information and the consumption of resources in minimal amounts. This is more applicable in cities, as effective communication networks facilitate the involvement of and decision-making within public policy actions. Greater effectiveness in functioning can be ensured through optimizing digital mediums, which in their turn facilitate large-scale participation and citizen involvement. With the help of integrated technologies, cities are not only able to enhance the delivery of information but also facilitate the active engagement of citizens within decision-making. This synergy of speed and efficiency ultimately creates a harmonious urban setting in which communication is efficient and effective, hence creating sustainable solutions to contemporary urban challenges (Yigitcanlar et al., 2009).

1.4.2. Global Connectivity

Global interconnectedness is now the theme of choice in the global digital communication landscape, actually transforming communication at a global scale. The advent of the internet and advances in mobile technology have made possible a networked society in which individuals can communicate and collaborate over vast distances almost instantaneously. This connectivity has not only narrowed geographical spaces but amplified the flow of information, culture, and ideas. But while the advantages of global connectivity are unequally distributed, they frequently expose inequalities in terms of access and resource distribution. When marginalized areas are attempting to connect to the digital world, the gaps to be filled need to be noted. With the historical background of externally imposed infrastructures, nations, especially in Africa, have specific challenges to overcome in order to provide equal participation in the global digital economy (Little et al., 2006).

Moreover, new technologies, and above all the introduction of 5G networks, offer opportunities as well as pitfalls in world connectivity. The potential applications of 5G, ranging from autonomous vehicles to smart cities, require ultra-low latency and super-fast transmission. This technological progression compels operators to reframe existing business models while scaling up investments in cutting-edge network capabilities. But they must be invested while keeping an eye on the cost-benefit ratio, since the deployment of 5G applications succeeds only with marketplace readiness and demand pull (Cuevas et al., 2017). With operators busy positioning themselves strategically in emerging markets, the interconnected models of digital communication will guide the articulation of innovative solutions that emphasize effectiveness and ease of use by the end user, ultimately giving rise to a more connected world. Additionally, the role of the African diaspora cannot be overstated in the context of international interconnectedness. The diaspora is a bridge and channel for knowledge, expertise, and resources which can empower local movements within various parts of the world. By drawing on experience and knowledge acquired outside the continent, the African diaspora can facilitate partnerships which focus on local agendas and develop technological capability bases. As observed, the key to ending the existing disparities depends on finding equilibrium between internal partnership and external partnership (Little et al., 2006). Through a culturally responsive practice, stakeholders can help ensure that projects are not imposed but are instead created with consideration of the needs of the served populations. Finally, strengthening this interconnectedness will open the doors to long-term growth, innovation, and resilience in this ever-changing world of digital communication (Figure 1.10).

Figure 1.10. *Global connectivity.*

Source: https://images.stockcake.com/public/f/e/4/fe495b7a-bb51-42e0-b217-169a23e79495_large/global-connectivity-concept-stockcake.jpg.

1.4.3. Cost-Effectiveness

In digital communication, cost-effectiveness is a key parameter to examine the execution and capital viability of many technologies. With the rapid growth of digital infrastructures, companies are in tighter pressure to utilize their resources to the maximum to offer quality services with higher standards. This problem becomes particularly evident in wireless communications, where traditional methods may not give the best combination between the performance and cost. For instance, the integration of the new techniques such as hybrid beamforming is an excellent cost-saving opportunity over digital beamforming without any loss of transmission quality. By optimizing beamforming technology, firms can spend financial and technological investments judiciously, and therefore reach enhanced operating efficiency within digital communications systems. Moreover, the shift towards cloud-based architectures in the form of cloud radio access networks (C-RAN) highlights cost-effectiveness as an important aspect of digital communication solutions. Digital processing is centralized in a C-RAN configuration, reserving user access points for analog signaling. This framework not only keeps the hardware load on individual remote radio heads (RRHs) to a bare minimum but also streamlines power consumption network-wide. Cost-effectiveness is achieved by lowering the operational costs and enhancing the network capacity, as analyzed through optimization of digital beamforming in conjunction with fronthaul quantization practices. Continuous improvements in cost indicators allow organizations to make changes to their systems so that performance is never compromised for fiscal viability. Periodic evaluation of such practices is crucial to maintaining competitiveness in a dynamic digital landscape. Lastly, the quest for cost-effectiveness in electronic communication transcends purely economic considerations; it requires a comprehensive understanding of technological advancement and its implications on service delivery. Through an examination of the economic viability of new technologies and their operational reactivity, stakeholders are able to make informed decisions to optimize cost-effectiveness and network capability. For instance, the technology integration of hybrids beamforming within a C-RAN setting enables firms to achieve economies of scale and make efficient technological investments. With continued research and practical applications, the findings from systematic optimization here lay the groundwork for ongoing innovation in digital communication and reinforce cost versus performance as a fundamental tenet of the discipline (Kim et al., 2017; Miller et al., 2013).

1.5. CHALLENGES AND LIMITATIONS

In the online communication space, there are several challenges and limitations that impede effective connectivity and information exchange. A significant challenge is posed by the socio-economic disparities that impinge upon access to technology, particularly among marginalized populations. For instance, in a research study of a newly resettled Karen refugee community in the UK, it was seen that the main inhibitors of their communication were cost, education, literacy, and age (Castells et al., 2012). These conditions, in turn, affect not only the ease with which individuals are able to gain access to digital communication resources but also whether and how they can maintain connections with their global diaspora and engage in local or political life. Such barriers create severe communication gaps that prevent the opportunity for social integration as well as the transmission of critical information. This renders it critical to understand and thus overcome such challenges so as to promote fair access to digital communication technology. Coupled with the heterogeneity of digital communication is the rapidly changing pace of technology, which is more likely to overshadow the infrastructure and regulatory frameworks put in place to underwrite it. As the demand for multimedia services with high data rate increases, the existing radio frequency (RF) communication spectrum is deemed inadequate to support such demands (Chowdhury et al., 2018). The adoption of optical wireless communication (OWC) technology is a bright prospect but comes with a need for high investment in infrastructure and the education of users in order to be well-adopted. Most of the possible users might lack the ability to understand these complex technologies, opening a knowledge gap that restricts access to the digital communication world. This fast rate of technological change, coupled with a lack of awareness and availability, thus becomes a significant limiting factor in the actualization of the full potential of digital communication tools in diverse groups of people. Moreover, the features of digital communication also encompass issues with respect to privacy, security, and disinformation. As communication increasingly depends on digital media, they are open to various threats in terms of cybersecurity, including data theft and identity stealing. In addition, the proliferation of misinformation erodes the credibility of information sharing, such that it becomes more challenging to build well-informed communities. The impact of such challenges can be particularly deadly on those groups seeking to mobilize politically or maintain cultural identity, further exacerbating their communication issues. It is hence in such a situation critical that digital communication campaigns are not only technologically advanced but also have effective steps and training programs. Addressing these multifaceted challenges and constraints is critical in creating a safer and more inclusive space for digital communication.

1.5.1. Digital Divide and Accessibility Issues

In the virtual environment, the problem of the digital divide is a major barrier to equal access to information and technology. The divide is not just defined in terms of differences in access to the internet and hardware but also as differences in digital literacy and competencies across different social groups. Socioeconomic status, geographic location, and access to education play a significant role in determining one's ability to make effective use of digital channels. For instance, rural societies may lack the supporting infrastructure to enable high-speed connectivity to the internet, which slows their entry into a rapidly digitizing economy. Through such a mechanism, the tech gap contributes to the exacerbation of social inequalities, with disadvantaged communities facing further obstacles that limit their ability to tap digital opportunities for education, employment, and politics. Closing these gaps is important in creating a more inclusive society where all groups are able to enjoy digital communication. Closing the digital divide requires various stakeholders to concentrate on building accessibility and pushing for digital inclusivity. This requires an in-depth understanding of the kinds of barriers faced across various segments of communities, particularly by the elderly and disabled individuals. Technological initiatives that seek to address these issues have emerged, focusing on the potential of mobile technology to provide accessible and affordable solutions to desktop computing (Carlsson et al., 2025). By focusing on mobile learning and creating digital literacy initiatives specifically for such audiences, we can equip people to thrive in the digital world. Moreover, initiatives must also address the collaborative potential of technology, facilitating effective communication and engagement within communities. Therefore, stakeholders can ensure that not only is every individual a participant in the digital world but also a contributor who can shape its destiny. While with growing recognition of how critical digital inclusion is, there are still serious challenges in translating vision into action. There have also been numerous national initiatives aimed at filling the digital divides, but numerous communities still lack the necessary resources and support frameworks to achieve genuine engagement with technology (Craig et al., 2015). One of the biggest challenges to success is user-centered design and developing support structures that keep on offering help, particularly to vulnerable communities. Community centers that offer access to resources and training have the capacity to lead digital competencies and facilitate digital engagement for people with accessibility challenges. However difficult it is to implement such schemes, the identification of the key success factors in digital participation provides stakeholders with an effective blueprint for planning change. Finally, a digitally inclusive society is contingent on the unified efforts to overcome these challenges and tap the revolutionizing power of technology for the common good of all members of society (Figure 1.11).

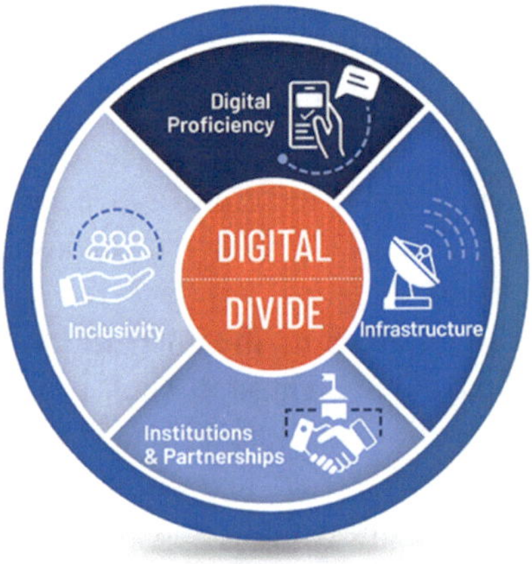

Figure 1.11. *Digital divide.*

Source: https://www.analog.com/en/_/media/analog/en/signals/bridging-the-digital-divide-challenge/digitaldivide-fourcomponents.jpg?rev=fb489fdc9190489699e200ba21290db4&sc_lang=en&la=en&h=432&w=584&hash=2800301FDC02D93589A6F93E0A765CD9.

1.5.2. Privacy and Security Concerns

The development of communication technology has undoubtedly transformed the mode of communication in organizations and between individuals, but it has also brought forth significant privacy and security concerns. The increasingly common and sophisticated nature of cyber-enabled crimes necessitates that law enforcement agencies and organizations upgrade their investigatory tools in the digital space. As these actors become more involved in the digital space through digital forensics, privacy rights are under threat of being violated. Unregulated research is highly detrimental to people's rights, posing ethical concerns regarding the use of personal information and the boundaries of lawful searches. Furthermore, firms are put at risk as their intellectual properties are subject to unregulated examination, undermining their autonomy and confidential information integrity. The tension between stringent investigative demands and moral constraints forms the backdrop to a central examination of privacy rights in online communication environments, seeking an even-handed resolution balancing the freedom of individuals with valid security needs (Ferguson et al., 2020). Resolving the complexities associated with privacy and security concerns in online communication requires an ethically sensitive framework that balances the interests of all stakeholders.

The evolution of digital forensics that aims to discover evidence pertaining to cyber offenses must be balanced with procedures that protect individuals' privacy and companies' rights to hold on to their intellectual property. An infrastructure for this, such as the proposed PRECEpt: Privacy-Respecting EthiCal framEwork, is called for to make sure that digital forensic procedures do not override the lawful right to privacy of an individual being investigated. As it is, the practices adopted have an unsettling imbalance, whereby security requirements override individual and organizational rights. A free consideration of such moral implications is important towards the development of guidelines derived using theory of justice, which can ultimately enable digital forensics to function in a fairer and more accountable setting (Ferguson et al., 2019). Aside from specific cases, the social implications of privacy and security problems with online communication extend to community trust and perceptions of justice. Inadequacy in openness and perceived unfairness in the implementation of cyber investigations can result in public distrust and resistance against law enforcement operations. When citizens feel their rights are being violated in the name of security, it not only discourages them from complying with the authorities but also prevents them from completely engaging in digital communication forums. It is important to instill a culture of respect towards ethical standards in digital forensics so that confidence in legal institutions can be regained and ensured that advancements in digital communication are not made at the expense of civil liberties. By adopting paradigms that place privacy at the center along with security, one can establish a stronger and more equitable digital culture that is respectful of justice values for all stakeholders involved (Ferguson et al., 2019).

1.5.3. Information Overload and Misinformation

Information overload has emerged as a defining feature of our computer age, fundamentally transforming the way individuals interact with content. When confronted with a record information flood served up on billions of channels, including social media, news sites, and online forums, a knack for legitimately processing and using this information has often been forfeited. This copious abundance of material, dubbed infobesity and data smog, is an irony in which the alleged benefit of information access is translated into mental blockage. As the world navigates the complexities of Floridi's Fourth Revolution, there is a need to address information overload, meaning that people and institutions need to learn to cope with this unstoppable flood through winnowing discerning facts amidst the din (Bawden et al., 2020).

Figure 1.12. *Misinformation of digital communication.*

Source: https://journalistsresource.org/wp-content/uploads/2017/01/Pixabay.jpg

Misinformation, an information age endemic threat, expands on digital communication's existing issues. While credibility of news and fabricated content have blended into a singular entity in an age where verification is a luxury few can afford, audiences are left stymied as to how to distinguish truthful sources from fictional tales. Institutions of conventional media have relied on perceived credibility to get and maintain their audience's attention; the digital age has, however, altered those dynamics as viewers crave authenticity more and more. With more discerning content consumers, media organizations must adapt by producing messages that resonate personally and thus restore trust in a time of ubiquitous skepticism. The new platform and program case studies suggest establishing authenticity can be a mainstay in combating disinformation and balancing audience expectations with credible journalism (Schulz et al., 2018). Despite all the issues brought about by information overload and misinformation, there are remedies that could counter such issues to build a wiser society. Building effective personal information management capabilities, as well as improving digital and media literacy, is necessary in order to empower people to be able to deal better with the complex information ecosystem. Strategies such as filtering, withdrawal, and mindful consumption work towards cultivating a more positive relationship with information, thus providing more informed decision-making. Additionally, the collective effort of the media towards enhancing the accuracy and transparency of their content can assist in re-establishing trust with the public. In this setting of electronic communication, the quest for concision and credibility will become paramount as society is faced with both information overload and misinformation (Bawden et al., 2020) (Figure 1.12).

1.6. FUTURE TRENDS IN DIGITAL COMMUNICATION

The fate of digital communication is deeply intertwined with technology development and how it reshapes human communication. With society evolving side-by-side with new technologies, the anticipated impact of new communication technology tends to obscure from its original intended function. For instance, rather than building a straight line of anticipated change, the convergence of new technologies produces complex social dynamics and meaning. Scholars argue that a more effective approach to understanding such trends is to embrace participatory, adaptive systems rather than linear oversimplifications. Through an emphasis on embedded innovation, participants in the digital sphere might enable advances in communication better suited to the nuanced behaviors and desires of users, thus further encouraging overall society adoption of digital structures (Anderson et al., 2008). This perspective highlights the necessity of a perpetual rethinking of the application and valuation of technologies in the grand plan of communication. The second trend shaping the future of internet communication is the increased utilization of Web 2.0 technologies, particularly in promotion and advertising. Recent scholarly writing identifies that Destination Marketing Organizations (DMOs) have accepted these technologies as powerful tools for promoting destinations, particularly during times of budget cuts. Web 2.0 supports interactive communication, with users able to interact with content and add their own stories. Yet, while the existing literature grows, it is still fragmented and lacks an overarching theoretical foundation through which to tackle key issues related to these technologies. There is a need for immediate more studies to compile these findings, complete the gaps, and develop more integrated strategies that optimize the impact of Web 2.0 in destination marketing (Mariani et al., 2020). An understanding of these complexities will be crucial to optimize the utilization of digital communication for uses that meet consumer requirements and build profound relationships between consumers and brands. Lastly, the trends in digital communication are not only reserved for promotion; they also significantly impact interpersonal communication methods. The more social media and other internet platforms people use, their communication behavior and style change accordingly. This transition is defined by the transition towards more visual and ephemeral forms of communication, such as images and short videos, which are appropriate to users' decreasing attention spans. Second, the introduction of artificial intelligence to online communication holds promise for personalized interaction, chatbots, and prediction engines that ensure maximum user interaction. However, as these trends continue to develop, it is still important to critically explore their implications for privacy, authenticity, and the quality of human relations. The interplay between technology and human action will still be a field for future research, as the digital world continues to develop and expand (Mariani et al., 2020) (Figure 1.13).

Figure 1.13. *Current trends in digital communication.*

Source: https://external-content.duckduckgo.com/iu/?u=https%3A%2F%2Fwww.mcgill.ca%2Fcontinuingstudies%2Ffiles%2Fcontinuingstudies%2Fcurrent-trends-in-digital-communication.jpg&f=1&nofb=1&ipt=23032b709f2d039a3d81146ec5ff59f3a85efab7a47bede46f6c8ee06203009d.

1.6.1. Advancements in Artificial Intelligence

The evolution of artificial intelligence (AI) has significantly transformed various sectors, including digital communication, through enhanced efficiency and innovation. One of the most notable advancements is the integration of machine learning algorithms that can analyze vast quantities of data in real-time, facilitating improved communication strategies across multiple platforms. These AI systems can derive insights from user interactions, allowing for more targeted content and personalized experiences, which are crucial in an era where user engagement and retention are paramount. As a result, organizations leverage AI to optimize their communication channels, ensuring that messages resonate effectively with diverse audiences, thereby fostering a stronger connection and enhancing user satisfaction. This transformation exemplifies the broader implications of AI as it continues to redefine how communication is conceptualized and executed in the digital landscape. In addition to enhancing communication strategies, AI plays a pivotal role in addressing challenges faced in cyber security within digital communication frameworks. The increasing frequency and sophistication of cyber threats necessitate robust defense mechanisms capable of responding to attacks in near real-time. Traditional methods of threat detection are no longer sufficient in this landscape, as they are often hindered by the sheer volume of potential threats. Artificial intelligence, combined with machine learning and advanced data mining techniques, empowers cyber security specialists to collect, analyze, and interpret attack patterns effectively, thus enabling proactive security measures. Understanding the nuances of cyber threat intelligence is crucial for organizations aiming to safeguard their digital communication environments, particularly as they integrate more AI-driven tools into their operations, ensuring that their communication remains secure and resilient against emerging threats (Conti et al., 2018). Furthermore, the intersection of AI and the creative arts presents exciting avenues for enhancing digital communication. Technologies such as AI-generated visual art and virtual reality applications are not only enriching artistic expression but also enabling new forms of communication and interaction. For instance, AIs capacity to create compelling visual narratives can serve therapeutic purposes for individuals with cognitive challenges, such as dementia, enhancing their communication skills and overall well-being. The innovative use of AI in this context highlights its potential to bridge gaps in understanding and foster social connection among diverse groups. As research progresses, the implications of AI within the creative domain are poised to influence not only artistic practices but also the broader conversation on how we perceive and interact with digital content, ultimately transforming the landscape of digital communication (Harwood et al., 2019).

1.6.2. The Role of Virtual and Augmented Reality

With the rapidly evolving domain of virtual communication, the role of virtual reality (VR) and augmented reality (AR) is assuming a much larger position. The two technologies offer immersive environments which facilitate new interaction and engagement opportunities beyond the traditional boundaries of communication. AR in particular is a bridge between the real world and the virtual world since it facilitates the overlay of digital information on to individuals' physical environments. This capability is enhanced by our understanding and exposure to our surroundings, as evidenced by its utilization in architectural presentation. With the capability of stakeholders to interact with 3D models through touch, AR provides enhanced spatial understanding, allowing improved decision-making in complex projects.

Figure 1.14. *The role of virtual and augmented reality in digital communication.*

Source: https://external-content.duckduckgo.com/iu/?u=https%3A%2F%2Fcdn.elearningindustry.com%2Fwp-content%2Fuploads%2F2024%2F07%2FShutterstock_2022847892.jpg&f=1&nofb=1&ipt=ac14aeed74a55dcca5aed3fdc11a864f09e10634e8d74c261d45a5f54020f5bb.

As these technologies evolve, their potential to enhance communication and collaboration between disciplines only becomes more apparent, and the importance of further research and the integration of AR and VR technologies increases (Baglioni et al., 2019). Education is another sector that is being drastically altered due to AR and VR technologies. Blended reality, where digital and physical interactions converge, is revolutionizing learning through the ability to gain deep knowledge and retention from interactive simulations. Mixed reality allows interaction with both the digital and physical worlds at once, providing collaborative learning experience. Scalability and accessibility for multiple users in a learning environment continue to be a challenge for these technologies. Problems arise about how to best deliver mixed reality simulations to achieve genuine co-presence, as necessary for maximum learning benefit from these technologies. Overcoming such problems is critical to developing effective, straightforward mixed reality experiences, thereby expanding their application in the learning environment and enhancing overall digital communication tactics (Birt et al., 2017). Furthermore, the intersection of VR and AR is not only in education and architecture but also in other industries, including healthcare, video games, and marketing. In medicine, for instance, AR software is used in training surgeons and educating patients to help practitioners visualize structures of the anatomy in real time, which actually improves the learning experience as well as patient safety. In marketing, these technologies provide new ways for brands to engage with consumers, providing interactive experiences that captivate audiences and generate brand loyalty. As the applications above demonstrate, the impacts of VR and AR extend far beyond visualization; they transform how information is presented and experienced. In accepting the potential of these technologies, industries can reshape their communication paradigms and effectively meet the evolving needs of their publics in the digital age (Baglioni et al., 2019) (Figure 1.14).

1.6.3. Predictions for the Next Decade

The coming decade in online communication holds change of a transformative nature, fueled by technology growth, availability of data, and user expectations. Advances like Artificial Intelligence and machine learning will make communication platforms more personalized and effective. As geotechnology evolves at high speed, decision-makers will possess better spatial data to shape policy formulation and planning. As noted in the trend analysis of the 1990s, the integration of geoinformation into digital communication provides necessary information for many sectors, particularly for urban planning and environmental management (Geertman et al., 2000). The capability will continue to grow, allowing for a more evidence-based problem-solving process that goes beyond boundary-crossing and facilitates more interdisciplinary collaborative work. Besides, the social climate will also experience unimaginable demographic shifts that will reshape communication lines. America's demographics will be further diversified by 2034, a factor that will need further inclusive and culturally nuanced lines of communication (Chung et al., 2008). By the transformation, digital channels will need to evolve in order to respond better to Varied audience preferences and language sophistication. This adaptability will not just enhance user engagement but also render communication pertinent and available across cultures. Therefore, the digital landscape of communication will place emphasis on inclusivity, making room for different voices to speak and be heard in the global dialogue. Finally, the technology landscape itself will further dissolve the line between virtual and actual experience. The development of immersive technologies like virtual and augmented reality will redefine how humans interact with other humans and with content. While these technologies progress, they will have the potential to enrich conventional modes of communication and deliver more interactive and participatory experiences. Web 3.0 trends include increased decentralization of information and greater emphasis on user power and content creation (Chung et al., 2008). In this evolving context, electronic communication will increasingly be an interactive dynamic arena, creating community and collaboration in addressing increasing complexities in user expectations and electronic ethics.

SUMMARY

This voyage of digital communication, we can observe that technological progress has revolutionized the manner in which we communicate and engage. The innovations in the medium and instrument of digital communication, such as social media and web collaboration tools, have enabled unprecedented levels of connectivity and engagement. This shift not only facilitates individual interactions but also enhances business collaboration across geographical distances. With these developments, individuals and institutions must accommodate new norms of communication that promote immediacy and accessibility, which will increasingly affect social structures as well as work-oriented dynamics. The implications of this shift call for ongoing research and adaptation as society adjusts to the unfolding context of communication. Also, integration of advanced technologies like the use of Artificial Neural Networks in computer tools demonstrates the complexity of modern communication systems. For instance, design of a single-phase DC-AC converter for applications in wind generator systems demonstrates how digital communication interfaces can enhance operational reliability in technical applications (Buswig et al., 2011). These advancements not only optimize technical effectiveness but also emphasize the central role played by effective digital communication in conveying sophisticated information and innovations. With ever more integrated systems depending on digital communication, understanding these technical advances becomes crucial to professionals who wish to utilize technology appropriately in their own areas of specialization, particularly those that rely on real-time information and communication effectiveness. Finally, a respect for the intricate nature of digital communication requires a respect for how its impact percolates over industries and employers, for example, how it impacts employment, for example, in the building construction industry (Hussein et al., 2018). The linkage between job satisfaction and communication underscores the necessity of fostering an organizational culture that embraces contributions and praises. Better communication channels increasing job satisfaction can lead to better performance and higher levels of commitment among employees. As communication technology evolves, stakeholders across different fields must be able to understand these processes so that technological advancements channel their way into tangible outputs of better individual and collective outcomes. Daring to make one's way through digital communication signifies an ongoing change, which compels its users to consistently innovate and evolve.

The world of digital communication has transformed comprehensively, a step towards more interactive and user-controlled content. The position of young people in shaping and engaging with digital media is perhaps one of the most essential aspects of this transformation. The report by the Center for Media Education specifically notes that there has been a new culture of media that aims not just at teens but allows them to become active members of the community as well. This is a significant shift towards active engagement over passive consumption wherein sites have become more interactive and user-involved. Thus, the interactivity of online communication has formed a dynamic relationship in which young people influence the online world while concurrently being influenced by it, reiterating the significance of understanding this two-way interaction in future digital communication strategies (Montgomery, 2001). Furthermore, integrating online platforms into socio-political spheres provides important information about the civic involvement of young generations. Research indicates that demographic, or in this case socioeconomic status, is related to online and offline civic activity. Younger generations use social media as a channel for political voice and mobilization, which is a significant departure from previous civic behavior. This change involves not just a reorganization in how political argument is facilitated but implies as well a potential reconfiguration of engagement practices for future generations. Understanding these changing inclinations can help teachers and policymakers foster environments conducive to active citizenship, as social media remain a key influencer of the political agenda (Smith et al., 2009). Lastly, the intersection of digital communication and learning spaces has extreme effects on instruction and learning. Since digital literacy is a skill that must be acquired, instructors must update their curriculums to include the principles of successful digital communication. This involves not only teaching learners how to apply various platforms but also transferring critical thinking skills necessary to evaluate the truthfulness of information online. Moreover, facilitating discussions of the ethics of digital communication can allow students to responsibly participate in these areas. By focusing on the identification of both the benefits and the difficulties inherent in digital media, educational settings can best prepare students for success in an increasingly interconnected world. In general, creating an informed and active digital citizenry is of most importance in order to reap the full potential of digital communication within educational and societal systems.

REFERENCES

1. Aksulu, A., Degbelo, A., Powell, A., Rajabifard, A., Jiang, B., Nosek, B. A., et al. (2019). Geospatial information infrastructures. https://core.ac.uk/download/322902166.pdf (accessed on 18th September 2025).
2. Alevizou, P., & Conole, G. (2010). A literature review of the use of Web 2.0 tools in higher education. https://core.ac.uk/download/30676612.pdf (accessed on 18th September 2025).
3. Allamandri, F., Campion, S., Centonza, A., Chernilov, M., et al. (2007). Service platform for converged interactive broadband broadcast and cellular wireless. https://core.ac.uk/download/333481.pdf (accessed on 18th September 2025).
4. Anderson, B., & Stoneman, P. (2008). Predicting the socio-technical future (and other myths). https://core.ac.uk/download/9063918.pdf (accessed on 18th September 2025).
5. Baglioni, L., Menconero, S., & Russo, M. (2019). Parametric surfaces for augmented architecture representation. https://core.ac.uk/download/188831772.pdf (accessed on 18th September 2025).
6. Baraulya, E. V., Ivanchenko, O. V., Mirgorodskaya, O. N., Putilina, E., et al. (2019). Marketing relations and communication infrastructure development in the banking sector based on big data mining. https://core.ac.uk/download/288638643.pdf (accessed on 18th September 2025).
7. Barolli, L., & Xhafa, F. (2011). JXTA-Overlay: A P2P platform for distributed, collaborative, and ubiquitous computing. https://core.ac.uk/download/187265156.pdf (accessed on 18th September 2025).
8. Batty, M. (2007). Planning support systems: Progress, predictions, and speculations on the shape of things to come. https://core.ac.uk/download/1683968.pdf (accessed on 18th September 2025).
9. Bawden, D., & Robinson, L. (2020). Information overload: An overview. https://core.ac.uk/download/286715468.pdf (accessed on 18th September 2025).
10. Birt, J. R., & Cowling, M. A. (2017). Visualizing mixed reality simulation for multiple users. https://core.ac.uk/download/196605626.pdf (accessed on 18th September 2025).
11. Bown, O. (2006). The extended importance of the social creation of value in evolutionary processes. http://ualresearchonline.arts.ac.uk/1530/1/bown_ecai06.pdf (accessed on 18th September 2025).
12. Boyd, R. W., Chen, C., Magana-Loaiza, O. S., Malik, M., et al. (2013). Rapid generation of light beams carrying orbital angular momentum.

https://core.ac.uk/download/223096887.pdf (accessed on 18th September 2025).

13. Brasher, A., & McAndrew, P. (2009). Automatic generation of audio content for open learning resources. https://core.ac.uk/download/41242.pdf (accessed on 18th September 2025).
14. Britchenko, I., Cherniavska, T., & Machashchik, P. (2018). New trends in development of services in the modern economy. https://core.ac.uk/download/323058294.pdf (accessed on 18th September 2025).
15. Carlsson, J., Patokorpi, E., & Tétard, F. (2025). A conceptual framework for mobile learning. https://core.ac.uk/download/pdf/6944626.pdf (accessed on 18th September 2025).
16. Castells, M., Castells, M., Castro, L. A., Chen, W., Cho, N. V., Dudley, S., Lockley, E. G., et al. (2012). Communication practices of the Karen in Sheffield: Seeking to navigate their three zones of displacement. https://core.ac.uk/download/9426476.pdf (accessed on 18th September 2025).
17. Chowdhury, M. Z., Hossan, M. T., Islam, A., Jang, Y. M., et al. (2018). A comparative survey of optical wireless technologies: Architectures and applications. http://arxiv.org/abs/1810.02594 (accessed on 18th September 2025).
18. Chung, J., & Wilkening, S. (2008). Museums & society 2034: Trends and potential futures. https://core.ac.uk/download/71340899.pdf (accessed on 18th September 2025).
19. Comrie, A., Cullen, D., & McDonald, D. (2010). Final report of work-with-IT: The JISC study into evolution of working practices. https://core.ac.uk/download/9033706.pdf (accessed on 18th September 2025).
20. Conti, M., Dargahi, T., & Dehghantanha, A. (2018). Cyber threat intelligence: Challenges and opportunities. http://arxiv.org/abs/1808.01162 (accessed on 18th September 2025).
21. Courtois, C., De Marez, L., Van Damme, K., Verbrugge, K., et al. (2015). What's APPening to news? A mixed-method audience-centred study on mobile news consumption. https://core.ac.uk/download/55765445.pdf (accessed on 18th September 2025).
22. Craig, M., Damodaran, L., Gilbertson, T., Olphert, C. W., et al. (2015). Digital inclusion–the vision, the challenges and the way forward. https://core.ac.uk/download/30650610.pdf (accessed on 18th September 2025).
23. Cuevas, M., Dohler, M., Laya, A., Lema, A., et al. (2017). Business case and technology analysis for 5G low latency applications. https://core.ac.uk/download/96762095.pdf (accessed on 18th September 2025).

24. Dal Falco, F., & Vassos, S. (2017). Museum experience design: A modern storytelling methodology. https://core.ac.uk/download/153363274.pdf (accessed on 18th September 2025).
25. Dijk, J. van, Heuvelman, A., & Peters, O. (2003). Interactive television or enhanced television? The Dutch users' interest in applications of ITV via set-top boxes. https://core.ac.uk/download/pdf/11463461.pdf (accessed on 18th September 2025).
26. Dobreva, M., Feliciati, P., & O'Dwyer, A. (2012). Introduction: User studies for digital library development. https://core.ac.uk/download/46601796.pdf (accessed on 18th September 2025).
27. Ferguson, I., Irons, A., Renaud, K., Wilford, S., et al. (2019). PRECEPT: A framework for ethical digital forensics investigations. https://core.ac.uk/download/287585744.pdf (accessed on 18th September 2025).
28. Ferguson, R. I., Irons, A., Renaud, K., Wilford, S., et al. (2020). PRECEPT: A framework for ethical digital forensics investigations. https://core.ac.uk/download/288653619.pdf (accessed on 18th September 2025).
29. Firestone, C., Lasica, J. D., & Kelly, P. (2007). The mobile generation: Global transformations at the cellular level. https://core.ac.uk/download/71339786.pdf (accessed on 18th September 2025).
30. Geertman, S., & Stillwell, J. (2000). Geoinformation, geotechnology, and geoplanning in the 1990s. https://core.ac.uk/download/51004.pdf (accessed on 18th September 2025).
31. Gómez Vázquez, L. M., & Soto Vélez, I. (2011). Los medios sociales como una herramienta estratégica para la comunicación corporativa. https://core.ac.uk/download/62896839.pdf (accessed on 18th September 2025).
32. Harwood, T., Maltby, J., & Mukaetova-Ladinska, E. (2019). Role of artificial intelligence (AI) art in care of ageing society: Focus on dementia. https://core.ac.uk/download/228191193.pdf (accessed on 18th September 2025).
33. Hochstenbach, P., Karstens Elbaek, M., Russell, R., Schmelz Pedersen, T., et al. (2008). DRIVER technology watch report. https://core.ac.uk/download/55769861.pdf (accessed on 18th September 2025).
34. Hussein, E. M. (2018). A study on the factors affecting job satisfaction amongst employees in construction industry workplace (Case Study Saudi Arabia). https://core.ac.uk/download/12006374.pdf (accessed on 18th September 2025).

35. Iosifidis, P. (2012). Mapping digital media: Digital television, the public interest, and European regulation. https://core.ac.uk/download/9559529.pdf (accessed on 18th September 2025).
36. Kim, J., Lee, I., Park, S. H., Shamai, S., et al. (2017). Joint design of digital and analog processing for downlink C-RAN with large-scale antenna arrays. http://arxiv.org/abs/1704.00455 (accessed on 18th September 2025).
37. Kulczycka, U. (2018). Analiza przestrzeni informacyjnej Biblioteki Uniwersytetu Gdańskiego. https://core.ac.uk/download/186630622.pdf (accessed on 18th September 2025).
38. Li, F. (2020). Team to Market (T2M): Creating high performance teams in the digital age. https://core.ac.uk/download/287606354.pdf (accessed on 18th September 2025).
39. Little, S. (2006). Meeting the information challenge: Exploring partnerships with Africa. https://core.ac.uk/download/82905854.pdf (accessed on 18th September 2025).
40. Mariani, M. (2020). Web 2.0 and destination marketing: Current trends and future directions. https://core.ac.uk/download/322487243.pdf (accessed on 18th September 2025).
41. Miller, R. G. (2013). ULS FY14 planning and budget report. https://core.ac.uk/download/16509440.pdf (accessed on 18th September 2025).
42. Minguella Canela, J., & Muguruza Blanco, A. (2017). Digital manufacturing: What are we able to print? https://core.ac.uk/download/157811997.pdf (accessed on 18th September 2025).
43. Montgomery, K. (2001). Teensites.com: A field guide to the new digital landscape. https://core.ac.uk/download/71339873.pdf (accessed on 18th September 2025).
44. Rexhepi, Z. (2014). Information sources in Kosovo–Traditional media vs. new media. https://core.ac.uk/download/187496991.pdf (accessed on 18th September 2025).
45. Rius-Riu, M., & Seale, J. (2001). An introduction to learning technology in tertiary education in the UK.
46. Rojas-Torrijos, J. L. (2020). Gamification of sports media coverage: An infotainment approach to Olympics and Football World Cups. https://core.ac.uk/download/299806676.pdf (accessed on 18th September 2025).
47. Sasvári, P. L. (2012). The effects of technology and innovation on society. https://core.ac.uk/download/11857629.pdf (accessed on 18th September 2025).

48. Schulz, G. W. (2018). Street cred: Are media consumers craving more "authenticity" in the digital age? https://core.ac.uk/download/211351234.pdf (accessed on 18th September 2025).
49. Smith, A., Brady, H., Schlozman, K. L., & Verba, S. (2009). The Internet and civic engagement. https://core.ac.uk/download/71352353.pdf (accessed on 18th September 2025).
50. Soukup, P. A. (2004). Walter J. Ong, S.J.: A retrospective. https://core.ac.uk/download/215444171.pdf (accessed on 18th September 2025).
51. Yee, J. (2004). A typographic dilemma: Reconciling the old with the new using a new cross-disciplinary typographic framework. https://core.ac.uk/download/9341172.pdf (accessed on 18th September 2025).
52. Yigitcanlar, T. (2009). Managing ubiquitous eco cities: The role of urban telecommunication infrastructure networks and convergence technologies. https://core.ac.uk/download/10895362.pdf (accessed on 18th September 2025).
53. Yonis Buswig, Y. M. (2011). Development of a DC-AC power conditioner for wind generator by using neural network. https://core.ac.uk/download/12007862.pdf (accessed on 18th September 2025).

CHAPTER 2

SIGNALS AND SPECTRA

Contents

INTRODUCTION

The discipline of signal processing is inherently inextricably linked with the study of spectral properties, creating an infrastructure that increases our knowledge of wide-ranging phenomena, from communications to sensor technologies. With advancements in technology, the need for efficient and reliable means of signal transmission and reception has grown exponentially. This has resulted in innovations in wide-ranging methodologies, ranging from sophisticated radar systems and frequency-domain optical technologies. For example, the behaviors of various waveforms of radar have been analyzed under scrutiny, and the effectiveness of techniques like Chirp and Multitones has been shown in well-optimized detection environments. A comprehensive research on radar waveforms revealed that these techniques provide significant precision in range profiling as well as detection abilities, proving their applications in the domain of software-defined radar prototypes (Denoulet et al., 2012). These developments highlight the significance of signal and spectral analysis in theoretical and applied settings alike, reflecting its essential place within several technological advancements. Furthermore, the merging of optical frequency combs, especially those produced by high-Q micro-resonators, is transforming several areas through noteworthy performance and efficiency improvements. By providing miniaturized device sizes with high energy efficiency, micro-combs have become a significant innovation within RF photonics. Their large optical bandwidths and high repetition rates make them suitable for precision and reliability-demanding applications like telecom and spectroscopy. Current progress has highlighted micro-combs ability to simplify operation, minimize size, and decrease power consumption, a paradigm shift in the current signal processing paradigms (Chu et al., 2018). With the evolution of these optical technologies, they introduce new frontiers of research and application—demonstrating the need for ongoing exploration in the context of signals and spectra. The investigation of signals and spectra is not restricted to conventional technologies but also encompasses novel applications that have the potential to reformulate future advancements. Scientists are more and more attracted to the promise of micro-combs and advanced radar methodologies due to their potential in addressing modern challenges in communication and sensing. Though there is visible progress, the area also has to contend with the continued challenges involved in implementing these technologies in real-time environments. These involve guaranteeing reliability, reducing costs, and increasing scalability of solutions. In this process, the synergy between old knowledge and new innovations will be key to redefining signal processing boundaries. Eventually, the ongoing review of matured and new techniques is necessary for enhancing our knowledge and utilization of signals and spectra in the ever-increasingly complex technology.

Definition of Signals

Signals are the basic constructs across many disciplines of science and engineering, embodying information transmission in the dimension of space and time. Broadly speaking, a signal is a function that transmits data by fluctuating parameters like amplitude, frequency, or phase over a specified duration. This fluctuation enables signals to represent complex information ranging from basic audio tones to complex seismic activities. The division of signals usually exists in two forms: continuous and discrete. Continuous signals change continuously over time, whereas discrete signals are symbolic of sampled data points, commonly used in digital processing systems. Clarification of these nuances forms the foundation of examining how signals influence their surroundings, more so when discussing non-linear systems and their effects on signal fidelity as well as processing efficiency. Stepping further into the realm of the nature of signals, one becomes aware of the significance of their spectral characteristics.

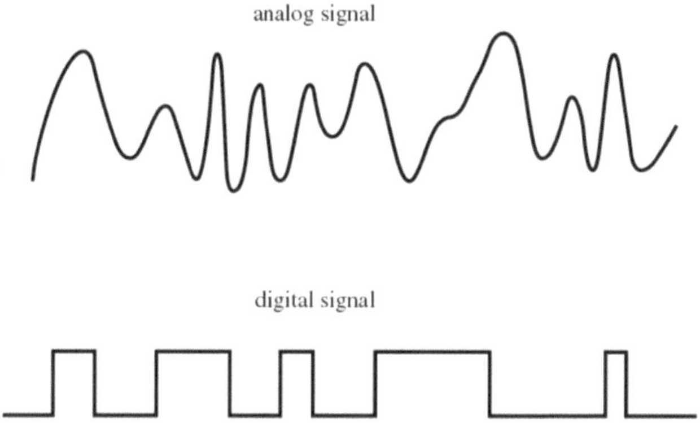

Figure 2.1. *Definition of signals in digital communication.*

Source: https://digestiblenotes.com/images/physics/signals.jpg.

The spectral description of a signal discloses the energy distribution of the signal between its various frequency components, enabling an understanding of its underlying patterns. Through methods like the Fourier transform, it is possible to break up complicated signals into their individual frequencies, rendering them understandable and making analysis easier. Such a decomposition is especially important in applications like audio engineering and telecommunications, where modifying or improving signals is common. Additionally, in seismic signal research, adaptive methods, as shown in some recent work, can significantly enhance non-stationary signal analysis using joint time-frequency representations, thus providing a more precise representation of a signals behavior under different conditions (Giaralis et al., 2007). The capacity to analyze and understand the spectrum of a signal can guide decisions that increase the predictability as well as dependability of information sent. New technologies have also served to highlight the importance of signals in spatial analysis, especially following the introduction of spherical harmonic transforms into signal processing. In various applications, ranging from image processing to cosmological research, the transformation of signals to spherical coordinates has provided deep insights into data compression and analysis. Specifically, a rapid algorithm for the directional correlation of band-limited signals has shown that rapid computation of spherical data is achievable, thereby significantly decreasing the processing time (Jacques et al., 2005). This novel methodology not only streamlines the management of large datasets but also reaffirms the signals' role as central to uncovering meaningful content from complicated systems. Finally, the study of signals, placed in the broader context of their spectra, lays out a rich field with far-reaching implications across disciplines with sound frameworks for both analysis and interpretation in the assessment of dynamic events (Figure 2.1).

Definition of Spectra

Origins of spectra are obtained from the examination of signals in various domains, namely both time and frequency. Spectra can thus be used to describe a representation of the different frequency components of a signal and provide insight into their structure and behavior. In signal processing, the spectral analysis finds periodicity in data, which is invaluable in applications ranging from telecommunication to audio engineering. Besides, spectra characterization can be varied depending on the nature of the signals being analyzed; for instance, non-stationary and stationary signals possess different spectral features that yield information of a different character. This is crucial during the development of analytical tools to process spectral information effectively and in detail as a foundation for efficient signal processing techniques. The idea of spectral analysis goes beyond static imagery, providing for an intricate description of signals from the perspective of temporal evolution. As an example, non-stationary signals, which evolve over time, need to be analyzed using advanced analytical methods to accurately convey their dynamics. Along these lines, recent techniques have proposed sophisticated methods like the Quaternion Fourier Transform, allowing for the representation of complicated signals in terms of quaternion-valued entities. The innovation allows for the presentation of a scalar quaternion-valued spectral density, a form that amplifies the interpretative potential of spectral information. Such presentations enable comprehensive analysis of frequency-dependent behavior of signals, including their unpolarized and polarized components designed to describe complex behaviors typical of real-world problems, thus transforming the field of spectral analysis (Bihan et al., 2017). In addition, the uses of spectral analysis can be illustrated through its application in designing artificial signals that meet specified design spectra, particularly in engineering and seismology. The evolutionary power spectrum (EPS) is a concept that dynamically describes the change in spectral content of a signal over time. For example, in the creation of compatible non-stationary earthquake accelerograms with reference to existing design spectrums, a least-squares optimization problem is used to match the computationally derived EPS with a target design spectrum. This particular use of spectral definition indicates the significance of spectra as more than abstract theory; it is of essential importance in risk assessment and engineering design, where predictions have to be confidently made. Lastly, the relationship between theory models and practice applications reveals the inherent significance of spectra in understanding and manipulating complex signal behaviors (Giaralis et al., 2009) (Figure 2.2).

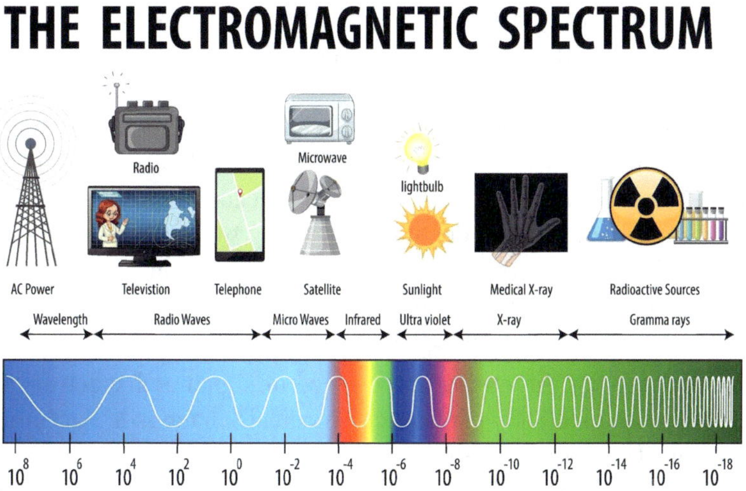

Figure 2.2. *Definition of spectra.*

Source: https://external-content.duckduckgo.com/iu/?u=https%3A%2F%2Fstatic.vecteezy.com%2Fsystem%2Fresources%2Fpreviews%2F001%2F868%2F617-%2Flarge_2x%2Fscience-electromagnetic-spectrum-diagram-free-vector.jpg&f=1&nofb=1&ipt=dca4cce0b1079b48e942d8c2b4abc05aa1e216426c5674e9c9180e74e430f852.

Importance of Studying Signals and Spectra

Signal and spectrum analysis plays a very important role in a broad variety of practical scientific areas and is a fundamental part of data analysis and interpretation. Signals simply mean the raw data gathered from various phenomena, whereas spectra provide a wide platform for understanding the behavior and properties of the signals. Such is the symbiosis between spectrum analysis and scientific investigation, particularly in astrophysics. Spectral analysis can deconstruct complex stellar behavior and elemental composition in celestial objects. The Advanced Compton Telescope (ACT) is an excellent example of the significance of spectral investigation in gamma-ray astronomy. With the assistance of high-resolution spectroscopy of supernovae nuclear emissions, scientists can decipher the processes that govern the production of elements in the universe, thus offering critical data to our knowledge of astrophysical phenomena.

Therefore, analyzing the signals and spectra not only enhances our universe knowledge but fundamentally drives science innovation. Outside of astrophysics, signal and spectrum analysis is critically important to medical diagnostics and imaging. Medical imaging by nuclear magnetic resonance is based mostly on signal and spectral principles, enhancing our ability to image anatomical structures and detect disease successfully. Hyperpolarization methods, particularly in nanodiamond methods, mirror how signals are overlapping with leading edge technologies. As explained, optical dynamic nuclear polarization methods have also significantly amplified the nuclear magnetic resonance signals, boosting the fidelity of imaging techniques. The finding of 13C nuclei hyperpolarization using nanodiamonds not only improves the quality of imaging but also opens new paths for research in biomedicine and materials science (Aguilar et al., 2020).

This underscores the utmost importance of signal and spectrum learning because they immediately influence medical diagnostic accuracy and advancement. The interdisciplinary applications of signals and spectra also bring about our collective knowledge among disciplines, ushering in future technologies. The application of spectral analysis in telecommunications and environmental science provides one good example of this fact. In environmental monitoring, for example, spectral analysis can detect pollutants and quantify ecological change and thus contribute directly to conservation and public health initiatives. In telecommunications too, advance relies on the signal and spectrum technology to optimize data transmission and improve communication devices. Through the study of the spectrum properties of signals, scientists and engineers can build improved systems and ultimately enable more connectivity in our increasingly digital lives. Thus, the importance of signal and spectrum study extends beyond individual fields, inducing developments that have wide-ranging implications for society at large.

2.1. TYPES OF SIGNALS

The signals can be broadly categorized as continuous and discrete types, with each being appropriate for conducting various functions in various areas of science and engineering. The continuous signals, as they are expressed in the form of functions occurring over a continuum of an interval, are typically found in analog systems where information transmission occurs unbroken. For instance, audio signals, which can reproduce the subtle nuances of sound waves, fall under this category as they denote continuous phenomena. Discrete signals, on the other hand, consist of individual values, typically derived from sampling continuous signals at discrete periods. Such quantized versions are employed in digital systems where processing and storage efficiency is paramount. These rudimentary classifications are essential in comprehension as they also define the analysis techniques used in signal processing, determining methodologies that penetrate telecommunications, environmental observation, and so on. The role of non-stationary signals, particularly in seismic activity, thus enhances our horizon of knowledge with respect to types of signals and their implications in real-life applications. Non-stationary signals possess time-varying properties, making them challenging to analyze and necessitating advanced modeling techniques. For example, artificial accelerograms' development that adhere to specified design spectra such as those used in the European aseismic code provisions (EC8) is important in structural assessment in earthquake engineering. By representing these accelerograms as simulations of a non-stationary random process, scientists can effectively describe their frequency domains through an evolutionary power spectrum (EPS). These models involve employing optimization techniques in order to match mismatches between target and generated spectra towards ultimately presenting improved seismic design methods. This underscores the need to integrate sophisticated analytical platforms in dealing with complex types of signals, especially ones in contact with dynamic systems. Acoustic signaling is a fascinating class of signals, particularly when one examines the diel patterns in many species. Research with melon-headed whales has shown that their acoustic activities are tuned to accompany various segments of the day, illuminating the role biological rhythms play in influencing signaling strategies.

There are adjustments in the spectral properties of their echolocation clicks—from frequency modulation to disparate levels of intensity—that serve as a response to environmental noise conditions varying between day and night. These adjustments serve critical roles in effective foraging and social interaction within these marine environments, illustrating the inherent plasticity of acoustic signals to environmental influence. The ability of organisms to control their signaling based on changing environmental situations proves the complex dynamic equilibrium among signal types and their corresponding contexts, showing how crucial context is in defining signals and their spectra in natural and engineered systems.

2.1.1. Continuous Versus Discrete Signals

In signal processing, the distinction between discrete and continuous signals is inherent. Continuous signals are defined over a continuum of time and take on an infinite set of values within a boundary. This aspect allows them to better represent real-world processes with nuances discrete signals may fail to capture. Analog sound signals, for instance, are continuous and allow the replication of sound waves with fidelity. In contrast, discrete signals are only specified at discrete points, leading to a representation that is perhaps easier to analyze and manage with the computer.

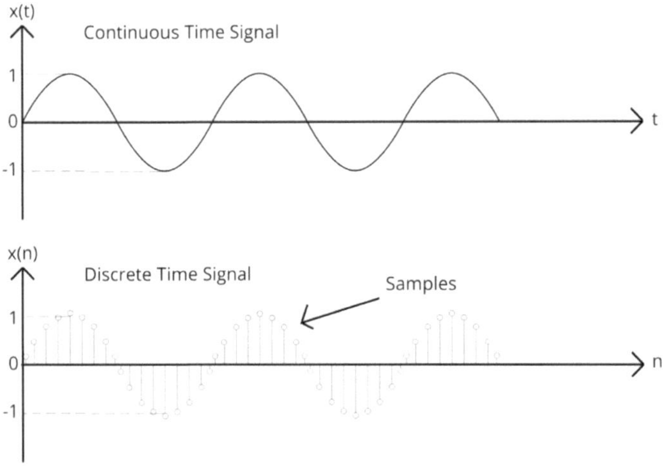

Figure 2.3. *Continuous versus discrete signals.*

Source: https://external-content.duckduckgo.com/iu/?u=https%3A%2F%2Felectronicsprojects.in%2Fwp-content%2Fuploads%2F2022%2F12%2FContinuous-time-signal-and-discrete-time-signal-Image-1.jpg&f=1&nofb=1&ipt=3f4b563db16d1a75a107cc0deaa2f1aa735f33531faf2ee186a03a3da1795dc0.

The sampling theorem points out that continuous signals can be accurately represented by discrete signals if adequately sampled at a rate higher than twice the highest frequency of the continuous signal, which highlights the relationship between these two types of signals. The implications of these differences are clearly seen in practical applications. Discrete signals are ubiquitous in digital systems, where information are processed in binary formats to enhance the efficiency and integrity of information transfer. For instance, when it comes to the detection of gravitational waves, researchers developed methods to look for periodic signals in data from resonant detectors. Employing a year's worth of data, the research aimed to find signs of gravitational waves, exemplifying the practical usefulness of discrete representation in signal-saturated conditions (Moleti et al., 2000). Discrete signals make it possible to use advanced algorithms that are capable of eliminating noise and improving signal identification, further illustrating the significance of discrete representation in modern technology. Conversely, continuous signals find their essential application in areas where high-resolution representation is imperative. For optical communications, for example, advancements in silicon nanophotonic integrated circuits have been remarkable in processing continuous mid-infrared signals. Such platforms enable varied applications including environmental monitoring and medical diagnostics in which accurate and instantaneous representation of the signal is critical (Spott et al., 2012). The efforts at converting optical signals from the continuous form to a more convenient detection form demonstrate a pivotal signal format change. In attempts to meet the deficit of dealing with continuous signal handling—like the cooling requirements of some photodetectors—researchers keep finding ways that keep pace with technological demands, thus ensuring the sustained relevance of both the continuous and discrete signal paradigms in contemporary applications (Figure 2.3).

2.1.2. Deterministic Versus Random Signals

The comparison between deterministic and random signals is foundational in understanding the characteristics of various signal types and their spectral properties. Deterministic signals are characterized by their predictable nature, which allows for precise mathematical modeling. For instance, sinusoidal signals are typical examples of deterministic signals due to their predictable frequency, amplitude, and phase. These signals can be described completely using mathematical equations and are essential in applications requiring stability and reliability, such as in control systems and communications. On the other hand, random signals, which exhibit inherent unpredictability, challenge traditional analytical methods. Analyzing random signals often necessitates statistical approaches that assess the average behavior over time, making them crucial in fields like telecommunications where noise and unpredictability are prevalent. Moreover, understanding the implications of deterministic versus random signals necessitates a closer look at their spectral analysis. The spectrum of a deterministic signal typically reveals sharp peaks at specific frequencies, reflecting the signals harmonic content and periodicity. In contrast, random signals exhibit broader spectral characteristics, often represented by continuous spectra or power spectral density functions. The Fisher-Shannon representation plane highlights the complex relationship between different types of signals and their information properties, further elucidating how random processes can exhibit fractal behavior and multifractality as implied by the Bandt and Pompe methodology. These insights into spectral characteristics illuminate significant differences in how deterministic and random signals are processed and interpreted in various applications, thereby impacting fields such as signal processing and diagnostics. In physiological contexts, the distinction between deterministic and random signals takes on even greater significance, particularly when analyzing biological rhythms such as the human heartbeat. A healthy human heartbeat may follow patterns that display multifractal characteristics, indicating a complex interplay of order and randomness. This complexity is crucial for maintaining physiological stability, as evidenced by the loss of multifractality in conditions such as congestive heart failure (Barabasi et al., 1999). The implications of these findings extend beyond mere theoretical interest; they suggest that a thorough understanding of the deterministic components of physiological signals, as well as the underlying random behaviors, can lead to better diagnostic tools and therapeutic strategies. Thus, exploring the balance between deterministic and random signals not only enhances our understanding of signals and spectra but also offers practical applications in health and technology (Olivares et al., 2015) (Figure 2.4).

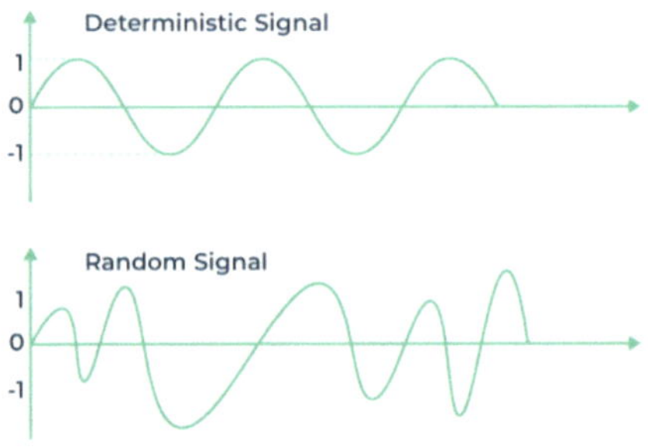

Figure 2.4. *Deterministic versus random signals.*

Source: https://external-content.duckduckgo.com/iu/?u=https%3A%2F%2Fmedia.geeksforgeeks.org%2Fwp-content%2Fuploads%2F20240105124609%2FRandom.png&f=1&nofb=1&ipt=9f86180a3f8a349c7f5d520b1170325e2c106ef269ae1338f302dc3eb753a318.

2.1.3. Analog Versus Digital Signals

In signal processing, the idea that analog and digital signals differ is fundamental. Analog signals are smooth waveforms that represent physical values in terms of varying voltage levels, thus they are more vulnerable to noise and interference. This variation is best witnessed in natural phenomena, for example, sound and light, which can be detected through instruments like microphones and cameras. These signals are linearly related to the source that they represent. Digital signals, however, are quantized information, mostly in the binary format for information transmission. This quantization ensures cleaner transmission and storage without degradation over long distances. Since technology is moving towards more advanced systems requiring accuracy and dependability, the application of digital signals is a popular choice. Digital systems are generally able to utilize advanced error-correction techniques to ensure maximum signal integrity and are hence critical in modern applications. Applications of analog and digital signals can be applied in many areas, especially in instrumentation and monitoring systems. For example, in material characterization, techniques such as ultrasonic testing employ analog signals to quantify parameters such as velocity and attenuation (Vary et al., 2025). This method is particularly useful in material integrity assessment since the continuous signals' analog nature allows for precise measurement. On the other hand, digital monitoring systems have emerged for more specialized applications, like capturing strain gauge signals from aerodynamic tests. These are capable of both time and frequency domain to provide complete insight into the working conditions' structural integrity (Smalley et al., 2025). This dualism of analysis describes the way that the two types of signals are used for different purposes, where analog signals are applied to fine detail in real-time measurements and digital signals introduce robust analysis structures to data post-processing. Additionally, the interaction between analog and digital signals produces opportunities for new technologies and ways of approaching things. Hybrid systems combining both types of signals, for instance, can exploit each. These systems are capable of converting real-time analog signals into digital form, providing a bridge that supports the best attributes of both continuous and discrete data handling. This can greatly improve applications in telecommunications, where both bandwidth and fidelity are paramount. In addition, improvements in signal processing algorithms add another level of complexity, enabling enhanced interpretation and manipulation of both analog and digital signals. The development of such technologies' history helps emphasize the ongoing relevance of learning about analog and digital signal differences, as they are the foundation on which today's communication and acquisition systems are built. Hence, the subject matter of such signal types remains central to the broader signal and spectrum debate.

2.2. SIGNAL REPRESENTATION

Representation of signals is a fundamental aspect of data interpretation and analysis in all areas of science. It has become a central topic in modern scientific and engineering use because ever stronger representations are required to extract useful information from complicated signals. Electrical current characterization in variable speed drives provides a good representation of the need for sophisticated analysis techniques. Through motor current signal analysis, researchers have developed ways to quantify the operation of electrical machines in an effective way, illustrating how load variation and mechanical defects influence the properties of signal features. Such findings point to understanding representations of signals as a way to facilitate fault detection at higher levels of reliability and improved accuracy, as shown in the research on conventional spectrum analysis combined with advanced techniques like amplitude modulation bispectrum representation (Ball et al., 2011). Apart from that, with more advanced technology used, the signals become increasingly sophisticated and call for more advanced analysis methods. Traditional spectral analysis is valuable but perhaps not enough to detect subtle movement in the signals, especially in varying environments. New studies have introduced new methods such as adaptive chirplet decomposition and the Wigner-Ville transform that offer a clearer understanding of non-stationary signals. Such techniques enable scientists to capture temporal evolution of signal frequencies and hence assist in the identification of structural responses during seismic events.

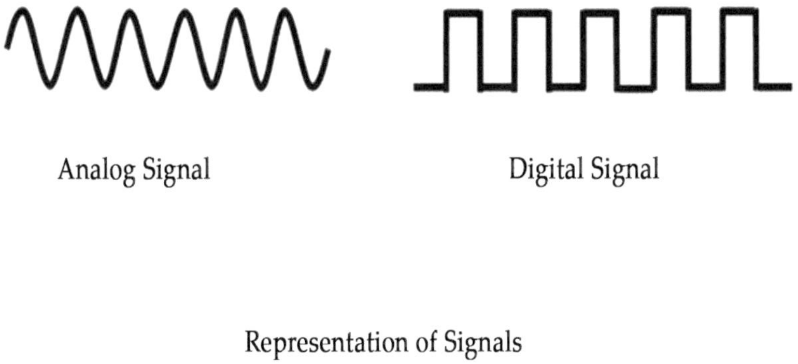

Figure 2.5. *Signal representation.*

Source: https://redaksi.pens.ac.id/wp-content/uploads/2020/03/1-4.jpg.

By applying these techniques to actual seismic accelerograms, scientists have been in a position to track variations in frequency content, which offers crucial information regarding the structural well-being of buildings under stress (Giaralis et al., 2007). This improves the dialogue in signal representation, illustrating the intersection of complex signal processing techniques with everyday engineering challenges. Development in signal representation techniques not only improves fault detection but also our understanding of the phenomena that dictate signal behavior. This type of analytical improvement is crucial in numerous fields such as mechanical engineering and earthquake engineering. The limitations of classical approaches lead to the examination of adaptive approaches able to cope with the complexity of non-linear signal dynamics in real-world applications. As a case in point, detection of structural response to earthquakes is a compelling illustration of how advanced signal representation can affect architectural design reliability and robustness. As the methods are more and more refined by researchers, they enhance our capability to predict and mitigate the occurrence of structural collapse, identifying the importance of sophisticated signal representation in furthering our understanding of complex systems and preserving the life of built conditions across various settings (Figure 2.5).

2.2.1. Time-Domain Representation

Time-domain signal description is a pillar of signal processing, particularly in the field of studying non-stationary signals. Through time-domain description, monitoring the behavior of a signal with respect to time is possible, allowing crucial information on such occurrences as sudden change and transient events. By undergoing the conversion of signals to the time domain, scientists can have a better view of how the amplitude of signals varies with time and understand its behavior in detail. This is particularly applicable in seismic engineering, where the time-dependent nature of ground motion is an important consideration for structural responses. In addition, of especial importance has been the introduction of adaptive techniques, enhancing extraction of valuable information from time-domain representations, uncovering the nuances of frequency content with the evolution of seismic events and emphasizing the relevance of time-domain analysis to practical applications (Giaralis et al., 2007). While exploring higher-level techniques for representation of time-evolving attributes in signals, the merit of time-domain representation finds further support. Techniques such as harmonic wavelet transforms have emerged as strong techniques that are capable of maintaining the innate non-stationarity of measured signals. This capability of molding recorded accelerograms into conformity with design requirements, seen in the case of aseismic design practice, is founded on this capability. By focusing on the instantaneous energy and mean instantaneous frequency derived from these time-domain expressions, researchers can ensure that the transformed signals are realistic and useful for structural inspections. This approach not only helps in achieving compliance with regulation such as that detailed in the European code provisions (EC8) but also provides a further insight into the signals that operate on structural response during seismic activity (Giaralis et al., 2010). Time-domain representation carries over into practical utility beyond theoretical analysis, particularly in observation and mitigation of the effects of seismic activity on buildings. Using joint time-frequency representations that are derived from time-domain analysis, the engineers can precisely track the change in the response of the structure over seismic events. This is useful because it enables the designers to identify non-linear behavior associated with damage or shifting natural frequencies during an earthquake. Consequently, the intersection of time-domain data with structural integrity assessments is centrally involved in seismic force design development of structures, and through observation and tracking by careful scrutiny, the engineered response can be calibrated toward present actual signal behaviors, demonstrating the dominant role of time-domain representations in ongoing interaction between application and theory in seismic engineering (Giaralis et al., 2007) (Figure 2.6).

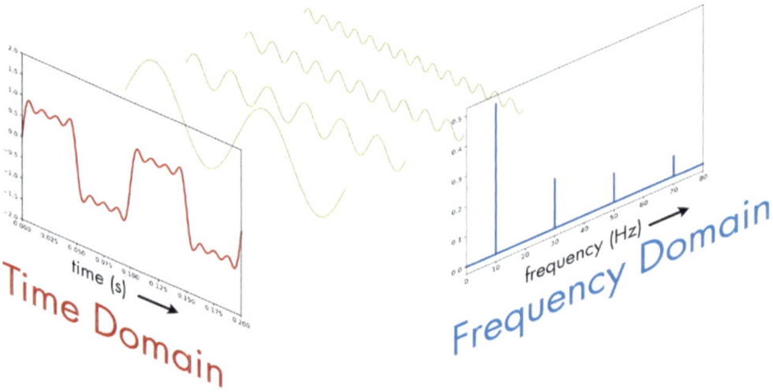

Figure 2.6. *Time-domain and frequency-domain representation.*

Source: https://encrypted-tbn0.gstatic.com/images?q=tbn:ANd9GcQfvfzZ_6snEb9wIV3TDiCMkl6n2Ie5jl2sWg&s.

2.2.2. Frequency-Domain Representation

Frequency-domain representation is a key concept of spectrum and signal research, offering a different perspective from conventional time-domain analysis. The process essentially involves the separation of signals into their respective frequencies, which may provide information about the intrinsic characteristics that would not be apparent through simple time-domain observation. The Fourier Transform is the basis in this system, and it provides the time-domain signal for conversion to its frequency components. This conversion allows the identification of dominant frequencies, spectral details, and possible drift in the signal, giving us greater understanding of complex systems. Frequency-domain analysis is particularly valuable in applications such as telephony and audio processing, where manipulation of specific ranges of frequencies is essential to achieving best performance and minimizing noise. In addition, frequency-domain techniques are utilized in structural health monitoring, especially seismic engineering. Employing the frequency content of signals to analyze them, engineers can ascertain the dynamic behavior of structures subjected to various vibrational loads. For example, time-frequency analysis methods such as the Wigner-Ville transform provide essential information about how the frequencies of a system vary during earthquake occurrences. The capability is necessary for the capture of instantaneous frequency changes that would indicate damage to the structure or non-linearity, ultimately enhancing the robustness and safety of infrastructure. Techniques discussed in the literature indicate the benefits of these analyses, the effectiveness of which in the detection of natural frequency changes relevant to structural response under conditions of loading contributes to enhancing predictive maintenance methods applied across engineering fields (Giaralis et al., 2007). Its adaptability is also exemplified in the case of aseismic design, whereby transformed accelerograms are subjected to testing for building strength. The use of harmonic wavelet-based methods allows engineers to create design spectrum-compatible time histories without compromising the non-stationary characteristics of record measurements. This is particularly significant in seismic design methods, where insight into the time-dependent energy content and frequency composition of ground motions is important. The comparison of the mean instantaneous frequency and instantaneous energy of changed records with original accelerograms enables specialists to test designs under true seismic conditions. Such utilization of frequency-domain analysis as part of the practice in engineering, therefore, not only improves the design process but also the seismic resistance of structures, highlighting its pivotal role in current structural engineering (Giaralis et al., 2010).

2.2.3. Complex Exponential Representation

In signal processing and signal analysis, complex exponential form is an essential tool for signal description and their corresponding spectra. It employs complex numbers to express sinusoidal functions to simplify otherwise difficult computations. By employing Eulers formula that links trigonometric functions to complex exponentials, it becomes easy for analysts to dissect signals into individual frequencies. This decomposition is central to the study of Fourier transforms, whereby processing of signals in the time and frequency domains is facilitated, and signal behavior is more comprehensively understood. The elegance of representation in terms of complex exponentials not only simplifies mathematical equations but also provides the basis for the study of the dynamics of signals in terms of changes with time, and is therefore an essential tool in modern signal processing. In addition, complex exponential form lies at the heart of the study of numerous physical phenomena in terms of signal properties and spectral characteristics. The study of coherence and decoherence in quantum systems, for instance, illustrates how the use of complex representations enables insight into temporal dynamics. Experimental data supporting irreversible quantum decoherence in nematic liquid crystals, as seen in recent work, highlights the relevance of eigen-selection effects in this model (Abragam et al., 2013). In using complex exponentials to represent perturbed states in the quantum system, researchers can examine the quantum interference and thermalization interaction. This approach, aside from illuminating experimental non-idealities, quantifies their effect on signal decay, rendering theoretical expectations concerning signal behaviors and dynamics justified. The relevance of complex exponential representation does not only lie in theoretical explanations but also in actual applications within other branches of science. The study of persistence phenomena in turbulent flows is an example of how fluctuating fields are characterized by complex signals in various aspects. The findings on probability of persistence reveal unexpected dynamics in velocity and temperature fluctuations within convective boundary layers to show how complex representations convey information on their statistical characteristics. As shown from the recent research, the PDFs of persistence have varying power-law behavior at particular scales that correspond to inherent self-similar mechanisms (Banerjee et al., 2020). These observations point out the general applicability of complex exponential form to the explanation of events in a wide range of fields from quantum mechanics to fluid dynamics, and reiterating its significance in signal and spectrum analysis.

2.3. SPECTRAL ANALYSIS

The field of spectral analysis is highly significant in signal interpretation and signal processing across most fields of science. By analyzing the frequency makeup of a signal, scientists gain knowledge of the intrinsic attributes and patterns of the physical processes that generate signals. Spectral analysis, in general, allows for the detection of dominant frequencies that might signal patterns or anomalies, crucial in fields such as engineering and communications. Fourier transform methods, for instance, decompose signals into their frequency components, providing a visual representation of how signal energy is partitioned across different frequencies. This adaptability in application reflects how spectral analysis helps in better comprehension of time-domain as well as frequency-domain features in signals. During the recent past, advancements in graph signal processing (GSP) have revolutionized signal analysis in non-standard data structures, such as those in sensor networks or social networks. This new technique enables image representation as graph signals in which pixels are connected by edges that reflect their structural relationships. Utilizing image processing with GSP techniques, tasks such as compression, restoration, and filtering can be executed more efficiently and accurately. For example, spectral-domain signal manipulation allows sophisticated processing methods that exploit the unique nature of image patches. The exploration and integration of GSP methods show the continued advancement of spectral analysis in its expanding relevance and growing usefulness in today's data-centric worlds (Cheung et al., 2018). In addition, spectral analysis finds application in engineering disciplines, particularly in failure detection and classification of mechanical breakdowns. Techniques developed to identify incipient failures in turbomachinery represent real-world applications of spectral analysis in real-life applications. Dynamic measurement characteristics can be effectively described by engineers using the time domain and spectral methods, thus improving fault-detecting capability. Statistical classification techniques, such as the use of moments, allow for discrimination between various failure modes in terms of signal characteristics. Introducing new coherence functions, such as hyper-coherence, adds additional diagnostic potential, echoing the experimental approaches currently under research in the field (Coffin et al., 2025). The convergence of these theoretical and applied advances speaks to the ongoing relevance of spectral analysis in achieving optimal performance and ensuring safety within engineering systems (Figure 2.7).

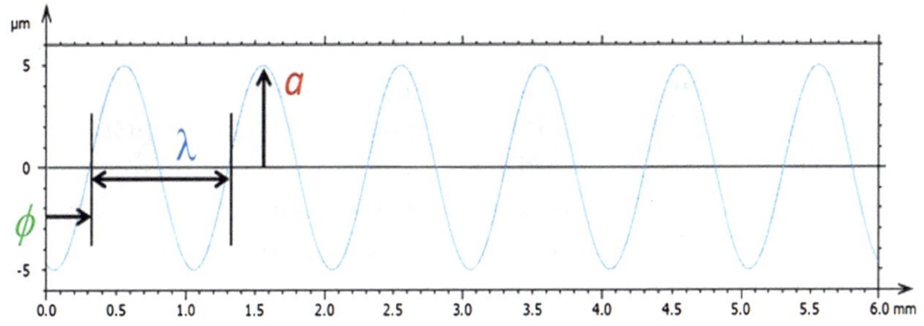

Figure 2.7. *Spectral analysis.*

Source: https://www.digitalsurf.com//uploads/2023/11/parameters-that-define-a-sine-wave.jpg.

2.3.1. Fourier Transform Fundamentals

The Fourier Transform truly revolutionizes the analysis of signals by transforming time-domain information into frequency-domain expressions. This mathematical tool enables the decomposition of complex signals into their frequencies, providing data on harmonic as well as non-harmonic terms. By applying the Fourier Transform, scientists can perceive patterns and periodicities in signals that are not readily apparent in the time domain. Such a decomposition is particularly useful in implementations like telecommunications and audio processing, where frequency content is important to optimize signal transmit and sound quality. The inherent linearity of the Fourier Transform facilitates broad applications, but one must be aware of constraints, particularly when the signals are non-stationary, with time-varying frequency content. As the signals become increasingly more complex, especially in areas such as earthquake engineering and structural dynamics, the traditional Fourier Transform falters. As a result of the need to analyze non-stationary signals, which change in frequency, techniques such as wavelet transforms and empirical mode decomposition were introduced. These allow time-frequency analysis to be brought together, maintaining better changing natures of a signal than the Fourier Transform. For instance, Giaralis et al. (2007) observes that advanced techniques like the harmonic wavelet transform provide seismically more sensitive diagnoses of structural responses so that engineers can assess structural integrity under dynamic conditions. These developments reflect the ongoing evolution of signal processing technologies, evidencing the necessity to update analytical methods in order to keep pace with the complexity of modern applications. The relevance of Fourier Transform principles is not merely theoretical knowledge, but they are applied practically, particularly in structural health monitoring. The contribution of such techniques based on the Fourier Transform, like the principle of mean instantaneous frequency, is of utmost significance in distinguishing normal behavior of the structure from distress signals. In seismic event analysis, the ability to record changes in effective natural frequencies is an important factor in predicting potential structural failure. According to Burrus et al. (2007), employing instruments such as the harmonic wavelet transform under such circumstances enables better functionality in interpreting responses in ground motion and structures. This use of Fourier principles to real-world problem-solving not only reinforces the importance of frequency analysis but also shows the need for advanced methods within the greater universe of signals and spectra.

2.3.2. Applications of Spectral Analysis

Spectral analysis is now a major technology in most scientific and engineering disciplines because it has the ability to decompose high-level signals into their constituent frequencies. Decomposing intricate signals into their constituent frequencies simplifies studying underlying processes by revealing underlying patterns and interdependence in the data. For instance, spectral analysis is utilized in environmental science for monitoring amounts of pollutants in water and air samples, enabling scientists to quickly analyze and quantify harmful chemicals.

Advanced techniques such as Independent Component Analysis (ICA) have been extremely promising in accurately reconstructing spectra from mixtures of complex ecotoxicants, like aromatic and polyaromatic hydrocarbons, as demonstrated by recent studies (Kraskov et al., 2010).

These applications, apart from providing valuable information regarding environmental health, are also crucial in building regulatory policies to combat pollution. Apart from environmental applications, spectral analysis is also of vital significance in audio processing and source separation, which are of vital concern in fields such as telecommunications and music technology. Machine listeners need to distinguish independent sound sources from complicated mixtures, an activity where conventional processing techniques are unlikely to be effective. Utilizing probabilistic modeling paradigms, researchers have developed sophisticated separation systems emulating the human auditory capacity based on independent component analysis and spectral template-based models (Dempster et al., 2010). These methods allow individual sounds to be identified more accurately in a complex audio environment, improving sound quality and clarity in many applications.

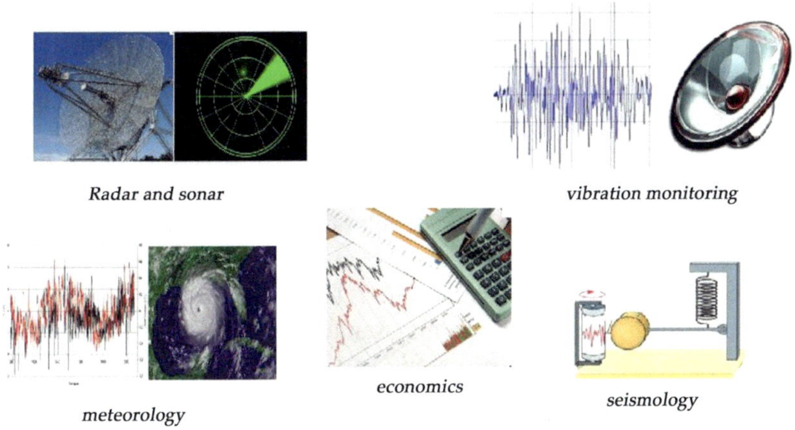

Figure 2.8. *Applications of spectral analysis.*

Source: https://hmco.enpc.fr/UNIT_ComplexSystems/~gen/5_SpectralAnalysis/5_SpectralAnalysis_web.publi/web/res/some_application.jpg.

Consistent improvement in this field underscores the revolutionary impact of spectral analysis, which continues to improve to respond to the complexity of modern hearing disorders. Spectral analysis also finds its uses in the biomedical field, specifically in medical imaging and diagnosis. Techniques such as Magnetic Resonance Imaging (MRI) rely heavily on spectral analysis to decode the spatial distribution of different tissues based on their unique spectral signatures. Spectral analysis of data allows physicians to identify abnormalities and assess the health status of internal organs non-destructively. Advances in spectral analysis methods have enhanced the precision of such systems, which has improved diagnostic outcomes as well as patient care. As argued in contemporary discussions regarding the union of several techniques of analysis, joint use of spectral analysis and traditional imaging will unveil hidden secrets previously unknown, thus further establishing the foundations for diagnostic advancements and medical technology (Kraskov et al., 2010). This shows the extensive application areas of spectral analysis in the improvement of not only the environment and sound technologies but also crucial aspects of human life and safety (Figure 2.8).

2.3.3. Limitations of Spectral Analysis

The study of spectral analysis always puts forward diverse constraints that must be acknowledged so that an accurate presentation of its application can be ensured. One obvious challenge is spectrally mixed data, best described with remote sensing applications when dissimilar materials can coexist within the same pixel. The overlap allows the spectral interpretation to be flawed, with single spectral signatures unattainable accurately. Traditional methods like principal component analysis (PCA) and independent component analysis (ICA) have been applied to manage these complexities. However, only they are found to fall short since they rely on only linear assumptions that could lead to data misinterpretation. New developments, such as spectral unmixing (SU), have been established to complement these analyses by demixing the mixed spectra into its constituent parts. Despite these developments, the fundamental nature of spectral mixtures remains a significant constraint in the context of spectral analysis (Brun et al., 2012). Another significant constraint concerning spectral analysis is that it depends heavily on possession of properly calibrated instruments and quality of data, which has been collected. Calibration and noise errors have the potential to significantly contort spectral readings and consequently result in misleading inferences about the material properties being researched. Further, variability sources like environmental factors and instrumental drifts may introduce non-consistent variability from measurements, thus complicating data interpretation. These issues call for rigorous pre-processing and post-processing steps to maintain the data in check. While techniques like regularization frameworks are capable of mitigating some of the sources of noise effects, they further contribute to complexities of model choice and hyperparameter tuning (Giovannelli et al., 2009). This close interplay between precision and variability underscores the ongoing challenges faced in the area of spectral analysis. Aside from these technical difficulties, the intelligibility of the findings of spectral analysis is weighed down by the commonly very high dimensionality of the data. Spectral data sets tend to have a large number of variables, and this will lead to issues such as overfitting and redundant information obscuring instead of accentuating the findings. As dimensions increase, standard statistical approaches may become less interpretable and generalizable, decreasing their utility in real-world applications. Unsupervised techniques can mitigate such problems by reducing dimensionality while preserving much information. Complexity of hyperparameter selection, however, still remains a concern because incorrect choices will lead to poor analytical outcomes. Hence, while spectral analysis is a powerful instrument in numerous industries, the limitations of the method require ongoing research and development to enhance its lifespan and viability (Brun et al., 2012; Giovannelli et al., 2009).

2.4. SIGNAL PROCESSING METHODS

The area of signal processing techniques encompasses intense interaction between mathematical concepts and practical applications in a broad range of fields, most notably the field of nondestructive evaluation (NDE). In doing so, the ability to retrieve meaningful information from complex waveforms and images comes into play. The inclusion of advanced resolution-enhancement algorithms and defect detection is a must because it makes proper interpretation of the data created by techniques such as ultrasonic testing and eddy current analysis possible. A good software platform, the NDE Wave & Image Processor Version 2.0 being a case in point, represents this integration through a common interface upon which different signal- and image-processing techniques are available to be used. This approach not only enhances individual competence but also enables joint work by scientists and engineers in the field of NDE to be directed toward better analysis of inbuilt complexities in spectra and signals (Roth et al., 2025). Also, signal processing approaches can be employed for remote sensing in maritime environments. One outstanding achievement in this area is illustrated by the two-frequency microwave radar technique aimed at the remote sensing of ocean surface currents and wave spectra. This innovative technique relies on the electromagnetic scattering resonance principle and can effectively utilize energy from gravity waves. Multiplexing signals with various frequencies makes it possible for researchers to capture the complete long-wave ocean spectrum, thereby gaining valuable information on marine conditions. Development of specialized scatterometers incorporating coherent radar capability further enhances precision and accuracy in measurements, allowing for real-time assessment of surface currents as well as wave patterns.

The technique not only adds strength to the horizon of signal processing but also suggests its contribution to environmental monitoring (Eng et al., 1984). In addition, advancements in signal processing have initiated the development of advanced algorithms facilitating data collection and analysis in numerous disciplines. Utilization of wavelet-based techniques, for instance, has revolutionized de-noising and filtering operations crucial in signal clarity improvement in one-dimensional and two-dimensional settings. These methods in NDE are utilized to achieve maximum quality of signal while at the same time facilitating recovery of critical information on objects under test concerning material integrity. These enhancements are particularly necessary in applications like spectroscopy, where the need for precise and accurate data is highest. The integration of powerful signal processing techniques with data analysis methods demonstrates the need to develop comprehensive systems that can handle various modalities of data optimally. As technology keeps progressing, the sophistication of such techniques is bound to tackle increasingly complex problems, thereby enhancing our understanding of spectra and signals (Roth et al., 2025) (Figure 2.9).

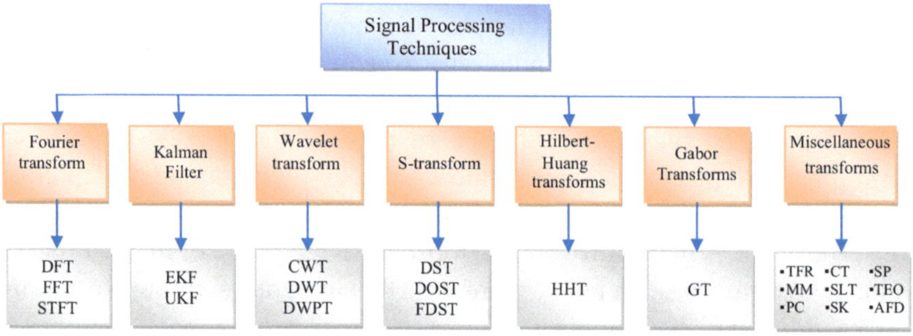

Figure 2.9. *Signal processing techniques.*

Source: https://ars.els-cdn.com/content/image/1-s2.0-S1364032115007157-gr3.jpg.

2.4.1. Filtering Techniques

The effectiveness of filtering techniques in signal processing is fundamentally rooted in the ability to enhance data quality through the removal of noise and unwanted components. Using a range of algorithms and strategies, such techniques facilitate extraction of valuable information from complex waveforms. In areas such as nondestructive evaluation (NDE), advanced software systems have been designed to integrate multiple filtering capabilities, thereby streamlining the process of analysis. For instance, the NDE Wave &

Image Processor Version 2.0 incorporates a broad spectrum of signal- and image-processing capabilities, including advanced wavelet-based filtering and de-noising. The software demonstrates how novel filtering techniques are able to improve signal resolution and interpretability of spectroscopic and imaging information towards more accurate judgments in numerous fields of science (Roth et al., 2025). Of the various methods of filtering, frequency domain filtering is of particular significance, especially in overlapping signals. Although in cases where time-overlapping linear frequency modulated (LFM) signals are found, ordinary methodologies such as time windowing cannot suffice because they cannot resolve the overlapping temporal features. Therefore, application of the fractional Fourier transform (FrFT) is a viable alternative, where signals can be efficiently demixed in time and frequency domains. By acting at the point of intersection between these two domains, the FrFT provides an elevated technique of demixing individual LFM signals, thus providing improved signal-to-noise ratio (SNR) under investigation. The technique has been effective in actual experiments, particularly in ultrasonic technologies that have complex signal environments where accurate resolution is critical (Cowell et al., 2010). The development and use of such filter techniques indicate the advancing trend of signal processing in contemporary research. As increasing demands for higher fidelity in data analysis take place, increasing demands are also placed on powerful and adaptable filtering techniques. The ongoing improvement of algorithms and integrating them into user-friendly software platforms ensures that scientists and engineers can effectively manage vast quantities of information while creating indispensable insights. Furthermore, ongoing studies of hybrid approaches that combine conventional filtering methods with emerging concepts such as machine learning will continue testing the boundaries of what is possible in signal processing. Lastly, the emergence of filtering techniques will come to play an important role in advancing our understanding of critical phenomena across various fields of science, confirming their role within the context of signals and spectra.

2.4.2. Modulation Techniques

Modulation techniques are critical in signal processing to enable effective information transfer through various channels of communication. Modulation entails varying a carrier signal's amplitude, frequency, or phase to represent data in a manner suitable for the transmission medium. This process results in the best bandwidth utilization and enhancement of the signal integrity. In radio communications, for instance, techniques like Frequency Modulation (FM) and Amplitude Modulation (AM) are utilized for various applications, striking a balance between signal integrity and distance of transmission.

Additionally, advances in digital modulation techniques, such as Quadrature Amplitude Modulation (QAM) and Phase Shift Keying (PSK), have enabled the realization of increased speed and more reliable data transmission in modern networks. These modulation techniques are fundamental to both terrestrial and satellite communications, delivering high fidelity under various environments. Application of modulation techniques can greatly affect the performance of observation experiments, particularly in the cosmological context of probing intrinsic phenomena.

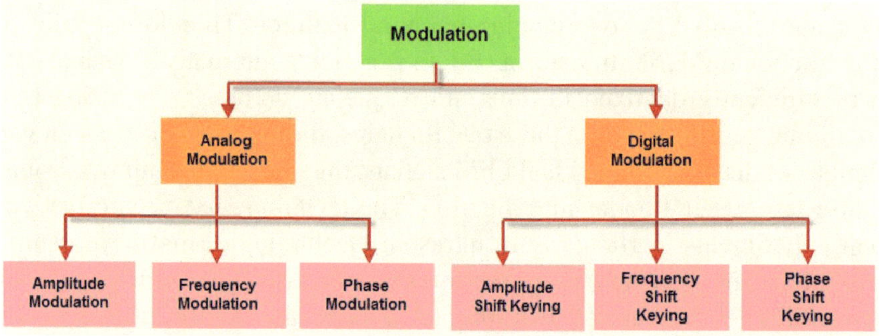

Figure 2.10. *Modulation techniques.*

Source: https://www.watelectronics.com/wp-content/uploads/9-25-2014-12-57-47-PM.jpg.

For example, modulation strategies are significant in maximizing the ground-based experiment science return of experiments such as the Clover experiment to measure the B-mode polarization of the Cosmic Microwave Background (CMB). Simulations show that fast modulation can reduce low-frequency detector noise very effectively, while providing redundancy on the instrument's polarization sensitivity projections on the observed sky. But it is important to note that while modulation can counteract certain kinds of noise, it might not address well issues associated with atmospheric variations or systematic errors that are experienced while acquiring data. This realization underscores the importance of maximizing modulation methods to improve the quality of signals in astrophysical measurements (Amblard et al., 2009). Moreover, the invention of complicated modulation techniques extends beyond traditional applications to new domains of interest, such as artificial neural networks to analyze speech intelligibility. The Speech Transmission Index (STI) is a significant parameter utilized for evaluating sound transmission channels, and traditional measurement approaches can be imprecise. Novel approaches utilize neural networks to estimate STI from running speech, effectively employing modulation concepts to obtain relevant parameters from transmitted signals. By training on diversified speech samples, these networks are found to be capable of accomplishing challenging non-linear mappings and feature extraction, resulting in a method that challenges classical measurements. This shows the manner in which modulation techniques are yet to be developed and mixed with emerging technologies, broadening the boundaries of signal analysis and providing enhancement over several communication systems (Cox et al., 2003) (Figure 2.10).

2.4.3. Sampling Theorem

The Sampling Theorem, which is the foundation of signal processing, dictates the requirements by which a continuous signal can be precisely characterized by its samples. The theorem is of fundamental importance since it makes possible the connection of the representation of signals in the continuous and discrete realms, thus facilitating efficient transmission and processing of data. According to the theorem, if a signal is band-limited, it can be reconstructed exactly from the samples if the sampling rate is two times the highest frequency present in the signal, also known as the Nyquist rate. This idea pervades numerous technologies, including digital audio and digital video, but finds its implications in more broad contexts including communication theory and data compression methods, suggesting the intersection of signal theory with engineering practice. Mathematical foundations of the Sampling Theorem have been explored with extensive depth proving its use in sophisticated signal analysis. Applications of the theorem encompass band-limited signal generation and analysis in different domains, for example, spherical harmonics in multidimensional space. Effective sampling methods are obtained in fields like cosmic microwave background (CMB) analysis by understanding the interactions between band-limited signals and spatial filters, leading to substantial advances in computational algorithms. Techniques developed for directional correlation of scalar band-limited signals on a sphere, for instance, capitalize on the sampling theorem to enable maximum signal processing. These developments demonstrate the value of the theorem in maximizing data extraction and representation from complex datasets, thereby enhancing the fidelity of signal interpretation within various scientific applications (Jacques et al., 2005). Moreover, the implications of the Sampling Theorem extend beyond immediate applications and call for further research in signal representation and transmission techniques. The indication that an ultraviolet cutoff, as described in existing theory, suggests a peak information density and new avenues of thought in communications theory. These guidelines have been replicated by researchers using fresh techniques aligned with the mathematical theory of the Sampling Theorem, thereby linking abstract theory to concrete practical applications. As research continues to be performed into the connection between high-frequency signals and their inherent boundaries, the information accrued does not only augment prevailing methodologies but also promotes fresh computing practices. This ongoing discussion captures the continuous interaction between practice and theory, pointing to the foundational role played by the Sampling Theorem in the evolution of signal processing techniques (Kempf et al., 2004) (Figure 2.11).

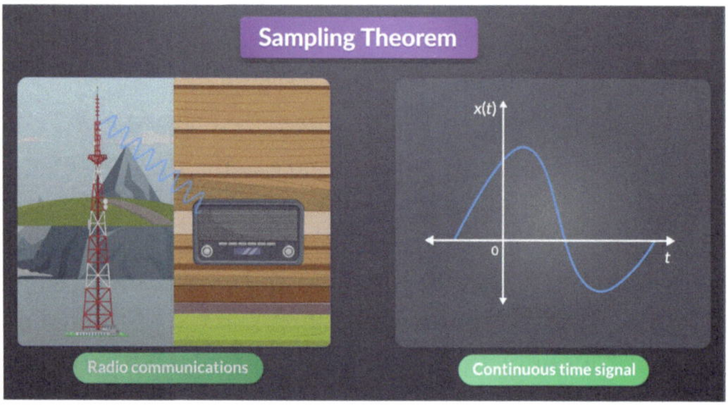

Figure 2.11. *Sampling theorem.*

Source: https://www.jove.com/files/media/science-education/science-education-thumbs/16076.jpg.

2.5. APPLICATIONS OF SIGNALS AND SPECTRA

Signal and spectrum research has triggered revolutionary innovation across numerous fields of science, most prominently metrology and telecommunications. The invention of optical frequency combs and, more precisely, those generated through high-Q micro-resonators has played a key role in promoting the accuracy and efficiency of measurement. Micro-combs offer researchers the chance to employ broad optical bandwidths with small size and energy requirements, characteristic features ideally suited to present-day technological demands. Worthy of attention are their application in RF photonics, which is better than the conventional approach as they provide enhanced functionalities. This technology for signals revolution not only optimizes existing systems but also predicts novel applications in quantum optics and spectroscopy, which marks the beginning of more accurate and multi-functionalized equipment in scientific research and communication networks (Chu et al., 2018). Use of signals in agricultural technology is an excellent example of how spectral analysis has been employed to enhance food safety and quality. Techniques such as pulsed electric field (PEF) processing have enhanced extraction and transport of soluble sugar and water in plant tissues significantly. Scientists are able to observe distinct differences in water mobility in treated plants through the use of time domain NMR relaxometry. For instance, the PEF treatment alters the transverse relaxation peaks and indicates higher permeability of carrot tissue. Such an alteration results in greater extractability of sugar and net retention of nutrients and therefore poses remarkable benefits to food companies. The implications extend beyond extraction efficiency; they indicate the potential for optimal nutrient composition toward enhanced health end results with reduced waste, offering a new synergistic blend of signals theory and practical agricultural applications (Aguilo-Aguayo et al., 2014). For diagnostics in medicine, applications of signals and spectra are of over-riding significance in non-invasive diagnosis and disease monitoring. Techniques such as spectroscopy enable medical personnel to probe biomolecules within body fluids to obtain data on metabolic and disease conditions. Physicians can also identify specific patterns of molecules with diseases from inspection of the spectra, thus enabling them to make an early diagnosis and personalize medication for the patient. The technique not only enhances patient outcomes but also facilitates the diagnostic process, enabling it to operate faster and at reduced costs. Furthermore, progress in signal processing algorithms has also increased the precision of these approaches to be implemented in real time and enhance decision-making within clinical settings. As signal and spectrum integration across medical technology continues to improve, its influence on healthcare delivery will be profound and far-reaching.

2.5.1. Telecommunications

The telecommunication environment is being shaped by significant transformations introduced through the advancement in signal processing and spectral management. The transformations are characterized by the implementation of advanced technology that increases the efficiency, reliability, and capacity of communication networks. One of the most important advancements is the development of optical frequency combs generated from high-Q micro-resonators. The technology provides a compact solution for telecommunications problems. These micro-combs are highly energy-efficient and have a broad optical bandwidth, making them applicable in a broad spectrum of applications from metrology to quantum optics. Surprisingly, their use in RF photonics is significant, as it brings advantages like reduced size and power usage compared to traditional processes, therefore pushing the area of telecommunications into new spheres (Chu et al., 2018). The modern studies in this area emphasize the need for incessant innovation to address the growing demands of modern communication systems. In addition, the performance metrics of telecommunication systems are being determined by advanced methodologies that can accurately analyze the performances of various waveforms of signals. In a recent study, an unbiased approach was presented for measuring radar waveforms performance, that is, Chirp and Multitones, which are a fundamental component of telecommunication. The research used an ultrawide band software-defined radar prototype, with outcomes that are comparatively as efficient as present state-of-the-art in this area. Experiments like these not only emphasize the applicability of theoretical simulations but also indicate the practical importance of such technologies to real-world use (Denoulet et al., 2012). The results through such comparisons can be utilized for designing future telecommunication hardware, in particular for enhancing detection and signal quality ranges. Overall, the integration of new advanced signal processing techniques with spectral management is revolutionizing the telecommunication sector. Micro-comb integration with RF photonics is a break-through in itself, allowing us to have enhanced functionalities while mitigating size and power problems. Also, the rigorous analysis of waveforms in radar is a testament to improvement in telecommunications technology, which promises improved performance in various uses. Since researchers are continually exploring what these technologies can offer, future telecommunications with increased efficiency and capability to address an even wider variety of communication needs appears promising. The advancement in this field exemplifies dynamic interaction between technological innovation and real application, whereby telecommunications manages to keep abreast of the pace of technology progress (Figure 2.12).

Figure 2.12. *Telecommunications.*

Source: https://upload.wikimedia.org/wikipedia/commons/thumb/4/40/Erdfunkstelle_Raisting_2.jpg/1200px-Erdfunkstelle_Raisting_2.jpg.

2.5.2. Audio and Music Technology

The evolution of music and audio technology has significantly influenced how sound is perceived, processed, and utilized across applications. Math modeling of sound in terms of spectra and signals lies at the center of this revolution. These concepts provide us with a model to work with for modeling sound waves based on their frequency, amplitude, and time characteristics. With advancements in technology, scientists are designing new ways of processing these signals, and the quality of audio is being enhanced along with more sophisticated techniques of music production. The complex connection between the human auditory system and audio signals is the platform on which numerous improvements are being made, as technology aims to mimic or improve the human sense in sound applications. In accordance, modern audio technologies not only increase artistic imagination but also make communication and information receipt more effective in a broad range of fields. Auditory spectrum analysis application is one of the greatest advances in music processing techniques. By modeling the human auditory system response to sound, scientists can develop more accurate approaches for onset detection and general sound analysis. Recent studies have demonstrated that algorithms based on auditory spectra have been successful in improving detection performance in musical tasks. For example, techniques using spectral flux and phase slope cues in combination demonstrate high precision, as reported in a study of various categories of pitched instruments. The study showed that techniques using auditory representations far surpass conventional methods, with an F-measure of 82.6% in onset detection (Benetos et al., 2010). These advances reflect an evolution toward methods closer to the manner in which human hearing processes, and generally better user experience within music technology. Further, the research on spectral features for audio classification, that is, discrimination between music and speech, is another driver of audio technology development. Recent advances have focused on the design of novel spectral features for better distinction between music and speech, based on biological models of perception. For instance, averaged cepstrum-based techniques have shown promise in identifying areas of significant cepstral changes, thereby improving classification accuracy. By removing noise and emphasizing salient spectral characteristics, these new features have also shown better performance in a number of test scenarios and compared to previously established classification schemes. Dynamic time warping, deployed in tandem with these spectral analyses, also confirms the merits of this approach, cementing the importance of spectral features in advancing audio and music technology (Jiang et al., 2008). These developments are an example of technology moving towards and merging with human perception, maintaining a trend towards increased auditory experience.

2.5.3. Medical Imaging

The medical imaging technology has made much progress with the driving force of technological development and increased understanding of signals and spectra of biologic tissue. Medical imaging encompasses a series of methodologies, from X-rays, MRI, and ultrasound, each conveying various pieces of information on anatomy and pathology. Among them, techniques like Raman spectroscopy have proven to be useful tools to discern specific molecular signatures in tissues to allow detection of disease like cancer. For instance, imaging Raman spectroscopy allows a rapid alternative to traditional imaging methods by using the principles of integral field spectroscopy (IFS). This advancement allows for rapid Raman image acquisition without needing time-consuming scanning protocols, thereby enhancing diagnostic precision and speed in the clinical setting (Darvin et al., 2018).

Figure 2.13. *Medical imaging.*

Source: https://www.postdicom.com/images/blog-posts/social-media-images/diagnostic-medical-imaging-social.png.

The integration of signal and imaging informatics has also transformed medical imaging, enabling clinicians to obtain useful information from sophisticated data sets. Particularly, the use of machine learning techniques has allowed novel solutions to dissecting the huge amounts of imaging data generated in the clinical environment. This not only augmented image interpretation but has also been used in predictive analytics, maximizing patient outcome through individualized medicine. The 2017 literature on medical informatics identifies key contributions to the field along these lines and highlights how sensor and signal technology has enhanced the management of imaging data. With the integration of informatics, clinical science, and engineering, scientists have developed a platform that enhances the use of imaging techniques, making it a robust platform on which future innovations in medical diagnostics are established (Deserno et al., 2018). With advancements in medical imaging, convergence between emerging technologies and analytical strategies will likely redefine the scope of patient treatment. Combining spectroscopy with advanced imaging techniques will enhance the visualization and measurement of biophysical and biochemical processes in real-time, leading to enhanced diagnoses and treatment recommendations. Furthermore, advances in imaging informatics will facilitate the creation of large databases to support multi-modality data aggregation. This vast dataset not only enriches clinical research but also helps in the enhancement of machine learning algorithms, which can ultimately lead to breakthrough advances in both diagnostic and therapeutic objectives. Thus, the cooperation among signals, spectra, and medical imaging is bound to play a crucial role in shaping the future of medicine, ushering in a paradigm shift towards more precise and effective patient management protocols (Figure 2.13).

2.6. RECENT ADVANCES IN SIGNAL PROCESSING

Signal processing as an area of research has experienced exponential growth, particularly within the field of spectral analysis, following the development and technological advancement and computational techniques. Among the new frontiers is the expansion of spectral unmixing (SU) techniques developed to represent the composition of a mixture of signals. The approach utilizes multi-dimensional data sets, for example, acquired through spectral imaging, and applies them to solve for the reflectance spectrum of individual materials in a viewed scene. This limitation has important implications for earth science applications, enabling scientists to accurately estimate and compare the ratios of different materials within pixels that are blended. These methodology advances circumvent the deficiencies of previous statistical algorithms like principal component analysis (PCA) and independent component analysis (ICA), which are not well-suited to deal with the sophisticated interdependencies among abundances. Recent studies demonstrate the way in which SU is capable of effectively extracting valuable information from hyperspectral images, further helping us to interpret complicated environments as mentioned in Brun et al. (2012). Importantly, the construction of novel decomposition techniques has revealed new frontiers in time-varying signal processing and extended the areas of signal processing application to other fields. For instance, the latest research utilizes adaptive chirplet decomposition and empirical mode decomposition to address non-stationary signals such as seismic data. By combining them with the use of the Wigner-Ville transform and the Hilbert transform, researchers can generate joint time-frequency representations that take into consideration the dynamic pattern of seismic events. This new approach facilitates greater insight into frequency content evolution over seismic events and enables the evaluation of structural responses to these excitations. The ability to track mean instantaneous frequencies of significant structural responses puts into perspective the potential use of such innovations towards monitoring and detecting potential damages in constructed infrastructures, in a direction towards more responsive and intelligent structural health monitoring systems as proposed by Giaralis et al. (2007). The merge of such fresh developments initiates a revolutionary epoch in signal processing, with higher precision in the handling of complex datasets for various purposes. The integration of multidimensional spectral analysis and adaptive decomposition methods is a far-reaching trend towards greater accuracy and informality in the interpretation of signals. With researchers further developing such techniques, numerous fields, from environmental monitoring to structural engineering, will benefit from increased analytical efficiency. As a result, technological developments in signal processing not only enhance theoretical understanding but also provide real-world solutions to real issues. The trend points to the central role of unremitting innovation within the discipline in illustrating the scope for influencing scientific research and engineering practice significantly. By continued study and application of these advances, the discipline is well situated to tackle the complexities of modern signal analysis and offer further enhancement of technology and society in general.

2.6.1. Machine Learning for Signal Processing

Combining signal processing with machine learning has transformed traditional practices to allow more intricate analysis and interpretation of data. Signal processing previously relied heavily on algorithms hand-designed, which were typically encumbered by their inability to adapt to multiple environments for the data. Machine learning, particularly generative models, introduces an attitude shift through learning automatically from data solutions rather than dependent rules. For instance, the product-of-filters (PoF) model has been proposed as a new framework that decomposes audio spectra's into sparse linear representations of filters in the log-spectral domain. This model takes leverage from conventional homomorphic filtering but enhances it with the assistance of statistical inference-based learned decomposition. Such advancements not only make processing of audio signals easier but also enhance the likelihood of extension to other fields of the signal processing domain (Hoffman et al., 2014). Machine learning techniques are implemented to enhance efficiency and accuracy in processing signals in real-world applications. For instance, for water pipe management, machine learning techniques are used to solve key issues like leak detection. Conventional approaches have mostly faltered on both efficiency and accuracy, which has led to a shift towards intelligent solutions. Utilizing wireless sensor networks (WSNs) with machine learning algorithms makes it possible to capture and forward data essential in detecting leaks in real-time.

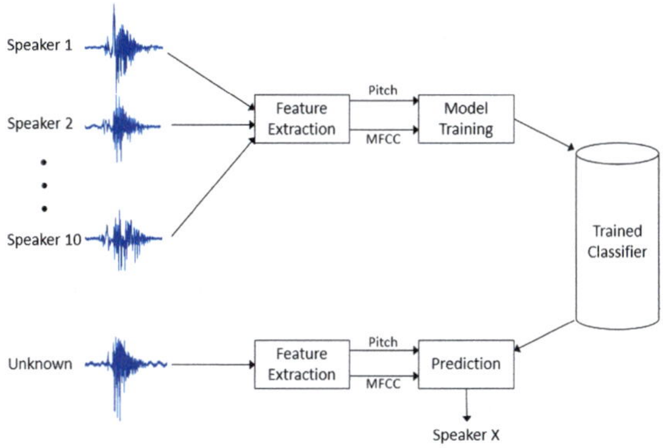

Figure 2.14. *Machine learning in signal processing.*

Source: https://encrypted-tbn0.gstatic.com/images?q=tbn:ANd9GcRBLFnIArvVz-ku-vjUerM__ObcLoWRo0a47uw&s.

By adopting techniques like support vector machines (SVM) and signal feature extraction, the systems can differentiate between water loss patterns, thus optimizing the efficiency and usage of water resources. The blending of machine learning with traditional signal processing techniques not only proves the adaptability of the techniques but also underscores their applicability in tackling real-world problems (Bøe et al., 2019). The scope of machine learning to influence signal processing is broad, extending to use beyond audio signal analysis and leak detection. As machine learning models continue to advance, they exhibit more sophisticated functions in fields such as image processing, telecommunications, and environmental monitoring. For instance, in telecommunication, machine learning algorithms can optimize data transmission through real-time adjustment of parameters based on network conditions. Similarly, in environmental monitoring, models can scan data from different sources to identify salient trends or anomalies and enable improved decision-making. As the marriage of machine learning and signal processing continues to mature, research and development work in progress has the potential to provide even greater efficiency and innovation. This ongoing synergy between signal processing and machine learning signals the beginning of a new age when data-driven knowledge shall transform industries and enhance our understanding of complex phenomena in the realms of spectra and signals (Figure 2.14).

2.6.2. Real-Time Signal Processing

Real-time signal processing is one of the main aspects of modern technology, especially in the understanding and interpretation of dynamic data streams. It refers to the processing and processing of signals in real time while receiving them and providing results immediately that are critically important in most applications, such as telecommunications, audio processing, and geophysical explorations. With techniques like the Fast Fourier Transform (FFT), it is possible to quickly transform signals to the frequency domain so that major spectral components can be determined. However, non-stationarity of real-time data is a significant challenge, especially in geophysics where environmental conditions change erratically. New techniques must be invented for accurate analysis and interpretation, beyond the traditional methods that break down when confronted with such varying conditions, ultimately deciding the destiny of signal interpretation under the paradigm of spectra analysis. Among the new techniques to confront non-stationary signals is the Empirical Mode Decomposition (EMD). The technique has gained momentum for its ability to decompose intricate signals into intrinsic mode functions (IMFs), with IMFs being fundamental oscillatory modes. EMD preserves real-time signal processing variability through the ability to extract instantaneous frequencies without having to assume the character of the signals. This adaptability is particularly significant in technologies such as the magnetotelluric (MT) method, where the instantaneous processing of electromagnetic fields is essential to determining subsurface geology. With the application of EMD, researchers can now build stable impedance tensors from non-stationary MT data, demonstrating the efficiency of the technique in relation to standard processing methods like BIRRP. Not only is accuracy improved, but visualization and filtering of noise as well, both of which are significant in dynamic settings. Energy leakage is another pertinent issue for real-time signal processing, particularly with the application of discrete wavelet transforms (DWT). This effect, when left uncontrolled, can lead to biased results and misinterpretation in spectral analysis. Researchers have noted that DWT, especially when applied on higher-order wavelet functions such as Daubechies, carries inherent time resolution and overlap in bands trade-offs, which exacerbate the issue of leakage. On this count, new approaches such as resampling processes have been proposed to neutralize the impact of energy leakage, rendering the purity of spectral outputs improved. By tackling such a limitation head-on in real-time schemes, the field can then improve its methodologies, and thus better characterize and interpret signals in a wide variety of applications. Such concurrence of methodologies is a pivotal point of signal processing discipline, that signals continuous innovation is the solution to effective management of composite signals as well as their spectra (Figure 2.15).

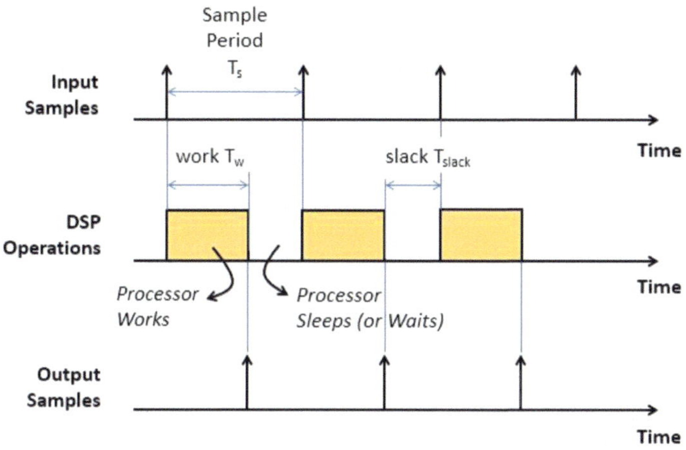

Figure 2.15. *Real-time signal processing.*

Source: https://schaumont.dyn.wpi.edu/ece4703b20/_images/realtimedsp.jpg.

2.6.3. Quantum Signal Processing

Quantum Signal Processing is an innovative departure from how we process information across a wide spectrum of applications. Employing the phenomenon of quantum mechanics, quantum signal processing attempts to enhance the accuracy and effectiveness of signal processing techniques. The combination of quantum phenomena allows the reevaluation of classical signal processing techniques on the basic level, especially the manner in which we perceive periodic orbits and their quantization. The application of methods such as decimated linear predictors, decimated Pade approximants, and decimated signal diagonalization, as given in Berry et al. (1999), illustrates the trend towards higher accuracy semiclassical spectra. The techniques not only improve numerical stability in calculations but also assist in bridging classical signal processing with its quantum counterpart, emphasizing the demands for innovative solutions in an increasingly technologically advanced environment.

Furthermore, Quantum Signal Processing has significant implications for the implementation of next-generation technologies, including quantum computing and communications. Control over single photons, for instance, has exceptional potential in quantum networks and cryptography. The research outlined in Bustard et al. (2015) demonstrates the promise of manipulating single photons through modified quantum memories for unprecedented spectral control. This flexibility of parameter tuning like wavelength and bandwidth is the key to component performance optimization in quantum optical technology. Quantum signal processing not only improves existing platforms but guides us towards establishing firm quantum systems through the ability to integrate different components easily by harnessing efficient spectral conversion. This pivotal intersection of quantum mechanics and signal processing is a symbol of vast potential of quantum effects being applied to tackle sophisticated real-world issues. In short, the research on Quantum Signal Processing is at the core of making real the inherent advantage of quantum mechanics in advanced signal manipulation. The better the scientists make the equipment employed in quantizing signals, the brighter the future appears for this pursuit, with hidden doors to making known the theoretical and factual implications of quantum effects on signal processing. The conclusions derived from novel analysis techniques, for example, those of Berry et al. (1999) and Bustard et al. (2015), identify the two-fold importance of precision and control over spectra in optimally applying quantum technologies. As this interdiscipline matures, it holds the promise not just to advance the current technologies but to create new paradigms for quantum information science, thereby charting the course of future studies and applications in the field of signals and spectra.

SUMMARY

The observation of signals and spectra shows the extensive impact that these concepts have across a broad scope of science. The intricate dynamics between signal processing and interpretation of spectral data facilitate better communication, accurate diagnosis, and technological advancement. As illustrated in this essay, the concepts behind signals and spectra form basic frameworks for the understanding of complex events in fields such as physics, engineering, and medicine. Continuous development of analytical techniques, such as Near Infrared Diffuse Reflectance Spectroscopy (NIR-DRS), highlights the necessity of constant research and development in this area. With well-established models with incredible determination coefficients, such as $R^2 = 0.9878$, the marriage of theory and practice in signal analysis can benefit a multitude of industries and applications, raising the bar for what has previously been achieved (Corsetti et al., 2014). Besides this, the contribution made by funding and collaborative research programs cannot be downplayed in terms of developing knowledge on signals and spectra. Research undertaken by the Northern Research Partnership (NRP) and the Scottish Sensor Systems Centre (SSSC), for instance, reflects the manner in which collaboration in education facilitates innovation and accommodates knowledge exchange. The impact of such joint ventures is indeed observed in the practical applications of sophisticated analytical models, assisting in improved as well as more efficient methods of measurement and prediction. With industries increasingly adopting these technologies, the potential for new breakthroughs in diagnostics as well as communications comes into play, opening up further research in the field of signal processing and spectral analysis. The commitment to joint projects therefore remains critical to the complete exploitation of the potential of scientific progress in these areas (Corsetti et al., 2014). The destiny of signals and spectra is therefore entangled with technology advancement and interdisciplinarity. With the improvement of better analytical techniques and methods, scientists can attain higher levels of accuracy and applicability of results in applied scenarios. The success of established models, like the NIR-DRS method of quantifying Shengmax inside I content, only serves to reinforce the applied importance of theoretical understanding, confirming that research on this subject is not only of academic value but also of essential importance for implementation. While the science community continues to grapple with the complexities of data, continuous synthesis of signals and spectra will be of absolute significance in advancing our understanding and skill in a wide range of areas. Therefore, the foray into this field holds the potential to reveal more insights and breakthrough technologies that can greatly enhance quality of life as well as operational effectiveness across the globe (Cheng et al., 2018).

In the study of intricate interactions of signals and spectra, consideration has to be given to the very fundamentals that govern their interaction. One most controversial area of contention is whether the clarity and dependability of signal transmission directly influence the final spectral analysis. Efficiency of communication in communication systems always hinges on the extent to which they can withstand noise, an intrusive interference in the signal transmission. When signals are contaminated by noise, such as in the Johnson (like) noise-based key-distribution scheme by Kish, the purity of the signals is more vulnerable to external attack. By understanding the part played by noise in this case, one is closer to understanding the weaknesses that can be open to communication systems and the necessity for robust security measures to guard against attack by an opponent (Yariv et al., 2006). Further, a close survey of primary distribution strategies reveals a thread of security vulnerabilities that critically impact the effectiveness of signal transmission. Analysis of advanced communication methods points out the susceptibility of such systems to hacking through passive attack methods. These methods manipulate intrinsic signal properties, such as the transient response to voltage variations, to extract sensitive data. An opponent's ability to determine complete knowledge of exchanged keys represents a formidable challenge to security systems. Realization of effects of such vulnerabilities in signal system design is, thus, indispensable, demanding better safeguards that can neutralize contemplated dangers and risks exhibited in applied practice (Yariv et al., 2006). In addition, such discussions over security are further substantiated by considering an analysis of chaotic modulation methods, where the use of synchronization among chaotic systems has been proposed as a way of enhancing security in communications methods. However, the approach has been shown to have existing key weaknesses, where it can be subjected to severe scrutiny and exploitation through high-pass filtering and low-order system synchronization. Such weaknesses highlight the significance of the stiffness of assessment in the deployment and design of secure communications. Through the identification of such weaknesses, researchers and practitioners can try to design systems that focus on signal fidelity as well as add robust security mechanisms intended to withstand sophisticated attack methods. To discuss these points is to facilitate the right perspective on the convergence of signals, spectra, and security in communication systems, with technological advancements always meaning enhancements and never jeopardizing system integrity.

REFERENCES

1. Abragam, A., De Gennes, P., Goldman, M., Segnorile, H. H., Oppenheim, A. V., Zamar, R. C., et al. (2013). Quantum irreversible decoherence behaviour in open quantum systems with few degrees of freedom. Application to 1H NMR reversion experiments in nematic liquid crystals. https://core.ac.uk/download/159302485.pdf (accessed on 18th September 2025).
2. Abry, P., Antonini, M., Antonino-Daviu, J., Argoul, P., Auger, F., Douka, E., Chu, F. L., et al. (2008). On the energy leakage of discrete wavelet transform. https://core.ac.uk/download/50571.pdf (accessed on 18th September 2025).
3. Aguilar, A., Ajoy, A., Druga, E., Gierth, E., et al. (2020). Enhanced optical 13C hyperpolarization in diamond treated by high-temperature rapid thermal annealing. https://core.ac.uk/download/323083276.pdf (accessed on 18th September 2025).
4. Aguilo-Aguayo, I., Brunton, N., Downey, G., Keenan, D. F., et al. (2014). Observations on the water distribution and extractable sugar content in carrot slices after pulsed electric field treatment. https://core.ac.uk/download/45656361.pdf (accessed on 18th September 2025).
5. Alvarez, G., Bowong, S., Cuomo, K., Devaney, R., Montoya, F., Femat, R., et al. (2005). Security analysis of communication system based on the synchronization of different order chaotic systems. https://core.ac.uk/download/16391511.pdf (accessed on 18th September 2025).
6. Amblard, A., Challinor, A., Battistelli, E., Bischoff, C., Johnson, B. R., Bunn, E., Bussmann, R., et al. (2009). Impact of modulation on CMB B-mode polarization experiments. http://arxiv.org/abs/0809.4032 (accessed on 18th September 2025).
7. Ball, A., Gu, F., Haram, M., Wang, W., et al. (2011). An investigation of the electrical response of a variable speed motor drive for mechanical fault diagnosis. https://core.ac.uk/download/1653055.pdf (accessed on 18th September 2025).
8. Banerjee, T., Chowdhuri, S., & Kalmár-Nagy, T. (2020). Persistence analysis of velocity and temperature fluctuations in convective surface layer turbulence. http://arxiv.org/abs/2005.04958 (accessed on 18th September 2025).
9. Barabasi, A. L., Meneveau, C., Peng, C. K., Poon, C. S., Sugihara, G., Moody, G. B., et al. (1999). Multifractality in human heartbeat dynamics. http://arxiv.org/abs/cond-mat/9905329 (accessed on 18th September 2025).

10. Baumann-Pickering, S., Hildebrand, J. A., Roch, M. A., Schnitzler, H. U., et al. (2015). Acoustic behavior of melon-headed whales varies on a diel cycle. https://core.ac.uk/download/323064229.pdf (accessed on 18th September 2025).
11. Benetos, E., & Stylianou, Y. (2010). Auditory spectrum-based pitched instrument onset detection. https://core.ac.uk/download/9982863.pdf (accessed on 18th September 2025).
12. Berry, M. V., Bogomolny, E. B., Cvitanovic, P., et al. (1999). Decimation and harmonic inversion of periodic orbit signals. http://arxiv.org/abs/chao-dyn/9912031 (accessed on 18th September 2025).
13. Bøe, H. J., Falkum, E., Gjerstad, C. L., Lystad, R. P., et al. (2019). Water pipeline leakage detection based on machine learning and wireless sensor networks. https://core.ac.uk/download/294619373.pdf (accessed on 18th September 2025).
14. Brun, N., & Dobigeon, N. (2012). Spectral mixture analysis of EELS spectrum-images. https://core.ac.uk/download/12043583.pdf (accessed on 18th September 2025).
15. Burrus, C. S., Cohen, L., Mallat, S., & Newland, D. E. (2007). Numerical treatment of seismic accelerograms and of inelastic seismic structural responses using harmonic wavelets. https://core.ac.uk/download/2708118.pdf (accessed on 18th September 2025).
16. Bustard, P. J., England, D. G., Fisher, K. A. G., MacLean, J. P. W., et al. (2015). Quantum optical signal processing in diamond. http://arxiv.org/abs/1509.05098 (accessed on 18th September 2025).
17. Chen, J., Heincke, B., Jegen-Kulcsar, M., Moorkamp, M., et al. (2011). Using empirical mode decomposition (EMD) for the processing of marine MT data. https://core.ac.uk/download/11902855.pdf (accessed on 18th September 2025).
18. Cheng, G., Kuang, H., Qian, X., Su, Y., et al. (2018). Rapid determination of saponins in the honey-fried processing of *Rhizoma Cimicifugae* by near infrared diffuse reflectance spectroscopy. https://core.ac.uk/download/323080465.pdf (accessed on 18th September 2025).
19. Cheung, G., Magli, E., Ng, M., Tanaka, Y., et al. (2018). Graph spectral image processing. http://arxiv.org/abs/1801.04749 (accessed on 18th September 2025).
20. Chu, S. T., Little, B. E., Mitchell, A., Morandotti, R., et al. (2018). Harnessing optical micro-combs for microwave photonics. http://arxiv.org/abs/1710.08611 (accessed on 18th September 2025).

21. Coffin, T., & Jong, J. Y. (2025). Signal detection techniques for diagnostic monitoring of space shuttle main engine turbomachinery. https://core.ac.uk/download/pdf/42840787.pdf (accessed on 18th September 2025).
22. Corsetti, S., Kiefer, J., McGloin, D., Zehentbauer, T. M., et al. (2014). Characterization of gasoline/ethanol blends by infrared and excess infrared spectroscopy. https://core.ac.uk/download/29151167.pdf (accessed on 18th September 2025).
23. Cowell, D. M. J., & Freear, S. (2010). Separation of overlapping linear frequency modulated (LFM) signals using the fractional Fourier transform.
24. Cox, T. J., & Li, F. F. (2003). Speech transmission index from running speech: A neural network approach. https://core.ac.uk/download/104410.pdf (accessed on 18th September 2025).
25. Darvin, M. E., Gersonde, I., Moralejo, B., Roth, F., et al. (2018). Nonscanning large-area Raman imaging for *ex vivo/in vivo* skin cancer discrimination. http://arxiv.org/abs/1810.03363 (accessed on 18th September 2025).
26. Dempster, A. P., Gelman, A., Wang, D. L., FitzGerald, D., Nocedal, J., Winn, J., Mandel, M. I., et al. (2010). Probabilistic modeling paradigms for audio source separation. https://core.ac.uk/download/30696438.pdf (accessed on 18th September 2025).
27. Denoulet, J., Garda, P., Le Kernec, J., Romain, O., et al. (2012). Empirical comparison of chirp and multitones on experimental UWB software defined radar prototype. https://core.ac.uk/download/42368116.pdf (accessed on 18th September 2025).
28. Deserno, T. M., Hsu, W., Kahn, C. E., Section Editors for the IMIA Yearbook Section on Sensor, et al. (2018). Sensor, signal, and imaging informatics in 2017. https://core.ac.uk/download/323077537.pdf (accessed on 18th September 2025).
29. Eng, W. P., & Schuler, D. L. (1984). An improved dual-frequency technique for the remote sensing of ocean currents and wave spectra. https://core.ac.uk/download/pdf/10372758.pdf (accessed on 18th September 2025).
30. Giaralis, A., & Spanos, P. D. (2009). Determination of design spectrum compatible evolutionary spectra via Monte Carlo peak factor estimation. https://core.ac.uk/download/2708109.pdf (accessed on 18th September 2025).
31. Giaralis, A., & Spanos, P. D. (2010). Derivation of Eurocode 8 spectrum-compatible time-histories from recorded seismic accelerograms via harmonic wavelets. https://core.ac.uk/download/2708110.pdf (accessed on 18th September 2025).

32. Giaralis, A., Baraniuk, R., Boashash, B., Burrus, C., Chen, J., Cohen, L., et al. (2007). Time-frequency representation of earthquake accelerograms and inelastic structural response records using the adaptive chirplet decomposition and empirical mode decomposition. https://core.ac.uk/download/2708117.pdf (accessed on 18th September 2025).
33. Giaralis, A., Politis, N. P., & Spanos, P. D. (2007). Algorithmic options for joint time-frequency analysis in structural dynamics applications. https://core.ac.uk/download/2708113.pdf (accessed on 18th September 2025).
34. Giovannelli, J. F., & Idier, J. (2009). Bayesian interpretation of periodograms. http://arxiv.org/abs/0908.3262 (accessed on 18th September 2025).
35. Hoffman, M. D., Liang, D., & Mysore, G. J. (2014). A generative product-of-filters model of audio. http://arxiv.org/abs/1312.5857 (accessed on 18th September 2025).
36. Jacques, L., Vandergheynst, P., Vielva, P., Risbo, T., & Wiaux, Y. (2005). Fast directional correlation on the sphere with steerable filters. https://core.ac.uk/download/147918039.pdf (accessed on 18th September 2025).
37. Jiang, M., Sadka, A. H., & Zhou, H. (2008). Feature extraction for speech and music discrimination. https://core.ac.uk/download/335077.pdf (accessed on 18th September 2025).
38. Kempf, A., & Zayed, A. I., et al. (2004). On fields with finite information density. http://arxiv.org/abs/hep-th/0404103 (accessed on 18th September 2025).
39. Klimenk, A., Zych, A., Zoglauer, A., Hoover, A., Phlips, B., et al. (2006). The Advanced Compton Telescope. https://core.ac.uk/download/72054708.pdf (accessed on 18th September 2025).
40. Kraskov, A., Anderson, B., Astakhov, V., et al. (2010). Independent components in spectroscopic analysis of complex mixtures. http://arxiv.org/abs/1009.0534 (accessed on 18th September 2025).
41. Le Bihan, N., Chainais, P., & Flamant, J. (2017). Spectral analysis of stationary random bivariate signals. http://arxiv.org/abs/1703.06417 (accessed on 18th September 2025).
42. Moleti, A., Cosmelli, C., La Posta, C., Coccia, E., Camilo, F., Ronga, F., Modestino, G., et al. (2000). Search for periodic gravitational wave sources with the Explorer detector. https://core.ac.uk/download/25301148.pdf (accessed on 18th September 2025).
43. Olivares, F., Plastino, A., & Rosso, O. A. (2015). Noise versus chaos in a causal Fisher-Shannon plane. http://arxiv.org/abs/1510.06950 (accessed on 18th September 2025).

44. Roth, D. J. (2025). Platform for post-processing waveform-based NDE. https://core.ac.uk/download/pdf/10556330.pdf (accessed on 18th September 2025).
45. Smalley, R. R. (2025). Microprocessor-based multichannel flutter monitor using dynamic strain gage signals. https://core.ac.uk/download/pdf/42883817.pdf (accessed on 18th September 2025).
46. Spott, A., Bristow, A. D., Jalali, B., Kuyken, B., Moore, D. S., Tien, E. K., et al. (2012). Bridging the mid-infrared-to-telecom gap with silicon nanophotonic spectral translation. https://core.ac.uk/download/55801285.pdf (accessed on 18th September 2025).
47. Vary, A. (2025). Computer signal processing for ultrasonic attenuation and velocity measurements for material property characterizations. https://core.ac.uk/download/pdf/42869708.pdf (accessed on 18th September 2025).
48. Yariv, A., Scheuer, J., & Kish, S. (2006). A classical key distribution system based on Johnson (like) noise–How secure? https://core.ac.uk/download/345075845.pdf (accessed on 18th September 2025).

CHAPTER 3

SOURCE CODING AND DATA COMPRESSION

Contents

INTRODUCTION

In digital communication, the efficiency of sending data depends greatly on source coding and data compression. These technologies are the key to optimizing storage and bandwidth occupied by digital information. With digital information increasing in this globalized world, how to keep it at a reasonable size and be easily accessible in the shortest time possible has become the priority. Source coding is actually the conversion of information into space-minimized form, while data compression is everything to do with minimizing redundancy in the data. By understanding and applying theories such as rate-distortion theory, engineers can design systems where efficiency need is weighed against preserving information quality. This early study provides the foundation for a thorough discussion of the principles and applications of these basic methods in modern computing and communications. A profound understanding of source coding and data compression has the potential to enhance various applications in practice from streaming media to data storage systems. The performance of such methods is typically illustrated by leading algorithms that embody the principles of best compression system design. For instance, many algorithms make use of rate-distortion theory in order to solve the inherent trade-off between theoretical optimality and limiting factors posed by transmission channels and storage devices. These compression techniques not only allow data to be stored in an efficient manner, but also enhance the rate with which data can be transmitted on networks. As mentioned before in other studies, such algorithms, as they are theoretically based but practically modified to address the diverse needs of users, prove their importance in contemporary society. Moreover, interaction between data compression techniques and information extraction has been of great concern in recent studies. New methods utilize compression algorithms to establish a quantifiable system for identifying the relationship between character strings, and these can find use in areas such as automatic language identification and authorship analysis. By utilizing information theory concepts, the new methods enhance the efficiency of existing data processing models such that it is now easier to comprehend and manage data sets. The variety of these techniques serves to underscore the necessity of developing new compression techniques that are not merely efficient but adaptable to various situations. Essentially, this introduction summarizes why source coding and data compression are relevant and lays the foundation for a detailed examination of their applications and techniques.

Definition of Source Coding

Source coding, a core principle in the field of information theory, is the process of converting information into a form applicable for the transmission and storage with maximum efficiency. Through the compression of data by eliminating redundancy but preserving its integrity and accuracy, this process compresses data to save space and conserve bandwidth. Through the use of methods that encode information with fewer bits than the original, source coding is responsible for maximizing bandwidth and storage capacities. Through the application of various algorithms and techniques, including Huffman coding and arithmetic coding, source coding permits information to be conveyed more economically through channels. Such methods not only enhance data transmission but also form an integral role in reducing the cost of data transmission and storage, proving crucial in modern computing and telecommunication. Fundamentally, source coding has a strong association with the theory of data compression. The synergy is illustrated in the way specific coding methods can make significant contributions to the effectiveness of compression algorithms. As an example, by incorporating principles borrowed from rate-distortion theory, it is easy to achieve optimal data compression that maximizes the trade-offs between data fidelity and the quantity of data transmitted (Effros et al., 1998). This intricate relationship between the coding, compression, and representation of data results in highly efficient systems with a spectrum of applications from database management at large to multimedia streaming. With the acceleration of data production in the age of the internet, the importance of source coding and its efficiency is greater than ever before. Moreover, the research on source coding techniques has resulted in a significant advancement in data classification and extraction. By applying techniques of data compression, it is feasible to develop techniques which describe and quantify the information contained in various types of data, for example, character strings and datasets of texts. These developments enable automatic language recognition and authorship determination, as well as enhance the operation of machine learning algorithms. Thus, source coding has advanced to being more than a mere technical necessity; it is now an anchor for innovative research and developments across different fields. With the ongoing exploration of information management complexities, the role of source coding in data compression continues to be at the forefront (Figure 3.1).

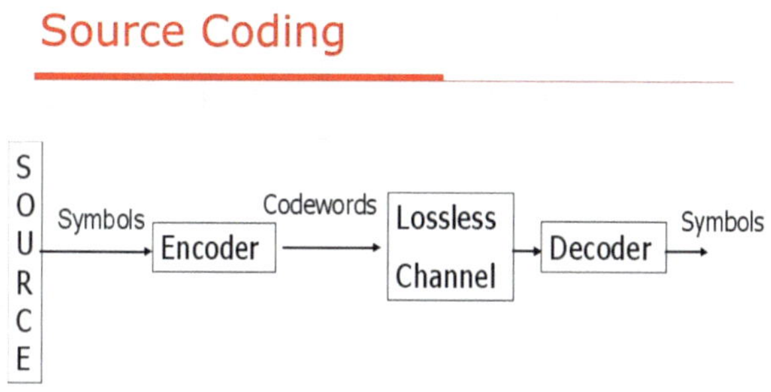

Figure 3.1. *Source coding.*

Source: https://image2.slideserve.com/5369517/source-coding-l.jpg.

Importance of Data Compression

Greater reliance on digital technology emphasizes the pivotal function data compression has in modern communication and information systems. With massive amounts of data being generated on a daily basis, effective methods for compression are necessary to assist in optimizing storage and bandwidth utilization. In the absence of compression, data transmission and storage would be unacceptable in commercial terms, entailing wastage and costs for individuals and organizations. For instance, in streaming media and cloud storage, file compression significantly reduces bandwidth and storage requirements, thus leading to more efficient data management and retrieval processes. This growing need also necessitates the development of advanced compression algorithms that accommodate different formats and purposes, supporting better performance with guaranteed data integrity.

Figure 3.2. *The importance of data compression.*

Source: https://fastercapital.co/i/Data-compression--NetVolume--Revolutionizing-Data-Compression-Techniques--The-Importance-of-Data-Compression.webp.

Moreover, data compression is one of the prime considerations for addressing energy constraints in a large number of systems, particularly cyber-physical systems that rely on sensor networks. Since the sensors continue to stream data, energy consumed for source gathering as well as compression will tend to reflect the energy consumed for transmitting the data. Appropriate usage of power not only optimizes the working period of energy-harvesting sensors but also supplies useful information without heavily restricting power supply. Efficient energy allocation policies must optimize compression schemes and transmission dynamics to deliver maximum performance under distortion and delay requirements. As demonstrated in studies, choosing effective management policies in such a scenario can significantly enhance stability of information flow among various sensors, thereby optimizing overall system performance (Castiglione et al., 2011). Beyond real-world efficiencies, data compression also plays a key role in deterring digital piracy and quality content assurance. The virtual arena facilitates easy duplication of high-fidelity content, which threatens unauthorized replication and copying. Compression techniques, such as digital watermarking, can readily secure copyright and ownership rights by inserting stealthy information into the data. Not only does this guarantee content integrity but also supports a range of applications from video broadcast monitoring to artistic rights management. Research advances show that trustworthy watermarking methods can survive types of attacks, granting original users continued access to quality content while keeping unauthorized usage at bay. Data compression technologies are thus not merely utilitarian; they form an integral part of ethical digital content delivery and management policies (Serdean et al., 2002) (Figure 3.2).

3.1. HISTORICAL BACKGROUND

Evolution of historical source coding and data compression is marked by a constantly changing interaction between advancing technology means and growing demands for efficient handling of data. In the early days, effort in this direction was more focused on optimally using limited bandwidth for communication, with telecommunication system requirements drawing much influence. Claude Shannon's pioneering effort of the 1940s established the age of modern information theory rooted in entropy and redundancy principles to explain data storage and transmission. Shannon's research on the maximum efficiency of data encoding paved the way for the creation of various source coding techniques that later led to methods aimed at reducing redundancy with a goal of maximizing informativeness of the content to be communicated. With growing computing power, the complexity of compression algorithms for data increased. Innovations in the 1970s and 1980s presented a number of techniques like Lempel-Ziv coding and Huffman coding that focused on lossless compression using variable-length encoding schemes. These innovations allowed for storage of larger data sets without affecting the integrity of the original data. The arrival of lossy techniques addressed the need for even greater reduction of data at the time, particularly in multimedia technologies. Notably, the evolution of formats like JPEG and MP3 profoundly changed the way images and audio files were being compressed and then distributed, reflecting the balance between the size and quality of digital representation. Such a change also reflect the harmony between techniques of compression and trends in machine learning, indicating an even broader trend for automating data processing operations. The union of data compression and artificial intelligence has far enhanced real-time data retrieval and analysis capabilities. Machine learning algorithms have of late been researched for use in augmenting traditional compression algorithms, as in the instance of Netflix video streaming and cloud storage services. By means of advanced frameworks, scientists are currently able to impose more efficient encoding and image restoration, as suggested in studies that implement 4D seismic data into inversion processes through machine learning. This incorporation is one such illustration of how data compression concepts can serve as building blocks in formulating new methods for complex data scenarios to lead towards better data integrity and access. Furthermore, recent efforts based on principles of information theory, have proven to be promising in application fields such as automatic language identification, a proposal for the all-pervasive nature and timeliness of source coding paradigms to various modern applications (Etienam et al., 2019).

3.1.1. Early Data Compression Techniques

The discovery of early compression algorithms has greatly influenced the digital information processing environment. Early compression algorithms were first necessitated by the necessity to optimize storage and transmission capacity. Simple algorithms, such as run-length encoding, were devised in order to reduce data size by employing the repeated character and its frequency instead of repeating characters. Even though primitive, the approach set the stage for highly sophisticated compression algorithms. As data formats evolved with the unfolding digital landscape, so did other approaches, including Huffman coding. Huffman coding applies variable-length coding based on data elements' probability of occurrence, thus allowing more dense representation of information. This kind of innovation was necessary in the days of low bandwidth and storage capacity, showcasing the initial realization of the need for efficient data management systems within the unfolding digital landscape. Concurrently, the move to digital media during the latter half of the 20th century facilitated further progress in compression algorithms. Video and audio information digitization saw an acute requirement for compression processes that could enable efficient storage and transmission without any perceivable loss of quality. One such innovation during this time was the development of transform coding techniques, that made use of mathematical transforms like the Discrete Cosine Transform (DCT) to represent data in a more efficient manner. This method breaks signals into frequency parts, enabling selective compression based on perception by human senses. As a result, fewer bits are used to represent less perceptual frequencies, maximizing the data representation as a whole. Research on these methods proposed the intricate involvement of human perception mechanisms and digital compression, a central issue in contemporary compression techniques (Serdean et al., 2002). Secondly, the development of watermarking technology was another turning point towards data compression creation. Early watermarking technology was essential in solving copyright and content protection problems in the digital world because it provided a way to embed recognizable information within digital media. This was important because digital media did away with the limitations on copying that were built into analog media, which always suffered degradation with each reproduction. Watermarking technology also became more secure even against other forms of attacks or manipulation to safeguard the integrity of the original content while still maintaining standards for compression. These are all reflective of an ongoing attempt at safeguarding intellectual property and indicate the increasingly complex interdependency between data compression, copyright, and digital rights management as additional evidence for the inevitable role of pioneering technologies in this context (Ingels et al., 2025) (Figure 3.3).

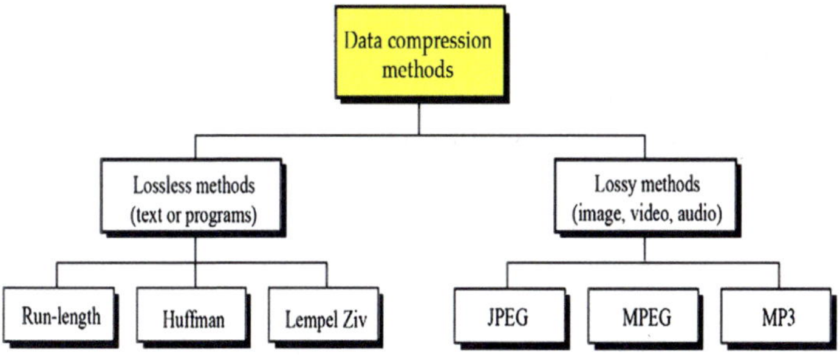

Figure 3.3. *Early methods of data compression.*

Source: https://encrypted-tbn0.gstatic.com/images?q=tbn:ANd9GcQWe0_G0I21Qw734Sp-al9k6awvV85oj4LePQ&s.

3.1.2. Development of Source Coding Techniques

The evolution of source coding techniques has been pivotal in the creative process of data compression, facilitating more efficient data transmission and storage. Early innovations in coding methods laid the groundwork for contemporary practices, enabling the reduction of data redundancy. By assessing various coding schemes, researchers evaluated their adaptability and efficiency, ultimately leading to the implementation of more advanced strategies that cater to specific types of data. Development within institutions like NASA has significantly influenced these techniques, where a focused effort has been made to craft coding methods that maximize compression while preserving data integrity. Such dedicated advancements underscore the necessity of source coding in optimizing data handling in an increasingly digital world, ultimately contributing to more robust data management solutions across a multitude of applications. In particular, NASAs endeavors toward developing efficient source coding techniques illustrate the practical applications and implications of such innovations. This organization undertook significant investments in both hardware and software implementations to create coding modules capable of high-speed, efficient data representation. As noted in recent projects, these coding techniques can adapt to local statistical variations, ensuring near-optimal code efficiency while protecting the accuracy of the transmitted data (Miller et al., 2025). The hardware components, like the high-speed very large scale integration (VLSI) coding chips, were tested at rates reaching 700 Mbits/s, showcasing the viability of these technological advancements for various high-data applications. Therefore, the work undertaken by NASA exemplifies a successful marriage between theoretical concepts and their realization, ultimately pushing the frontier of what is achievable in source coding. Moreover, the flexible nature of these developed techniques has bolstered their applicability beyond specific domains such as aerospace. The introduction of preprocessors into coding modules allows for optimized handling of diverse data forms, which can be externally adapted to meet the standards required by these advanced systems (Barbarien et al., 2009). As a result, the mechanisms utilized in NASAs coding methods extend to fields like telecommunications, image processing, and high-quality audio applications. The adaptability of the coding stages enhances their generalizability across various data forms, providing a comprehensive approach to source coding. Consequently, the ongoing refinement and development of source coding techniques remain critical not only for specialized sectors but also for broader technological advances that demand effective data compression solutions (Figure 3.4).

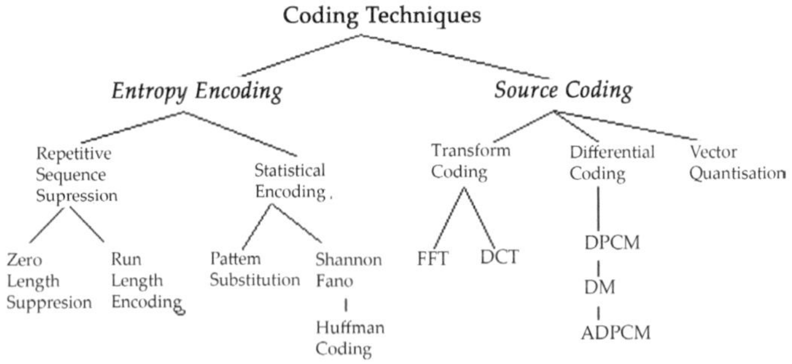

Figure 3.4. *Development of source coding techniques.*

Source: https://users.cs.cf.ac.uk/dave/Multimedia/coding.gif.

3.1.3. Key Milestones in the Field

The history of data compression and source coding is punctuated with some key milestones that have transformed the subject matter. Early on, Shannon's Information Theory being ushered in during the 1940s was a pivotal moment, enabling the quantification of information and defining the theoretical limits of data compression. This foundation allowed the development of several source coding techniques, of which the most important is Huffman coding, which maximized binary data representation through variable-length codes. These innovations allowed for the possibility of substantially compressing information without sacrificing important content. As technology progressed, effective compression techniques became increasingly needed in telecommunications and media storage, prompting further developments and systematic approaches to encoding data. Following the initial innovations, the 1970s and 1980s saw the creation of standardized compression algorithms further transforming the scene. Specifically, Lempel-Ziv (LZ) family of algorithms' introduction brought data compression through dictionary-based encoding schemes. This was particularly useful in uses such as file storage and transmission, where it helped redundancy in data to be handled easily. The commonly used LZW (Lempel-Ziv-Welch) algorithm was integrated into different technologies, such as GIF files and UNIX compress utilities, and its importance speaks for itself. In this period, realistic applications of compressive methods increased exponentially, pointing to how crucial adaptability and scalability are to meet the fast-changing digital environments. With the increase in data creation with the advent of the internet and electronic media, the demand for more efficient source coding grew in importance. This led to the development of complex compression standards, such as JPEG for image data and MP3 for music files, which offered a balance between quality and file space. Above all, technological advancements in data processing made it easy to apply these standards, as revealed by advancements in onboard data processing systems for satellites that optimized payload management and reduced data loss through access in a timely manner (Crawford et al., 2025). Similarly, the integration of advanced radar systems into geospace research demonstrated how new technologies optimized data collection and processing efficiency (Chau et al., 2019). Cumulatively, these milestones demonstrate the ongoing evolution of source coding and data compression, and the field's sensitivity to new technology and client demands.

3.2. PRINCIPLES OF SOURCE CODING

The source coding practices are founded mainly on the science of data representation in as little space as possible. Essentially, source coding aims at minimizing the information required to represent a source while maintaining the integrity and quality of the data. The operation involves realization of probability and statistical characteristics of the source, which are essential to determine how to compress the information. By techniques such as Huffman coding and arithmetic coding, redundancy in the source can be reduced to a high degree. These coding techniques leverage symbol frequency within a dataset, allowing more common symbols to be encoded in fewer bits than less common ones. This concept of tailoring representation based on the sources characteristics is pivotal in data storage and transmission maximization, and the comprehension of source coding constraints is therefore crucial in today's data-oriented age. Source coding deployment is therefore particularly prominent in contemporary communication systems, where data efficiency dictates performance and cost. In cutting-edge configurations such as cloud radio access networks (C-RANs), source coding concepts are central to maximizing spectral efficiency (SE) and energy efficiency (EE). The fronthaul—a highly critical transmission link between baseband units and remote radio heads—is often prone to capacity constraints that could adversely affect overall system performance. To bypass this challenge, techniques such as compression and quantization are a must. Such approaches not only alleviate the realities of scarce bandwidth but also ensure quality of service and user experience. The successful integration of source coding principles in C-RANs proves their generalizability in modern communications, exhibiting how new coding paradigms have the potential to transform traditional network operations (Lau et al., 2015). Apart from that, with wireless communications technology entering fifth-generation ground, source coding takes a more prominent role. The growing need for high rates and low latency guarantees the significance of advanced coding methods capable of meeting users' and emerging technology's demands. With the implementation of resource allocation optimization and massive coordinated processing, systems are enhanced through strategic utilization of source coding techniques. To address the complexity of modern data environments requires not just knowledge of coding mechanics but even an understanding of the overall networking environment, including software-defined networking and network function virtualization issues. At the intersection of these elements, the source coding principles emerge as the number one focus area, weighing the relationship between efficient data representation and robust communication networks (Barbarien et al., 2009).

3.2.1. Entropy and Information Theory

Entropy is the central information theory concept as a measure, numerical in character, for disorder or uncertainty within a data set. Physicists use entropy to measure the amount of disorder in a physical system, and information theorists use the same concept to investigate data compression and transmission. Increased entropy here refers to a greater degree of randomness of data points such that the data is more complicated and, maybe, less compressible. Decreased entropy implies increased redundancy and therefore increased scope for effective source coding techniques that can eliminate redundancy without compromising the integrity of the message. An understanding of entropy facilitates the best coding schemes, which are essential in effective transmission and storage, providing a solid foundation for source coding and data compression techniques growth. From the understanding of entropy, information theory also supports the viewpoint that various methods of data compression have a role to play in enabling more effective storage of information.

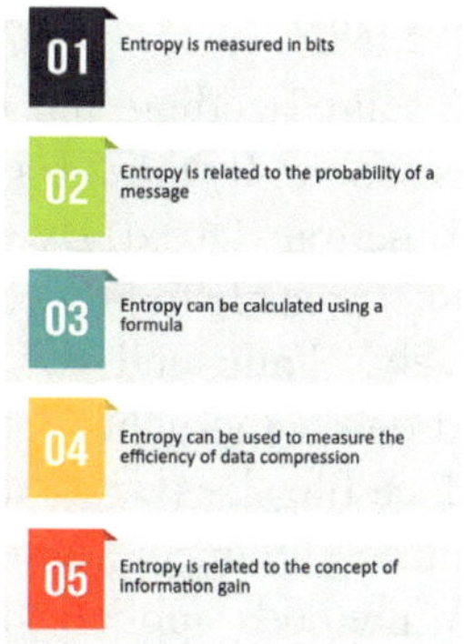

Figure 3.5. *Entropy in information theory.*

Source: https://fastercapital.co/i/Entropy--Understanding-Entropy--The-Role-of-E-in-Disorder--Entropy-in-Information-Theory.webp.

Employing methods that take advantage of redundancy present in low-entropy information, professionals can construct more streamlined information representations. Most traditional coding strategies strive to minimize bit representation, whereas new innovations take into account the addition of automation in extracting information from character strings. Studies indicate that novel techniques incorporating compression techniques can not only shrink the size of information but also calculate similarities and differences among a number of datasets. This versatility opens avenues in numerous areas, including language processing and classification of information, where precise measurement of the information content provides an avenue towards reliable classification and comprehension of sophisticated sets of information. The exchange of compression with practical use further exemplifies the critical importance of malleability to coding schemes. Since the digital domain keeps transforming constantly—most especially in cloud computing settings where data is continually being modified—ensuring that data is efficient as well as easy to access promptly becomes necessary. Research indicates that achieving balance between the efficacy of compression and ease of incorporating continuous updates is a challenging endeavor. Sophisticated coding methods that allow codewords to be reused without compromising on data representation compactness can neutralize the cost of large-updating. Therefore, this finding presents a compelling case for reevaluating traditional data storage schemes, and the necessity of codeword development schemes that compress as well as provide room for adapting to new data needs (Goyal et al., 2016) (Figure 3.5).

3.2.2. Redundancy Reduction

Redundancy elimination is a source coding and data compression technique building block. Redundancy is essentially the extra data that contributes nothing to the required information to be transferred. By eliminating this redundancy, many data compression methods are able to propel communication and storage systems much more effectively. Static and dynamic coding strategies, for instance, accommodate based on data structure so that encoding can be done optimally. These methods reduce file sizes without compromising the integrity of the original information, a necessity in environments with limited bandwidth and storage space. Reduction methods applied to redundancy affect not just the data, but the general system performance for systems that are committed to data processing and transmitting (Hirschberg et al., 1991). One of the prominent benefits of redundancy reduction is its ability to boost system speed and efficiency. With parallel algorithms, data compression can be performed simultaneously, which makes it faster. This is particularly relevant when huge volumes of data must be passed quickly, such as in telecommunications and streaming. Employing a combination of various compression methods can also enhance the efficiency of the compression process, achieving greater compression ratios at the expense of quality. Theoretical and empirical analysis of these algorithms guide us to analyze their effectiveness and lay the foundation for further research which may lead to optimalization of these techniques (Hirschberg et al., 1991). These developments not only ease data handling but also enable more complex systems to rely on instant data transfer. Moreover, techniques of redundancy reduction are also used in advanced fields, such as video signal processing, where bandwidth compression is a significant factor. In analog data digitized, redundancy might have numerous different forms and could require specialized steps to achieve the best compression. Through examination of the various data compression approaches, researchers can generate focused solutions that reverse the unique characteristics of varied types of data. This versatility not only improves efficiency in compression but also improves the overall user experience in situations such as video streaming, where uninterrupted playback is paramount. Continued research into compression methods for bandwidth shows a significant step towards maximizing use of available resources versus ever larger demands for high-quality data communication. These reductions in redundancy form a basis for continued advances in technology and communications policy.

3.2.3. Lossless Versus Lossy Compression

In source coding and data compression, lossless and lossy compression represent the essential distinction for understanding how digital information is optimized. Lossless compression codes, such as Huffman coding and Run-Length Encoding, strive to preserve the original quality of data in its pure state, such that the original file can be retrieved completely. On the other hand, lossy compression algorithms like JPEG for images and MP3 for audio on purpose throw away some of the information deemed less important, shrinking file size at the cost of continuously diminishing quality. This choice normally depends on the specific application; for instance, text document file archival storage might demand lossless systems as a top priority, while multimedia files, where slight degradation will be acceptable for storage and bandwidth considerations, can successfully employ lossy methods. The practical uses of deciding between lossy and lossless compression are more than theoretical uses. As an example, data storage in science often encounters high levels of information, which demands efficient compression techniques that can impact processing speed as well as storage capacity. Herein, techniques used for lossy compression—like SZ and ZFP—come in handy for compressing to optimized ratios with the level of data distortion under control, which is important in fields where measurement precision is paramount. Advances indicate an algorithm automatically selecting the best compressor to utilize based on data patterns allows forecasting the most fit compressor to utilize based on data attributes so that efficiency can be achieved with a minimal amount of distortion with heterogeneous datasets. This adaptability points toward the importance of dedicated compression methods in scientific studies, which generally involve cross-disciplinary research with diverse data requirements (Cappello et al., 2019). Moreover, the application of compression techniques has a central role in disseminating and utilizing geospatial data, illustrating the utility benefits of efficient data management. Computer maps, for instance, will have extensive geodata like necessary coordinates and thematic information vital to various applications, particularly in military and navigation uses. Very detailed maps' huge file sizes can cause slow downloading times and increased requirements for storage, making them hard to apply in internet environments. By employing advanced lossy compression methods, such as wavelet technology, image file sizes can be greatly reduced while maintaining sufficient quality for practical use. Not only does the compression make the loading process easier but also enhances the user experience on applications based on web-based picture libraries, thereby making information more accessible and facilitating effective decision-making (Sheriff et al., 2005) (Figure 3.6).

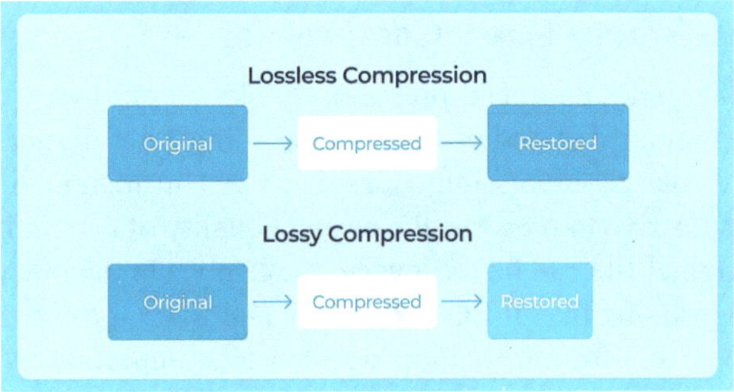

Figure 3.6. *Lossless versus lossy compression.*

Source: https://mlfk3cv5yvnx.i.optimole.com/cb:HA53.300ea/w:1024/h:535/q:mauto/f:best/https://www.ninjaone.com/wp-content/uploads/2024/04/N1-0921-Lossy-vs-Lossless-Compression-blog-image-1-2.jpg.

3.3. COMMON SOURCE CODING TECHNIQUES

In the field of compression of data, conventional source coding techniques are also the solution to compressing the data size without affecting its integrity. One of the best-known techniques is entropy coding, which maximizes the statistical performance of the source by assigning shorter codes to the more probable symbols and longer codes to the less probable symbols. This technique is particularly significant in application such as text compression, where recurring characters may efficiently be encoded, thereby achieving enormous space savings. The application of Huffman coding algorithms also depicts this technique, as they build variable-length codes dynamically based on symbol frequency to achieve maximum data representation. Nonetheless, conventional techniques do not manage to capture the diverse and complex nature of different communication systems, and therefore the need for the pursuit of novel paradigms that fit better the particular needs of different situations, like those in broadcast systems (Effros et al., 1999). Progress in universal source coding has also been notably helpful to the field of common coding methods, particularly for discrete stationary ergodic sources. The objective is to build coding schemes that perform well for a wide class of probable source distributions without having information regarding the specific details of the source. Recent developments indicate that one can achieve good universal noiseless variable-rate codes with small redundancy even for these challenging sources. For instance, the implementation of small maximum redundancy with relatively short block lengths verifies that such techniques are capable of tackling actual data compression cases like telecommunications and streaming. As the need for efficient transmission of data continues to increase, it becomes increasingly apparent how much these universal methods matter, offering higher flexibility and performance compared to traditional coding methods (Davisson et al., 1981). Furthermore, the formulation of source coding techniques calls for extension of thought beyond discrete coding schemes to their application over a range of broadcast systems. Classical models treat communication as a series of discrete transmitter-receiver pairs, limiting potential efficiencies obtainable from a systems level perspective. Taking a systems view of the broadcast system allows researchers to explore new coding paradigms that exploit the inter-receiver dependency among multiple receivers. Such a paradigm shift has promise to allow for improved compression ratios and more effective delivery of information, addressing the growing need for simultaneous dissemination of information on numerous channels typical of current multimedia and networked systems. adopting such an all-encompassing approach may not only enhance existing coding practices but also lay the groundwork for future advancements in source coding and data compression strategies (Effros et al., 1999).

3.3.1. Huffman Coding

The secret to effective data compression is disciplined elimination of redundancy, a precept realized in Huffman Coding. This algorithm, invented by David A. Huffman in the 1950s, is a classic example of entropy encoding, in which variable-length codes are supplied for input characters in proportion to their occurrence frequencies. The fundamental hypothesis is straightforward: increasing occurring characters are assigned shorter codes, leading to the data size as a whole being reduced. This principle not only optimizes space utilization in storage and transmissions but also optimizes the speed of processing data systems. The adaptability of Huffman Coding allows it to be applied to a myriad of applications, ranging from image file formats to text file processing. Its capacity for minimizing bits per symbol is in direct accordance with the broad goals of source coding and data compression in the sense that minimizing data with no loss is foremost (Hirschberg et al., 1991). As used in actual applications, Huffman Coding holds its own significance with special benefits, particularly in systems with a high redundancy rate. This is seen when traditional fixed-length coding schemes are compared, which can be space-wasting by giving equal lengths uniformly irrespective of the frequencies of the characters. With Huffman Coding, variable codeword lengths allow datasets to be more efficiently represented, supporting compression rates that are greater than static ones.

Additionally, the algorithm's ease of implementation in both static and dynamic forms increases its usefulness in several applications. Static Huffman Coding pre-arranges the code table, while dynamic Huffman Coding facilitates live adaptation against recurring changing input data. This flexibility is quite critical in the modern hectic digital era, where both volume and diversity of data created require robust and efficient compression strategies for speedy processing and delivery (Hirschberg et al., 1991). The integration of Huffman Coding with modern compression techniques proves its continued relevance in the field of data compression. For instance, its principles can integrate with sophisticated video compression formats such as H.264 to enable the ease of data transmission in streaming applications.

By leveraging residual information following first-time compression, such as in certain approaches with binary autoencoders, Huffman Coding can optimize the quality of received data further, ensuring bandwidth restriction without video loss. This double-layer approach not only enhances the strength of Huffman Coding in contemporary settings but also aligns it with ongoing research into greater efficiencies in data handling. The heritage of compression technologies testifies to the necessity of rudimentary algorithms like Huffman Coding, which is a scaffolding on which more sophisticated frameworks can be built, both historically used and applicable today for source coding and data compression (Figure 3.7).

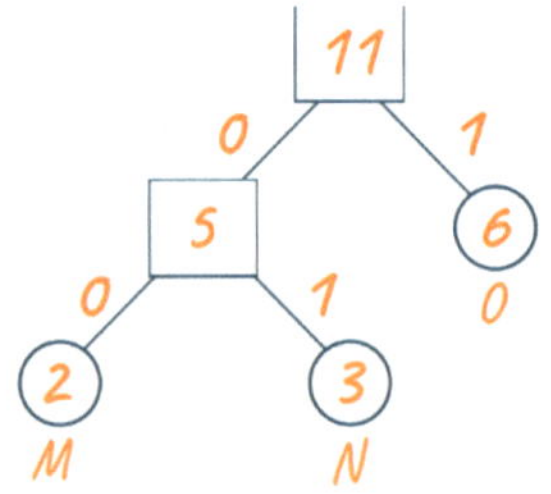

Figure 3.7. *Huffman coding.*

Source: https://favtutor.com/resources/images/uploads/Huffman_Coding_Python.png.

3.3.2. Arithmetic Coding

Arithmetic coding is a core technique in source coding and data compression that significantly differs from other standard methods like Huffman coding. Unlike Huffman, which codes symbols with varying binary codes based on their frequencies, arithmetic coding encodes the entire message to a single number between [0, 1]. The method is more efficient, particularly for sources whose probability distribution can be represented in a more complex manner. By representing the probability of each symbol occurrence, arithmetic coding can more effectively handle variable-length messages and thus minimize redundancy in the compressed information.

This inherent flexibility makes arithmetic coding eminently capable for a broad spectrum of applications, from text compression to multimedia encoding, and priceless for making data transmission and storage systems more efficient. The arithmetic coding is effective since it is adaptive and has the ability to utilize real-time frequency analysis during data processing.

The adaptability is also crucial when dealing with varied symbols with non-stationary probability distribution. Techniques such as the Directed Acyclic Word Graph (DAWG) are employed in modeling the source where symbol probability is approximated based on previously seen data. As explained, the probabilities of the next letters can be calculated based on the longest suffix existing in the encoded text, giving more precise approximations of the next symbol occurrence probability. Such a detailed model gives arithmetic coding a significant advantage, inclined to achieve superior compression ratios compared to independent-letter models and other non-adaptive techniques, showing its ability to enhance data compression levels effectively (Blumer et al., 2025).

In addition, arithmetic coding needs computational complexity and memory usage to be properly considered for implementation. Although it is superior in compression performance, practical implementation involves the use of efficient encoding and decoding algorithms to be feasible.

Parallel algorithms have since been developed to overcome the obstacles by enabling data to be coded simultaneously, improving system throughput while decreasing latency. By combining multiple compression algorithms into one, the overall compression rate can be dramatically improved, and effective aid in the high-speed transmission of data. In this way, not only does it prove arithmetic coding's scalability but also illustrate ongoing innovations in the area of data compression, pointing towards a promising line of development for research and application in both theoretical and practical fields (Hirschberg et al., 1991).

3.3.3. Run-Length Encoding

The concept of Run-Length Encoding (RLE) emerges as a suitable method within the context of source coding and compression of data, specifically for information that contain repeated patterns. With effective encoding of series of identical characters into a pattern that describes the character and frequency, RLE demonstrates a powerful use of space. For example, the string AAAABBBCCDAA is reduced to 4A3B2C1D2A, significantly reducing the amount of data that must be stored or transmitted. This method is particularly effective in uses like bitmap images or text files, where redundancy is present, allowing for valid decreases in file size without compromising the integrity of the original data. This compression technique is very useful in bandwidth-restricted environments, allowing for effective management of lots of data, hence optimizing overall efficiency in storage and transfer of data. One of the greatest advantages of RLE is its simplicity to implement and computational efficiency. Since the encoding process does not do multiple passes over the input data, it takes linear time and hence can be scaled. This characteristic of RLE makes it such that even for large datasets, encoding and decoding are still possible, hence enabling real-time applications. What, however, should be noted is that RLE is able to function best under specific circumstances. When data contain large stretches of repeating symbols, compression ratios can be enormous; but when the data is highly variable, RLE can make files larger and defeat its purpose. This limitation defines the necessity of selecting suitable compression algorithms based on the type of data being compressed, thereby increasing the complexity of the compression setting as identified in research such as various bandwidth compression techniques used in video signals (Ingels et al., 2025). An interesting part of data compression is discovered when examining the incorporation of RLE within more sophisticated contexts, such as the widely used Deflate compression algorithm. Deflate's application of prefix-free encodings and backreferences shows the direction of combining RLE with other compression technologies to maximize overall efficiency. As studied in recent research, mathematical formalizations of coding schemes within these models can maximize compression and decompression algorithms and actually address the primary issues of memory and runtime expense. These developments indicate a growing tendency toward enhancing compression algorithms by strict mathematical specifications and implementations, which consequently suggest new avenues for Run-Length Encoding applications optimization. By situating RLE in this larger context of data compression, or more particularly its successful implementation, the method's applicability to contemporary data issues becomes apparent, asserting itself firmly into the present discourse on source coding and efficiency (Appel et al., 2016) (Figure 3.8).

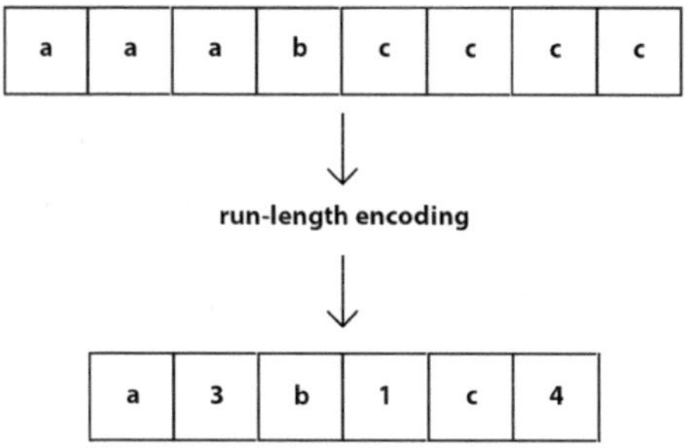

Figure 3.8. *Run-length encoding.*

Source: https://img.api.video/run_length_f9030faa12.png.

3.4. APPLICATIONS OF DATA COMPRESSION

Data compression is at the heart of many applications, particularly for telecommunications and web usage. By reducing the size of data being sent over networks, data compression significantly increases bandwidth as well as the general data transfer rate. For example, cellular communications benefit from the application of advanced source coding techniques to efficiently transmit large amounts of information. This efficiency is crucial, given that mobile data usage has been on the surge, placing pressure on infrastructures present. In addition, data compression techniques have a direct impact on user experience by enabling faster downloading and smoother streaming, thereby providing high-quality content with minimal or no delay. These advances point towards the integral role played by data compression in providing modern communications needs in an increasingly digitalizing world. Digital cinema data compression has revolutionized the production and distribution process. Hitherto dominated by film, the sector has shifted towards digitals, for which high-definition content needs to be transmitted. The integration of optical communication technology with sophisticated video compression algorithms enhances the process of film creation and exhibition. According to research, these advancements in optical networking substantiate the ability to carry bandwidth-intensive traffic required for digital cinema effectively. With the implementation of data compression techniques, video filmmakers have been successful in minimizing excess visual quality and as much storage capacity and bandwidth as required, thereby maximizing the overall efficiency of filmmaking. This revolutionizing effect reminds us of the importance of data compression in modern-day cinematography, enabling cinematographers to focus on creativity while simultaneously addressing technical matters. Data compression is also in use in many day-to-day applications, from multimedia storage to cloud computing. For instance, audio and video files are bulky, and data compression enables users to store and transfer files without loss of quality. In cloud storage, with perhaps limited disk space, high-quality data compression methods yield the optimum use of the user's storage space at the lowest possible cost. Besides, new compression algorithms continue to improve, supporting various types of files and operating satisfactorily on a broad variety of platforms. Therefore, data compression not only facilitates smooth user experiences but also enables data management system efficiencies. Briefly put, the applications of data compression extend far beyond technical parameters, influencing content creation, delivery, and consumption within our networked digital life (Figure 3.9).

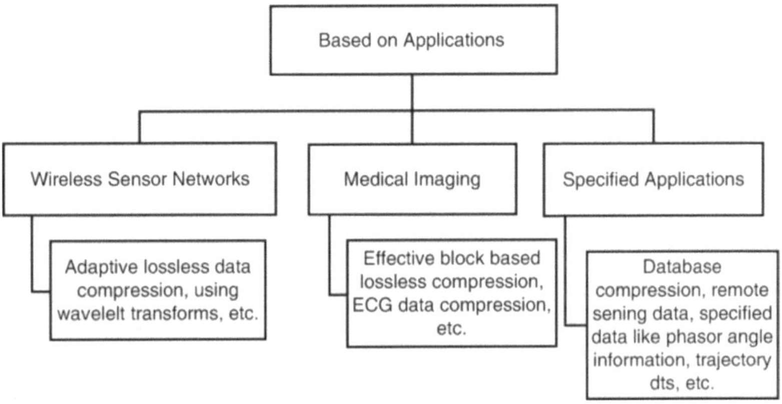

Figure 3.9. *Data compression and its application.*

Source: https://media.springernature.com/lw685/springer-static/image/chp%3A10.1007%2F978-3-030-12575-2_1/MediaObjects/472726_1_En_1_Fig5_HTML.png.

3.4.1. Multimedia Data Compression

The explosion of digital multimedia data demands efficient storage and transmission, and this is where the performance of multimedia data compression becomes vital. Text, audio, and video are varied types of multimedia data with varied behaviors and varying compression needs. This implies that varying procedures should be used based on the type of material to achieve the optimal balance between file size and quality. For instance, lossless compression schemes such as Huffman Coding and Arithmetic Coding are best suited for text and other data types where every bit of information matters. Conversely, lossy compression schemes such as those used with audio and video streams reduce file sizes while losing some loss of quality, which is handy for streaming and distribution purposes. Tarmizi et al. (1995) is also indicative of the importance of applying appropriate methodologies so that multimedia accessibility is maximized and storage space utilized efficiently. The study into the primary means of multimedia data compression presents a scene in which lossless and lossy solutions run parallel, each addressing different needs. Lossless compression makes it possible to exactly reconstruct the original data and therefore is essential for fidelity-based applications such as text files or archive photographs. On the other hand, lossy compression sacrifices some quality in order to deliver more reduction in file size. Techniques like Predictive and Frequency-Oriented coding are very efficient in compressing large amounts of audio and video data to acceptable quality. Contemporary standards such as JPEG for still image compression and MPEG for video compression are hybrid techniques that benefit from the use of both these techniques and therefore allow flexibility for use in multimedia applications. Tarmizi et al. (1995) emphasizes that the right method of compression must be selected so that viable compression ratios can be obtained as well as integrity of content. The implications of multimedia data compression are more than mere technical efficiency; they establish the usability and viability of accessing multimedia applications across various platforms. For instance, without effective compression techniques, the huge data needs of high-definition video or rich audio would render them not feasible for ordinary consumers. As noted in current studies, choosing a suitable compression technique will have to consider the user experience at the end, where slight losses in quality in lossy compression are readily accepted by the average viewer or listener. Thus, the deployment of suitable compression methods is not just a technical parameter but also about achieving maximum user engagement and interaction with web content. Additionally supports the proposition that compression of multimedia data is very important in increasing the spread and applicability of multimedia technology to the broader populace.

3.4.2. Data Transmission in Networks

The transfer of information in networks is at the core of facilitating communication between various platforms and technologies. With networks evolving further, efficient data transfer becomes increasingly important, particularly in the management of bandwidth and handling of resources. Source coding and data compression methods are the backbone of guaranteeing successful data transfer by compressing the data to be sent, hence conserving bandwidth as well as enhancing performance. This is especially so in today's scenarios, where high-speed data transfer is ever increasing in demand. Both lossless and lossy compression algorithms must be used appropriately depending on the application requirements, so that the integrity of the data is not compromised but its space during transit is minimized to the extent possible.

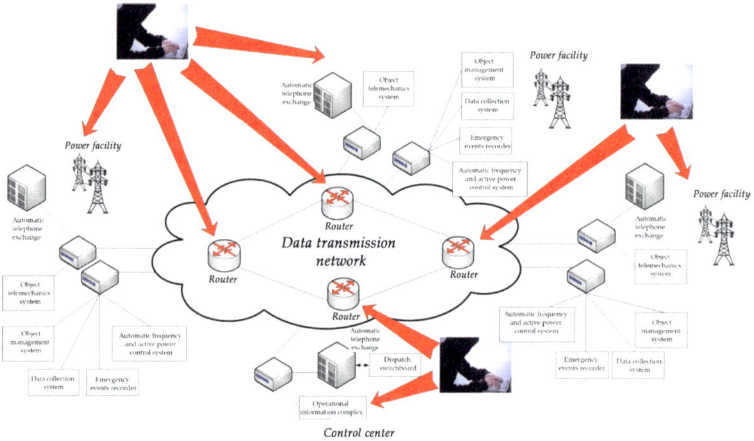

Figure 3.10. *Data transmission in networks.*

Source: https://www.mdpi.com/energies/energies-14-04755/article_deploy/html/images/energies-14-04755-g001.png.

In cloud radio access networks (C-RANs), issues concerning the transmission of data are even more pronounced, and those concerning the fronthaul interface, connecting the baseband unit to the remote radio head, are especially critical. Proper management of this link is required to ensure that the spectral efficiency (SE) and energy efficiency (EE) of the network are sustained (Lau et al., 2015). Compression algorithms are important here because they facilitate high rates of information processing without filling up the fronthaul's capacity. Quantization and extensive coordinated processing are required for compensating for bandwidth constraints. Through such processes, networks are able to provide quality of service while transmitting information in an efficient manner, emphasizing the need for source coding as well as compression in modern network designs. The dynamic nature of wireless networks calls for flexible data link layer protocols that adapt to changing traffic conditions while ensuring Quality of Service (QoS) reliability. Energy efficiency and adaptability across the entire protocol stack can play an important role in the efficiency of data transmission. By using feedback mechanisms, programs can tailor streams of data according to real-time situations such that bandwidth scheduling and mechanisms of error control can be improved (Havinga et al., 2000). This cross-connection-based approach facilitates a stronger network infrastructure capable of supporting various kinds of traffic in an efficient way. Hence, the compatibility of source coding, data compression, and adaptive network protocols plays a vital part in fulfilling the complex data transfer demands of continually expanding network systems (Figure 3.10).

3.4.3. Storage Optimization

The concept of storage optimization is a central aspect of the effectiveness of source coding and data compression algorithms, particularly as digital information accumulates at increasingly exponential levels. Sophisticated coding techniques are used by storage optimization to achieve the most space available through efficient utilization, reducing the storing and retrieving of information to a minimum. Complex networks, such as cloud radio access networks (C-RANs), can lead the way to new approaches to storage that maximize function to compensate for resource constraints. There is high energy and spectral efficiency in these networks, which enable one to provide numerous applications without expending extravagant amounts of money. The use of compression algorithms is also critical in addressing the constraints of scarcity in terms of capacity in fronthaul links, thus enhancing storage efficiency and user experience in such interconnected systems (Lau et al., 2015). Storage optimization is also not a matter of simply maximizing space but ensuring that the integrity and availability of information are maintained. Since software-as-a-service apps keep evolving, the demand for updates and changes to data stored is frequently realized. The challenge this presents is how one can ensure data remains compressed yet still provide access to efficient update and change. Studies indicate there is a tradeoff between update malleability and compression efficiency. Achieving both compressed states and operational simplicity is vital, particularly in the context of dynamic datasets where the latest version must be stored without creating unnecessary older versions (Goyal et al., 2016). Therefore, storage optimization is not merely compressing data but also ensuring it is easily editable and retrievable. In addition, the continuously increasing demand for efficient storage optimization is exacerbated by the enormous growing demand for data-driven decision-making across various industries. The use of high-performance compression algorithms is capable of significantly diminishing dataset size without compromising data quality. Optimizing resources for optimization becomes paramount in environments with large-scale data and tight bandwidth limitations. By analyzing constraints typical of storage systems, for example, memory limits and processing capability, optimizing solutions can be planned to enhance both efficiency and access. Considering this, emerging storage technologies need to consider capacity of high order and the need to support future user demands, paving the way for future source coding and data compression advancements (Lau et al., 2015; Goyal et al., 2016).

3.5. CHALLENGES IN DATA COMPRESSION

Data compression is a critical component in achieving the utmost efficiency in data transmission and storage. However, there are many complexities in the process, primarily because of ensuring compromises between data quality and compression ratio. With the ever-changing digital landscape, lossless data compression is increasingly needed, especially for operations handling sensitive information. To address these intricacies, scientists must remember not only the technological need of compression techniques but also where they are going to be applied. For instance, despite the fact that aggressive compression can minimize files, it tends to detrimentally degrade data integrity. This problem necessitates the creation of new approaches that maintain high levels of quality while effectively minimizing data footprints, a problem still existing in the data compression arena.

One of the significant challenges in data compression today is performance when handling huge amounts of data, specifically in applications such as cloud computing and wide-area wireless communication. Based on recent advancements in cloud radio access networks (C-RANs), the fronthaul—the transmission link from baseband units to remote radio heads—is typically under bandwidth pressure and high capacity demands (Lau et al., 2015). These constraints create sophisticated means of mitigating their impact on system performance, usually in the form of complex resource allocation and compression methods. Offering solutions to these problems entails relentless innovation to ensure that used data compression techniques do not yield significant loss of quality or increased operational costs, hence pursuit of more robust solutions is a point of interest for researchers. The problem of achieving effective parallelization in lossless data compression is yet another hindrance that decelerates processing without losing out on any quality. Recent research has also suggested approaches that include algorithms that focus on maximizing throughput from multiple computing resources in parallel, splitting input streams into blocks to enable parallel processing (Baron et al., 2015). While this method has the capability to significantly increase handling speed, it sacrifices compression quality when not applied judiciously. With the estimation of the source minimum description length and the close monitoring of the encoding of each parallel block, researchers can maintain an improved trade-off between performance and speed. Thus, solving this problem also relies on improvements in algorithms but also on the realization of the trade-offs inherent in the different compression methods, a problem of an n-dimensional nature in data compression.

3.5.1. Balancing Compression Ratio and Quality

Balancing quality and compression ratio is a very important operation in the field of data compression and source coding, particularly in applications where data must be handled with greater efficiency. The quality of a compression algorithm is typically measured according to its ability to compress data without affecting its fidelity. In applications such as multimedia processing and storage, making this balance difficult is increasingly becoming the case since greater compression ratios can lead to loss of quality at the expense of user experience. It is thus under this challenge that developers need to implement mechanisms that minimize such quality degradation while maximizing compression of data size, exemplified in advanced methodologies employing techniques such as context-adaptive coding and predictive modeling.

Figure 3.11. *Balancing compression ratio and quality.*

Source: https://fastercapital.co/i/Exploring-RAR-s-Impressive-Compression-Ratio--What-It-Means-for-Your-Files--Impact-of-Compression-Ratio-on-File-Quality-and-Integrity.webp.

Processor performance improvements in recent years have further affected the trade-off between compression and quality. Researchers have also explored the use of superscalar processors and Single Instruction, Multiple Data (SIMD) instructions as ways of speeding up compression. An example is the SIMD-BP128 scheme, which significantly accelerates compression without loss in quality. This method achieves roughly twice the speed of earlier techniques without reducing data compression by as much as 2 bits per integer, illustrating the way progress in computation can capitalize on synergy between efficiency and speed of data compression. Such advances are critical in situations where time pressure is no less critical than data quality being processed, especially in high-frequency data contexts. Besides, graph signal processing (GSP) application is an innovative approach to trade-off compression and quality, particularly for irregular data structures. Through performing digital image models as graphs' signals, researchers utilize graph spectral methods to efficiently compress and process pictures. The new paradigm can process images without destroying their original structure and leads to greatly enhanced compression quality. GSP-based methods, including those for image segmentation and restoration, indicate that trade-offs between compression and quality are not merely technical problems but also an opportunity to rethink perception and data handling in intricate networks. Thus, as the field progresses, this equilibrium is a fundamental concern for researchers and practitioners alike in seeking to promote not just effectiveness but also quality of digital information transmission (Figure 3.11).

3.5.2. Computational Complexity

The intricate interplay between computational complexity and source coding presents a fundamental paradigm for the understanding of data compression techniques. Source coding to efficiently represent data for storage and transmission inherently involves complexities in the feasibility and effectiveness of the compression algorithms. The computational requirements for processing, encoding, decoding, and correcting errors become paramount as the need to process large data sets grows in the digital era. The handling of such assignments is generally reflective of the constraints imposed by NP-hard problems, and researchers are forced to seek the best compression strategies that balance complexity against efficiency. As a result, algorithm design research is central to optimizing performance in data-driven applications, necessitating continuous research to develop approaches that minimize the computation burden while delivering optimal compression effectiveness. In studying applications to specific contexts, distributed video coding (DVC) is a case in point of the interrelation between computational complexity and effective data compression. The application by DVC of the Slepian–Wolf and Wyner–Ziv theorems illustrates how theory is applied in innovative solutions in novel areas like wireless video sensor networks (WVSNs). The structure of DVC exploits both the correlation among distributed sources and the computational paradigms that underlie data representation and transmission. Recent history of progress on DVC shows its potential for efficient coding schemes that resist diverse operating conditions. This stream of research not only addresses the intricacies of source coding but also makes advancements in applying complex coding structures to practical use, thereby propelling the efficacy of compression methods in applications where latency is low and dependability is high, promoting further the need for sophisticated computational methodologies within this realm (Fong et al., 2015). The functioning of the coding algorithms has a close relation with their computational complexity, as noted in the performance study of the MN-Gallager code on Galois fields. The code depicts the intricate relationship between autocorrelation, types of messages, and the channel conditions, which ultimately influence the efficiency of source-channel coding techniques. Results indicate that the Dynamical Block Posterior algorithm's performance is superior relative to traditional compression algorithms such as gzip, indicating the promise of high-capacity data transmission. Also, the observed evidence of reduced performance with higher correlations indicates the balance between theoretical advancement and practical limitations faced by engineers in real-time application. As coding techniques keep advancing, understanding how computational complexity affects the performance of algorithms will always remain important, as more and more demands for handling complex data keep on rising, illustrating the significance of sophisticated computational approaches to efficient data compression strategies (Figure 3.12).

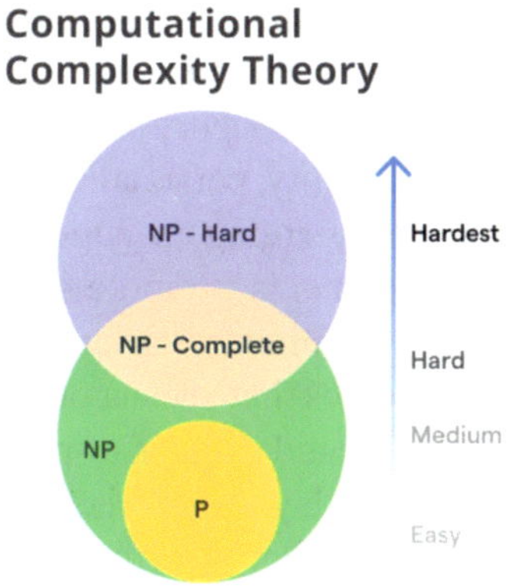

Figure 3.12. *Computational complexity.*

Source: https://cdn.botpenguin.com/assets/website/Computational_Complexity_Theory_cb0b0bf408.webp.

3.5.3. Evolving Data Formats

The evolution of data formats has significantly affected the use of data compression and source coding, particularly in the transmission and storage of information. As digital technologies evolve, consumers demand higher quality and more effective media experience, necessitating the invention of new data formats.

For instance, traditional film cinematography has evolved to digital cinema, which supports high-quality content that is less susceptible to degradation. This transition calls for strong compression techniques to manage the higher volume of data that is ushered in by high-definition video content. The new applications of optical communication technologies enable this bandwidth-hungry platform, efficient transmission at a cost to quality, which is a necessity for modern media consumption and production processes (Kaleem et al., 2016). Besides, the integration of various forms of media into combined platforms has brought about networked media, which has altered the understanding and utilization of data formats. With an estimated one billion individuals connected to the Internet and exponential increases in multimedia content, there has never been a larger demand for efficient source coding methods. Networked media is a diversity of modes—text, sound, images, and video—that are produced, kept alive, and retrieved from across connected networks. The diversity of these modes makes it challenging for data compression techniques, which need to cater to variable demands of speed and quality and operate effectively over a range of network types (Aggoun et al., 2008). These evolving needs underscore the need for ongoing research and development in data formats and compression algorithms to enable more compelling user experiences on different platforms. Aside from the challenge introduced by variable format, the future of data compression will also be shaped by developments in transport protocols and coding schemes. With consumers demanding high-quality, personalized multimedia experiences, new data delivery standards are emerging to support richer, more dynamic media. Not only do these developments promote improved quality, but they also encompass interactive and community-oriented applications, requiring versatile source coding methodologies that can accommodate evolving network conditions and user habits. With the aid of peer-to-peer networks, machine-to-machine communication, and cross-layer optimizations, next-generation data formats can be designed to better accommodate the evolving paradigm of digital media. The marriage of technological innovation and user-friendly design will also continue to propel the evolution of data formats and compression methods critical to addressing the demands of modern digital environments (Aggoun et al., 2008).

3.6. FUTURE TRENDS IN SOURCE CODING

Since the need for efficient data transmission grows stronger, the trends in source coding in the future are increasingly likely to coincide with advances in machine learning and artificial intelligence. Both technologies have the potential to revolutionize compression algorithms through the ability to adapt dynamically to data patterns in real time. Through machine learning, and particularly deep learning, it becomes possible to design models based on neural networks capable of learning optimal coding procedures from enormous datasets. This adaptability is of particular importance in high variability applications, such as multispectral imagery markets, where normal coding techniques would be found unsuitable. Such integration of advanced methodologies can lead to higher efficiencies, as quoted by available literature that emphasizes the use of proper compression techniques appropriate for individual datasets (Habibi et al., 2025). Thus, the future of source coding will likely see increased application of this technology synergy to meet the demands of diverse and dynamic data environments.

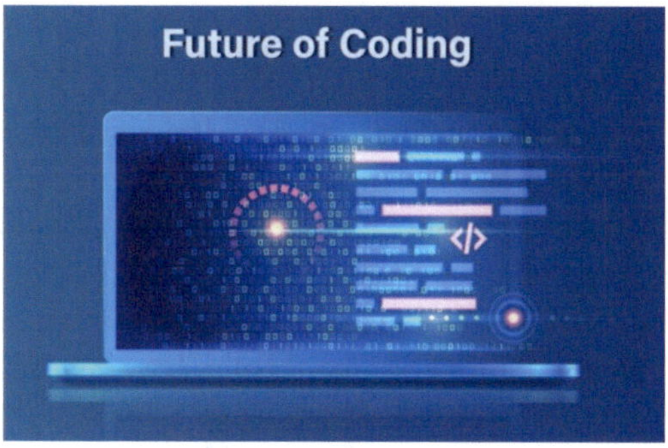

Figure 3.13. *Future trends in source coding.*

Source: https://static.habilelabs.io/future_of_coding_51b1fdc797.jpg.

Another source coding future trend is the concept of approximate sufficient statistics, allowing for more efficient data representation. Classical methods require rigorous data reconstruction, resulting in computational cost and perhaps gigantic code lengths, particularly in the processing of large volumes of independent samples. Recent studies suggest that using a Shannon-theoretic approach, one can achieve significant reductions in code lengths with the tolerance of some minute error margin in reconstructed data (Hayashi et al., 2017). This paradigm shift toward approximate solutions has a dual advantage: it significantly cuts down on the memory required for encoding without losing enough precision to diminish most real-world applications. As the digital age continues to advance and data creation increases, the development of such adaptive coding methods will be key to efficient data compression and transmission. Furthermore, the increasing complexity of data types such as those from IoT devices means source coding practices have to evolve. Traditional approaches may prove inappropriate for large and intricate data sets that originate from smart spaces, where the direction has to be towards scalable coding systems. Future trends may involve hybrid coding schemes that fuse both traditional entropy coding and new computational methods to offer a balanced approach to coping with diverse data formats. Also, the introduction of non-standard data types like video streams and multispectral data will widen the scope of traditional source coding methodologies. Ongoing research and development of robust compression systems with emphasis on adaptation and efficiency will be critical not only for the enhancement of compression ratios but also for reliability and data transfer speeds between platforms (Habibi et al., 2025). These trends are likely to radically alter the landscape of data compression and source coding in their realization (Figure 3.13).

3.6.1. Advances in Machine Learning Applications

The role of machine learning in the optimization of source coding and data compression has become increasingly significant, particularly with the development of joint source-channel coding techniques. Traditional approaches tend to interweave the source encoding and channel encoding processes, a process that has been proven effective in the past. But more recently, research has indicated that combined schemes are capable of delivering superior performance for incorporation in applications such as video communication, where error resilience and compression are required. With the application of machine learning techniques, researchers aim to create systems that learn adaptively to encode information without being explicitly programmed, outdoing the state of the art. This new technique not only makes the process simpler to encode but is also more adaptive in various channel conditions, thus embedding the changing capability within the field of machine learning (Akpabio et al., 2019). Yet another extremely interesting area of innovation arises from taking advantage of the concept of graph signal processing (GSP) for source coding and compression of data. In traditional data structures such as images, pixels are kept in a grid manner; but with GSP, it's possible to visualize data points in more complex form by representing relationships among pixels through graphs. By applying machine learning techniques to graph-based representation, it becomes possible to process and analyze images using their inherent structural information. This opens new avenues for the compression, restoration, filtering, and segmentation of images in which traditional methods would break down. Machine learning not just enhances the efficiency of processing but also the quality of result obtained through graph spectral techniques. These developments demonstrate the ability of machine learning to transform our understanding of source coding methods such that more advanced data processing can be performed (Cheung et al., 2018). The union of data compression and machine learning is also evolving, most significantly in the automation and optimization of coding techniques. With the integration of deep models for learning, researchers are studying end-to-end learning architectures that provide robust solutions to compressing and transmitting data. These models can learn automatic input data mappings, automatically learning to adapt to diverse compression needs without being told how. This versatility is especially helpful in variable environments where data feature tendencies may vary significantly. With machine learning continuing to influence source coding, the prospects for constructing highly efficient and resilient systems are enhanced, with interesting implications for several uses throughout communications and multimedia technology. Machine learning optimization represents a milestone in enhancing the capability of data compression algorithms.

3.6.2. Emerging Compression Algorithms

The landscape of compression algorithms is constantly evolving to adapt to increasing demands for effective data management between several applications, such as cloud computing and wireless communication. Emerging compression algorithms leverage advanced mathematical paradigms to enhance data redundancy elimination and minimize storage requirements. One of the novel developments entails introducing lossy data compression algorithms founded on constraint satisfaction gates. This new approach demonstrates that theoretical capability of these algorithms can tend exponentially towards Shannon's limit, and with superior performance with a larger variable number. These innovations not only provide higher efficiency but also practical utility in real-time operation, so it becomes feasible to apply them to scientific research as well as industry applications in data transmission and compression management (Shannon et al., 2005). In the case of modern transmission networks, particularly for cloud radio access networks (C-RANs), newly developed compression algorithms lead the way in optimizing the fronthaul connections between various network units. C-RANs can potentially reduce operational costs while increasing energy and spectral efficiency. Nevertheless, the fronthaul itself is the bottleneck due to capacity problems. Compression techniques are an essential part of alleviating these limitations as they make data transmission possible even with the inherent limitation in the network infrastructure. Recent studies elucidate how emerging techniques improve user quality of experience and system performance under constrained environments and supports the need for robust and adaptive compression techniques (Lau et al., 2015). The implications of such compression algorithm developments are broader than mere efficiency; they reflect a broader data processing paradigm shift. With more companies reliant on data-driven decision-making, the ability to process enormous amounts of data with low latency is essential. New algorithms don't just optimize capacity but also create new capabilities such as real-time analytics and self-processing. Research and implementation of these algorithms are necessary to make communications networks future-proofed and capable of meeting growing demands of a data-centric world. With continuous optimization of these compression methods, researchers and engineers can better address challenges posed by data growth size and complexity of transmission, positioning them in the development of source coding and data compression (Shannon et al., 2005; Lau et al., 2015) (Figure 3.14).

Figure 3.14. *Emerging compression algorithms.*

Source: https://casinobeats.com/wp-content/uploads/2022/07/Shutter-stock_1831742065.jpg.

3.6.3. Impact of Quantum Computing

The advent of quantum computing holds the promise of a new era for many computational models, not least the areas of source coding and data compression. While conventional computers process information in terms of bits, quantum computers use qubits, enabling them to perform complex calculations at unprecedented speed. This superior capability holds the promise of new compression techniques, since quantum algorithms can search vast collections of data more efficiently. A classic example is in the processing of scientific data where large quantities are generated by high-performance computing (HPC) software. As literature suggests, the significant reduction of data size is required to curb storage limitations and optimize data transfers, especially in the wake of current bandwidth bottlenecks (Cappello et al., 2019). Quantum computing is therefore set to revolutionize data compression into not only more rapid processing but even potentially improved compression ratios with new algorithms that take advantage of quantum mechanics. In designing compression algorithms, rate-distortion principles come to the forefront in addressing the compromise between data quality and compression ratio. Quantum algorithms are best at optimizing these parameters due to their inherent parallelism and ability to perform complex math operations. This advantage allows for real-time setting of compression techniques, such that the selection between available alternatives can be made automatically based on some properties of the data. For example, research into algorithms like SZ and ZFP demonstrates the applicability of such techniques using conventional computing. Quantum computing has the potential to enhance these techniques by making more accurate estimations of compression quality and adaptively changing the most appropriate compressor for various domains of data. As emphasized in recent studies, automatic optimization is able to achieve tremendous compression ratio enhancement—up to 70% in certain cases—without any loss of data integrity (Cappello et al., 2019). Further, the potential integration of quantum computing in data compression and source coding schemes can open up more resilient and efficient systems for managing data. With ever-growing environments surrounded by huge amounts of data, getting advanced compression tools is now more than ever an imperative requirement. Quantum-based methods not only enhance the processing burdens of handling vast data sets but also enable better utilization of network bandwidth, thus avoiding any data transfer bottlenecks. According to the study, programmable selection algorithms that utilized quantum capabilities were able to scale the computation demands of present-day applications with negligible overhead. Consequently, the impact of quantum computing on source coding and data compression goes beyond theoretical development in providing practical solutions that can respond to challenges posed by future and upcoming data landscapes.

SUMMARY

The field of source coding and data compression obviously shows the significance of these techniques to maximize data storage and transmission. With a review of various methodologies—lossless and lossy compression—it is apparent that a balance should be struck to a point of maximum between file size reduction and data integrity. With digital content saturating the internet, efficient coding techniques are a prerequisite. Techniques such as Huffman coding and arithmetic coding show effective methods employed to enhance compression ratios without losing vital information. The advances in source coding, aside from providing pragmatic benefits in terms of bandwidth utilization, facilitate rapid transmission of multimedia information, hence transforming the digital communication scenario. Furthermore, the impact of digital watermarking in combating piracy emphasizes the need for effective data protection methods in the age of unparalleled duplication convenience caused by technology. The solutions proposed demonstrate a multi-layered method towards intellectual property protection without undermining copyright integrity with the dividends of digital advances. The development of watermarking systems, more precisely wavelet transform-based systems, is a testament to the eternal struggle between innovation and security. These systems are warranted because the potential for generation and dissemination of digital copies without loss is an ambidextrous sword: while it encourages freedom of creation, it destroys the very texture of ownership in media. Lastly, the transition from theoretical models to practice, as illustrated by developments in rate-distortion theory, is a milestone in data compression. The convergence of theory and practice is indicative of ongoing improvement in algorithms designed to optimize performance without sacrificing efficiency. Notably, the advances in the design of multiple-model algorithms further improve the performance of data compression systems with significant performance increments. As technology advances, it is important that researchers stay alert, making sure the development of source coding methods not just meets the needs of the present but also prepares for the needs of the future in terms of data management and transmission, making the use of source coding as well as data compression all the more important in an increasingly advancing technological environment.

The role of source coding and data compression cannot be emphasized enough, especially in the data-centric world of today. This is the process by which data is converted into a smaller format, making it possible to store and transmit in an efficient manner without compromising the integrity of the information. Through the use of different compression mechanisms, digital file size can be significantly trimmed without losing quality, an essential for uses as diverse as online streaming to huge databases. For instance, geodata such as coordinates that are part of electronic maps make them more valuable in uses such as military strategy. But their volume, intrinsic to the data, can render them less convenient to use in network systems, triggering studies on the use of more advanced compression methods such as those using wavelet technology, which are more efficient than traditional methods (Sheriff et al., 2005). There are several schemes and approaches in data compression to accommodate different types of information and their specific requirements. Approaches vary from lossless compression, where the original information can be precisely recreated from the compressed version, to lossy compression, where there is some sacrifice on fidelity but more file size reduction is attained. The choice of compression scheme in some cases is also reliant on the context of application, specifically in weighing the trade-off between quality and efficiency. In video signal processing, for instance, bandwidth compression techniques play a crucial role in optimizing the dissemination of high-definition content without network resource overload. These approaches facilitate smoother user interfaces for multimedia or geospatial data retrieval, highlighting the multifaceted applicability of compression schemes within various digitally-mediated systems (Ingels et al., 2025). As a simple removal of files, the implications of effective source coding and compression also extend to novel applications in other areas, including cloud computing, telecommunications, and data processing. By capitalizing on compressed data, businesses can make their operating efficiencies more robust, reduce the costs of storing and transferring data, and offer greater user access to critical information. For example, the use of applications like Map Online, which uses advanced image compression technology for digital maps, illustrates how compression technology promotes user convenience in the form of online purchasing and keyword searching for maps. With the evolving nature of online information, source coding optimization and data compression techniques remain the momentum drivers of future progress and attaining optimal data communication systems (Sheriff et al., 2005).

REFERENCES

1. Adamski, M., Węgrzyn, M., Węgrzyn, A., Barkalov, A., Titarenko, L., et al. (2007). Measurements, models, systems and design. https://core.ac.uk/download/pdf/15312083.pdf (accessed on 18th September 2025).
2. Aggoun, A., Amon, P., Arbel, I., Chernilov, M., et al. (2008). Multimedia delivery in the future internet. https://core.ac.uk/download/335423.pdf (accessed on 18th September 2025).
3. Akpabio, I. (2019). Joint source-channel coding using machine learning techniques. https://core.ac.uk/download/211172878.pdf (accessed on 18th September 2025).
4. Aksyonoff, A., Büttcher, S., Jones, D. M., & Witten, I. H. (2015). Decoding billions of integers per second through vectorization. https://core.ac.uk/download/35146031.pdf (accessed on 18th September 2025).
5. Appel, A. W., McMillan, B., Huffman, D., Klein, G., Kelsey, J., Blanchette, J. C., Danielsson, N. A., et al. (2016). An implementation of Deflate in Coq. http://arxiv.org/abs/1609.01220 (accessed on 18th September 2025).
6. Barbarien, J., Coppens, S., De Cock, J., Evens, T., et al. (2009). Digital recording of performing arts: Formats and conversion. https://core.ac.uk/download/55805755.pdf (accessed on 18th September 2025).
7. Baron, D., & Krishnan, N. (2015). A universal parallel two-pass MDL context tree compression algorithm. http://arxiv.org/abs/1407.1514 (accessed on 18th September 2025).
8. Baronchelli, A., Caglioti, E., Loreto, V., Puglisi, A., et al. (2005). Artificial sequences and complexity measures. https://core.ac.uk/download/16409638.pdf (accessed on 18th September 2025).
9. Baxter, R. J., Davey, M. C., et al. (2003). Statistical mechanical aspects of joint source-channel coding. http://arxiv.org/abs/cond-mat/0303033 (accessed on 18th September 2025).
10. Blumer, A. (2025). Noiseless compression using non-Markov models. https://core.ac.uk/download/pdf/42828309.pdf (accessed on 18th September 2025).
11. Cappello, F., Chen, Z., Di, S., Liang, X., et al. (2019). Optimizing lossy compression rate-distortion from automatic online selection between SZ and ZFP. http://arxiv.org/abs/1806.08901 (accessed on 18th September 2025).
12. Castiglione, P., Erkip, E., Simeone, O., Zemen, T., et al. (2011). Energy management policies for energy-neutral source-channel coding. http://arxiv.org/abs/1103.4787 (accessed on 18th September 2025).

13. Chau, J. L., Coles, W. A., Hysell, D. L., Milla, M., et al. (2019). The case for combining a large low-band very high frequency transmitter with multiple receiving arrays for geospace research: A geospace radar. https://core.ac.uk/download/287622497.pdf (accessed on 18th September 2025).
14. Cheung, G., Magli, E., Ng, M., Tanaka, Y., et al. (2018). Graph spectral image processing. http://arxiv.org/abs/1801.04749 (accessed on 18th September 2025).
15. Crawford, P. R., & Kasulka, L. H. (2025). Science and Applications Space Platform (SASP) end-to-end data system study. https://core.ac.uk/download/pdf/42860980.pdf (accessed on 18th September 2025).
16. Davisson, L. D., McEliece, R. J., Pursley, M. B., Wallace, J., et al. (1981). Efficient universal noiseless source codes. https://core.ac.uk/download/9124076.pdf (accessed on 18th September 2025).
17. Effros, M. (1998). Optimal modeling for complex system design. https://core.ac.uk/download/4874479.pdf (accessed on 18th September 2025).
18. Effros, M., & Zhao, Q. (1999). Broadcast system source codes: A new paradigm for data compression. https://core.ac.uk/download/4875596.pdf (accessed on 18th September 2025).
19. Etienam, C. (2019). 4D seismic history matching incorporating unsupervised learning. http://arxiv.org/abs/1905.07469 (accessed on 18th September 2025).
20. Fong, A. C. M., Imran, N., & Seet, B. C. (2015). Distributed video coding for wireless video sensor networks: A review of the state-of-the-art architectures. https://core.ac.uk/download/46558677.pdf (accessed on 18th September 2025).
21. Goyal, V. K., Kusuma, J., & Varshney, L. R. (2016). Malleable coding for updatable cloud caching. https://open.bu.edu/bitstream/2144/39228/1/cutpaste_JSAC_v7.pdf (accessed on 18th September 2025).
22. Habibi, A. (2025). Study of on-board compression of Earth resources data. https://core.ac.uk/download/pdf/42885117.pdf (accessed on 18th September 2025).
23. Havinga, P. J. M. (2000). Energy-efficient wireless communication. https://core.ac.uk/download/pdf/11460969.pdf (accessed on 18th September 2025).
24. Hayashi, M., & Tan, V. Y. F. (2017). Minimum rates of approximate sufficient statistics. http://arxiv.org/abs/1612.02542 (accessed on 18th September 2025).

25. Hirschberg, D. S., & Stauffer, L. M. (1991). Parallel data compression. https://core.ac.uk/download/287623249.pdf (accessed on 18th September 2025).
26. Ingels, F. M., & Shahsavari, M. M. (2025). A study of data coding technology developments in the 1980–1985-time frame, Volume 2. https://core.ac.uk/download/pdf/42871670.pdf (accessed on 18th September 2025).
27. Kaleem, M., & Nazir, S. (2016). Optical network technologies for future digital cinema. https://core.ac.uk/download/293885083.pdf (accessed on 18th September 2025).
28. Kautz, J., Liu, M. Y., Sun, D., Tsai, Y., et al. (2017). Learning binary residual representations for domain-specific video streaming. http://arxiv.org/abs/1712.05087 (accessed on 18th September 2025).
29. Lau, V., Peng, M., Poor, H. V., Wang, T., et al. (2015). Fronthaul-constrained cloud radio access networks: Insights and challenges. http://arxiv.org/abs/1503.01187 (accessed on 18th September 2025).
30. Miller, W., & Rice, R. F. (2025). Advanced techniques and technology for efficient data storage, access, and transfer. https://core.ac.uk/download/pdf/42813245.pdf (accessed on 18th September 2025).
31. Mohd Tarmizi, M. (1995). Data compression for multimedia computing. https://core.ac.uk/download/268136539.pdf (accessed on 18th September 2025).
32. Serdean, C. V. (2002). Spread spectrum-based video watermarking algorithms for copyright protection. https://core.ac.uk/download/29816683.pdf (accessed on 18th September 2025).
33. Shannon, C. E., Papadimitriou, C. H., Huffman, D. A., Caire, G., Mézard, M., Gallager, R. C., Zecchina, R., et al. (2005). Lossy data compression with random gates. http://arxiv.org/abs/cond-mat/0504509 (accessed on 18th September 2025).
34. Sheriff, N. A., & Nik Ismail, N. I. (2005). Map online system using internet-based image catalogue. https://core.ac.uk/download/11780623.pdf (accessed on 18th September 2025).

CHAPTER 4

PULSE MODULATION TECHNIQUES

Contents

INTRODUCTION

In the context of modern telecommunication, pulse modulation techniques have evolved into techniques that are fundamental to effective transmission of information across various media. Techniques like On-Off Keying (OOK), Pulse Position Modulation (PPM), and Digital Pulse Interval Modulation (DPIM) constitute a foundation in establishing the performance of wireless as well as wired communication systems. In light of the ongoing demand for higher data rates, exploration of new and more sophisticated modulation schemes has never been more vital. The practice of pulse modulation optimization enables engineers to achieve the highest data capacity without adding errors and a compromise on signal integrity. For this reason, understanding these modulation schemes, their advantages, and their uses is very important for any electrical engineer or telecommunication practitioner. The efficiency of pulse modulation techniques is greatly subject to the environment in most applications, particularly in the case of optical communication systems. Interfering introduction of artificial light interferences (ALIs) poses a significant issue for indoor optical wireless communication channels where excessive optical power penalties render certain link configurations impractical. In a bid to address this, novel techniques such as discrete wavelet transform (DWT) based denoising have been proposed. This method not only reduces the level of complexity associated with traditional filtering processes but also enhances overall system performance, which is particularly beneficial to applications in pulse modulation systems. As technology advances, it is necessary to constantly assess ways through which various methods can be designed to meet rising issues so that communication can be guaranteed constantly across various environments. Additionally, the advent of coherent detection systems has revolutionized the roles of pulse modulation techniques even more, particularly in fiber-optic networks. The systems utilize DSP algorithms to mitigate the effects of optical fiber impairments, so as to improve spectral efficiency and enable multi-level modulation formats. As fiber technologies have evolved, the integration of coherent detection with sophisticated DSP techniques enables one to deal with linear and nonlinear distortions such as those generated by the Kerr effect. An understanding of the relationship between modulation techniques and these digital signal processing techniques is crucial for pushing the limits of high-capacity communication systems. By exploring the fertile ground of pulse modulation techniques, engineers and scientists are able to design solutions that meet the increasing requirements of modern communications networks and be kept in good working order under a variety of working conditions, as highlighted by the presentation of conventional and new compensation methods (Angelova et al., 2010; Amari et al., 2017).

Definition of Pulse Modulation

Pulse modulation comprises a number of techniques in which the data is converted into the time aspects of pulses. It basically alters the parameters of a pre-existing continuous wave, thereby making effective transmission of signals across a variety of mediums possible. Amplitude, frequency, or duration of the pulses are modified in pulse modulation in order to convey various amounts of information. This pulse modulation method is unlike the traditional analog methods of modulation in which the original signal is continuously modified. Converting the data into pulses enhances pulse modulation to signal integrity and offers rich practical advantages in terms of noise immunity and energy efficiency, thus becoming the best choice for modern communication systems. Most importantly, pulse modulation is known to have a multitude of sub-methods like pulse-width modulation (PWM), pulse-position modulation (PPM), and pulse-amplitude modulation (PAM). All the methods utilize different modulation principles to optimize signal processing for specific applications. An example is that PWM varies the pulse width to convey information, which is most effective in power control applications. PPM, instead, encodes information by altering the pulse position within a given time interval, such as in some optical communications applications. Meanwhile, PAM employs different pulse amplitudes to represent various levels of information in order to facilitate high data rates of transmission in cases where the amplitude can be discretely adjusted (Buswig et al., 2011). This type of sub-techniques demonstrates the applicability of pulse modulation to numerous fields of technology. Moreover, the integration of pulse modulation techniques has profound implications for breakthroughs in technologies, particularly in the field of optical communication and wireless transmission. For instance, pulse modulation enables advanced networking capability indoors through white light-emitting diodes (LEDs). Optical orthogonal codes (OOC) and multilevel signaling enable simultaneous access by many users in a network system, thus improving utilization of limited bandwidth (Brandt-Pearce et al., 2013). The ability to manage the peak-to-average ratio of power and dim level using pulse modulation not only enhances user experience but also powers the efficiency of lighting systems. Techniques of pulse modulation remain an essential function in ensuring high-speed, dependable communication since the techniques evolve (Figure 4.1).

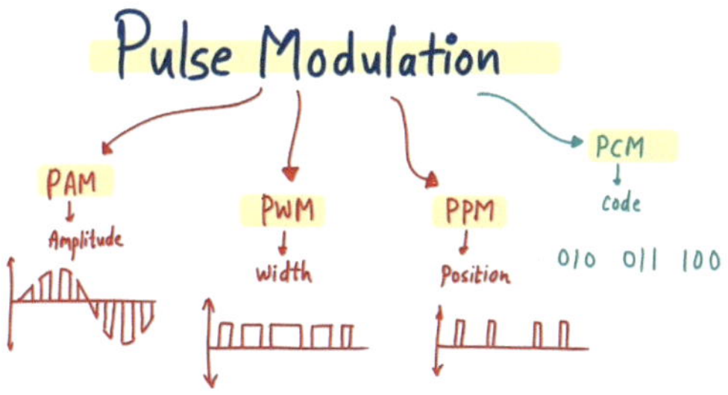

Figure 4.1. *Pulse modulation in digital communication.*

Source: https://i.ytimg.com/vi/2-YsquXnG8c/maxresdefault.jpg.

Importance in Communication Systems

In modern-day communications systems, signal transmission cannot be overstressed. Pulse modulation techniques are the means through which efficient transfer of digital data can be carried out over various media, enhancing bandwidth utilization and reducing interference significantly. Pulse modulation converts an analog signal to a discrete signal by pulse amplitude, frequency, or width modulation, thus fully utilizing the available spectrum. The evolution in these techniques is reflective of mounting demands for communication systems that are able to process high rates of data without compromising the integrity of information transmitted. This is particularly critical in cases where the signal-to-noise ratio is low, as efficient modulation can recover original information irrespective of distortions. Thus, pulse modulation continues to be the backbone of the emergence of communication technologies and their application across platforms.

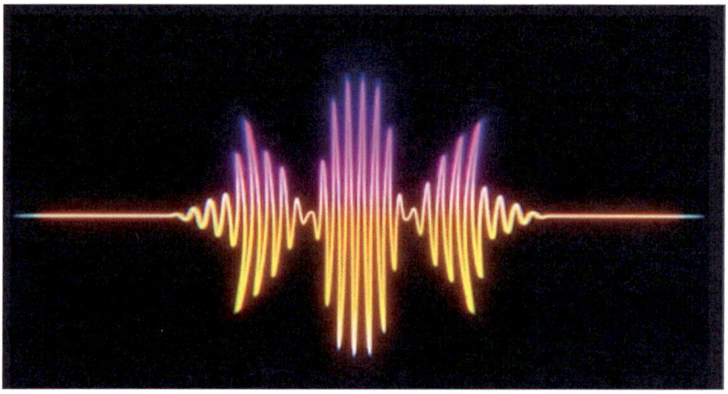

Figure 4.2. *Pulse modulation: The backbone of modern communications.*

Source: https://images.unsplash.com/photo-1724185773486-0b39642e607e?crop=entropy&cs=tinysrgb&fit=max&fm=jpg&ixid=M3w3MDgyMDR8MHwxf-HNlYXJjaHwxfHxQdWxzZSUyME1vZHVsYXRpb24lMkMlMjBDb21tdW5pY-2F0aW9uJTIwU3lzdGVtcyUyQyUyMERhdGElMjBUcmFuc21pc3Npb24lMkMlMjBTaWduYWwlMjBQcm9jZXNzaW5nfGVufDB8MHx8fDE3NDk5MjE1Mzd8MA&ixlib=rb-4.1.0&q=80&w=500&?utm_source=numberanalytics&utm_medium=referral.

Pulse modulation techniques' use in wireless communications is evidence of their suitability in today's networking. Under variable atmospheric conditions, as those studied in free-space optics (FSO), the effectiveness of pulse modulation is even more pronounced. The research highlights the manner in which certain modulation and coding techniques can significantly enhance performance via reduction of the effects of atmospheric turbulence and fog conditions, providing for stable signal transmission across challenging distances (Awan et al., 2009). These findings highlight the versatility of pulse modulation methods in providing for data integrity and reliability in changing communication conditions. Besides, advances in hybrid systems demonstrate the versatility of pulse modulation in addressing the particular demands of various applications from terrestrial broadcasting to satellite communications. Besides, eco-sustainability of communications systems is increasingly becoming the center of attention in ongoing work and research. As emphasized by previous research, the Triple Bottom Line (TBL) approach emphasizes the interplay between environmental, economic, and social aspects of technology deployment (Hemdi et al., 2017). Pulse modulation methods naturally assist in this endeavor through the optimization of energy efficiency; fewer resources are used without sacrificing high-quality communication. This efficiency becomes more important as the technology sector continues to grow, requiring environmentally friendly measures to offset possible effects on the environment. Consequently, the utilization of effective pulse modulation not only facilitates increased operation efficiency but also enhances the broader agenda for technological and telecommunication sustainability, testifying to the methods significance as major components of future communication systems (Figure 4.2).

Overview of Key Techniques

Pulse modulation techniques have come to be fundamental procedures in the digital communication area, effectively conveying information through pulse duration, amplitude, or position modulation depending on the input signal. Among different crucial techniques, Pulse Amplitude Modulation (PAM) is highly famous for its bandwidth efficiency and ease. PAM does this by varying the amplitude of the pulses to be able to represent more than a single bit per symbol, hence higher data transmission rates. This is particularly relevant in modern optical communication systems, where fast and efficient processing of data is needed. As PAM has spread, more advanced techniques have emerged in the form of Pulse Position Modulation (PPM) and Pulse Width Modulation (PWM), both of which possess specific strengths in specific uses where power reduction or reduced interference is vital to performance. Another significant technique is Pulse Code Modulation (PCM), which converts analog signal amplitudes into digital form through sampling and quantization. PCM is significant in telecommunications as well as digital audio since it supports the error detection and correction with various encoding techniques in order to promote more secure data transfer. The noise robustness and clarity of the technique have made it a central element in voice communications networks and broadcast media. PCM is also very compatible with digital signal processing architectures in order to enrich its capability under environments with diverse transmission conditions. With networks becoming more directed towards higher capacity requirements, PCMs role becomes more pertinent in proving how legacy systems can be reworked with the help of newer digital infrastructures in telecommunication. Beyond the fundamental techniques of pulse modulation, newer advancements in coherent detection systems call for solving fiber optic impairments, particularly in high-speed networks. The emergence of digital signal processing (DSP) offers solutions for treating linear distortions, crucial to signal integrity in WDM technology-based deployments. Recent research identifies the significance of effective fiber nonlinearity compensation (NLC) techniques, which counteract phenomena like the Kerr effect at greater transmission rates up to 400 Gbps and beyond. With increasing coherent modulation formats, they enable increased spectral efficiency towards meeting the demands of modern optical communications. Thus, ongoing development of NLC approaches is crucial in ensuring the sustainability of pulse modulation techniques in upcoming high-speed transmission systems, as noted by the construction of optical communication networks discussed in Amari et al. (2017).

4.1. TYPES OF PULSE MODULATION

The discipline of pulse modulation contains a range of different types, each serving a specialized purpose and having various advantages. Some of the most widely used are Pulse Width Modulation (PWM), Pulse Position Modulation (PPM), and Pulse Density Modulation (PDM). PWM varies the pulse width to convey information and is commonly utilized in motor control and dimming light-emitting diodes since it's energy efficient and contains minimal energy loss. By changing the length of time a signal remains in the on state to that of its off period, PWM effectively controls power delivery with little heat generation. This idea is paramount to apply when there is the necessity of having precise control of output power (Escolano et al., 2019). Beyond PWM, Pulse Position Modulation (PPM) introduces a new strategy that includes changing the pulse timing to convey data. In this method, the position of a pulse within a pre-specified time interval reflects the value of the information being transmitted. PPM finds great application in systems such as satellite communications, where bandwidth is crucial. By optimizing the use of the available spectrum with high robustness against noise, it ensures transmission of the signal in a quality manner. This technique also takes advantage of the advantages of modulation, overcoming limitations posed by inter-symbol interference and the potential for increased data rates (Alexander et al., 2015). Finally, Pulse Density Modulation (PDM) differs in its strategy, since it modulates pulse density rather than position or width. In PDM, the count of pulses averaged per fixed interval equates to the amplitude of the analog signal being converted. This technique beats others in applications such as audio processing, where fidelity has to be extremely high. Through guaranteeing minimal quantization errors with maximum transmission efficiency, PDM enhances the purity of the original signal and is therefore a widely used choice in digital-to-analog converters. All these pulse modulation types demonstrate different methods of ensuring maximum signal quality and transmission efficiency, which highlight the versatility and importance of modulation in modern communication systems (Figure 4.3).

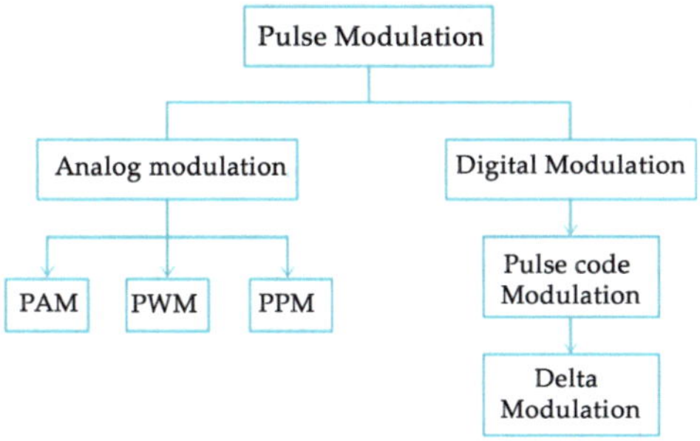

Figure 4.3. *Types of pulse modulation.*

Source: https://cn.edurev.in/ApplicationImages/Temp/b0677f2f-eebb-439e-8168-7f2ee59854fe_lg.jpg?w=360&dpr=2.6.

4.1.1. Pulse Amplitude Modulation (PAM)

Pulse Amplitude Modulation (PAM) is a simple technique of pulse modulation, in which the amplitude of individual pulses is modulated to send information. This technique of modulation is more effective in communication systems that require efficiency and dependability, particularly in transmitting digital information. PAM facilitates the encoding of multiple bits per pulse, which significantly increases the channel capacity. For instance, in a hybrid Amplitude Shape Modulation (h-ASM) scheme, PAM is combined with Pulse Shape Modulation (PSM) to achieve maximum bandwidth and reduce interference in challenging environments. The resultant system is optimized for high data rates and hence lowers the framework suitable for applications such as time-hopping ultra-wideband (TH-UWB) communication (Herceg et al., 2009). Thus, PAM's elasticity and flexibility render it an essential component of modern communication infrastructure that can address the increase in data transmission efficiency demand. One of the most interesting advantages of PAM is its potential for use in high-speed transmissions through a variety of media, such as optical fibers. Current advancements have brought out the promise of PAM for multi-gigabit transmission speeds, as evidenced by the fact that it has been utilized in plastic optical fibers (POFs). By employing intensity-modulated direct-detection systems with PAM, researchers have established the feasibility of as much as 10 Gbit/s transmission using 1-mm core diameter POFs. Such a breakthrough is of particular significance in domestic network applications and data communications, where cost savings and high-speed technologies rank highest (Charbonnier et al., 2011). The revolutionary potential of PAM in such settings accentuates its importance in a rapidly evolving technology world where requirements for more bandwidth and greater coverage are growing evermore. Besides, the error resilience of PAM under varying levels of noise renders it more robust and an improved possibility for real-world application. When PAM is applied in environments where additive white Gaussian noise (AWGN) is anticipated, accurate modulation and demodulation techniques can be used to minimize error probabilities, improving signal integrity. Combined with adaptive coding and modulation, PAM can be tuned to maximize performance based on existing channel conditions. This versatility positions PAM not just as a utilitarian choice for ordinary transmission uses but as a critical component in emerging communication systems that must process increasingly sophisticated streams of data. Thus, as the telecommunications field works to create more efficient and more powerful means of communication, Pulse Amplitude Modulation stands at the forefront of such advancement, projecting its enduring applicability to the background of pulse modulation methods (Figure 4.4).

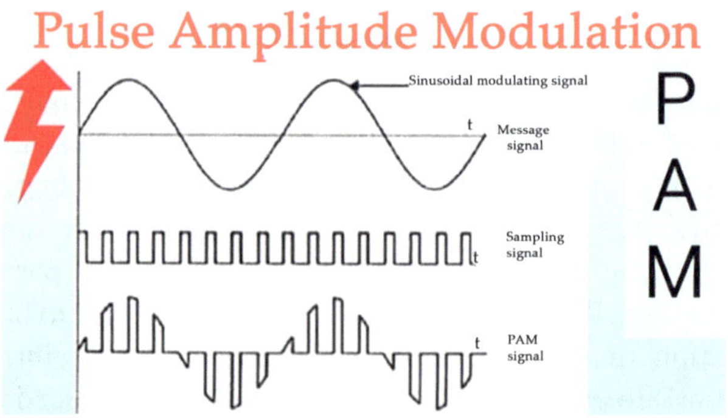

Figure 4.4. *Pulse amplitude modulation (PAM).*

Source: https://i.ytimg.com/vi/UkVSmGJk-vA/sddefault.jpg.

4.1.2. Pulse Width Modulation (PWM)

Pulse Width Modulation (PWM) is a critical power control and signal processing technique, evolving as a trending solution for controlling energy supplied to electrical loads. PWM works by adjusting the pulse width of a waveform to regulate the average power supplied to a load. PWM has been widely used in numerous applications, ranging from motor control to dimming of lights and power inverters, due to its efficiency of a high order and precise control over the output.

The most fundamental principle on which PWM is founded is that it can be used to represent analog signals in a digital format, thereby making it highly suitable for modern electronic systems that need performance as well as reliability. Apart from that, PWMs efficiency in power control is high towards reducing energy losses, vital in the current energy-conscious era, where efficient solutions are required to address rising demands for electricity. In the application of emerging power electronics, PWM techniques have evolved past straightforward applications, creating room for the development of high-class multilayer converters that allow for high-power applications. As identified in existing studies, multilevel converters are regarded as paragons of power-conversion systems that will accommodate complex industry demands. Multilevel converters utilize a number of voltage levels to enhance the quality of output with minimal harmonic distortion, which has been a usual drawback employing traditional single-level PWM techniques.

By embracing PWM and multilevel converter technology, engineers can achieve high performance in applications demanding high power and quality. The use of the latest modulation methods, such as phase-disposed and phase-shifted PWM, indicates the continuous development in the technology, which demonstrates PWMs versatility and adaptability in meeting high standards demanded by industry (García et al., 2009). The innovation of new PWM methods, such as a unipolar hybrid PWM system, indicates the continuous development of this modulation method.

Studies indicate that through this hybrid approach, which utilizes phase disposed PWM in conjunction with phase shifted PWM, power losses associated with converters of high performance are significantly reduced. Studies indicate that the approach put forth herein achieves the lowest overall power losses compared to traditional PWM approaches, thus enhancing system efficiency. These efficiency gains have real-world implications, extending the useful life of power semiconductor devices already in service and lowering the thermal stresses within the power systems. Furthermore, the high-quality voltage waveform produced by these advanced PWM methods ensures the integrity of the electrical output, which is critical in sensitive applications (Oghorada et al.,

2016). As a result, PWM remains to shape the future of energy management solutions to a very large degree, reaffirming its relevance in modern electrical engineering (Figure 4.5).

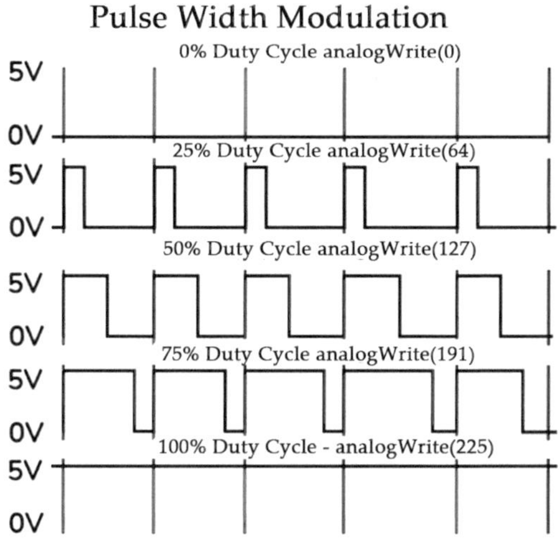

Figure 4.5. *Pulse width modulation (PWM).*

Source: https://media.geeksforgeeks.org/wp-content/uploads/20240505075642/Pulse-Width-Modulation.webp.

4.1.3. Pulse Position Modulation (PPM)

Of the signal modulation schemes, Pulse Position Modulation (PPM) is a significant enhancement, especially as a digital communication system technique. PPM transmits information by varying the position of a pulse within a given time period, thereby allowing several symbols to be transmitted in a single cycle. This process utilizes the timing of the pulse occurrence completely, and it benefits in terms of noise immunity as well as power usage. With the distribution of pulse positions into different time slots, PPM offers greater data rates compared to traditional modulation techniques such as Pulse Amplitude Modulation (PAM). Pulse position discrimination offers receivers improved performance characteristics that make PPM particularly suitable in situations where optical communications and remote sensing applications have a tight bandwidth constraint. Practical use of PPM over the years has resulted in considerable implications for system performance and design. One of the major advantages of PPM is that it is resistant to interference and fading, making it very dependable in hostile environments. This characteristic is particularly beneficial in systems where signal weakening is crucial, such as in wireless communications and satellite communications. Moreover, the simplicity of receiver implementation of PPM, in many instances being limited to timing recovery and synchronization, makes it preferable. With the telecommunication industry trying to keep up with the increasing needs for higher capacity and efficiency, techniques like PPM are being studied to meet this demand. These advances are preceded by future technologies that employ PPM for improved transmission capacity, which give rise to more efficient data transfer protocols in contemporary communication systems. Furthermore, integration of PPM with other modulation schemes, such as dual-header pulse interval modulation (DH-PIM) techniques, can further enhance system performance. By integrating PPM with detection techniques such as majority decision detection and retransmission, researchers were able to attain relatively huge gains in error rate performance, particularly under adverse transmission conditions. Studies discover that hybrid systems such as these significantly cut down on the slot error rate (SER), showing extremely enhanced signal reliability compared to simple single-symbol transmitting systems. Compatibility with complementary techniques offers a way to maximize pulse modulation usage in real conditions so that systems are able to cope with increasing complexity and demands of today's digital communications environments (Hemdi et al., 2017; Aminzadeh-Gohari et al., 2007) (Figure 4.6).

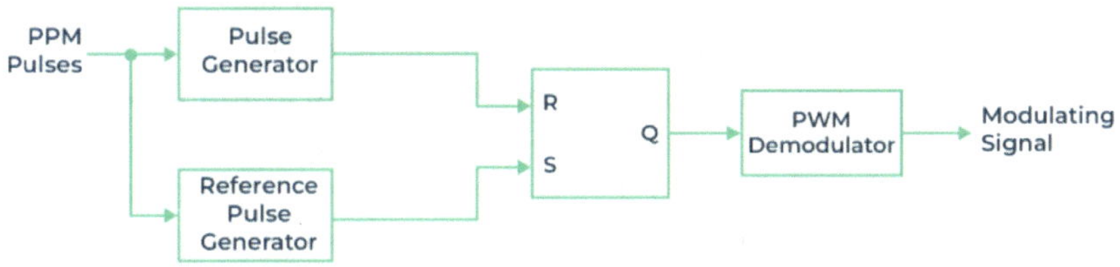

Figure 4.6. *Pulse position modulation (PPM).*

Source: https://media.geeksforgeeks.org/wp-content/uploads/20240104183735/PPM-Detector.png.

4.2. APPLICATIONS OF PULSE MODULATION

Applications of pulse modulation are vast and ubiquitous, and they have a profound effect on fields such as telecommunication, audio engineering, and renewable energy systems. By modulating signals in the form of pulses, such techniques enable cost-effective transmission of data on various platforms with minimal interference and maximal use of bandwidth. In telecommunication, for instance, pulse code modulation (PCM) is widely applied to digitize analog signals so that clearer voice and data can be transmitted over long distances. This digitization increases the robustness of the system against noise and allows multiplexing, whereby several signals could be transmitted in parallel over a single channel. Additionally, use of pulse modulation also finds congruence with growing concerns around sustainability since it enables technologies with minimal power consumption during data transmission, showing a growing concern with environmental aspects in engineering design (Hemdi et al., 2017). In sound engineering, pulse modulation has revolutionized sound synthesis by allowing more sophisticated manipulation and creation of sound. An exemplary instance is the application of pulse width modulation (PWM) in synthesizers, wherein adjustment of pulse widths creates rich harmonic frameworks and emotive soundscapes. This technique is particularly favored by electronic music genres, as it presents a range of timbral choices that readily complement other sound modulation methods. PWM is also instrumental in the adjustment of the output signal of devices like digital amplifiers, which assists in improving power efficiency and sound quality. The precise regulation of all pulses width offers audio engineers the capability to create distinctive listening experiences while achieving modern performance and sustainability requirements (Buswig et al., 2011). The implementation of pulse modulation in renewable energy systems, particularly wind power generation, shows its indispensable use in modern technological applications. Implementation of pulse modulation techniques facilitates the conversion of variable DC output from wind turbines into stable AC power for grid compatibility. Through the utilization of advanced controllers, such as Artificial Neural Networks, the systems stabilize the output voltage effectively, enhancing the reliability of renewable energy sources. Mathematical modeling and simulation techniques applied in these applications display the efficacy and simplicity of pulse modulation in optimizing energy conversion processes. With the direction of energy requirements shifting towards cleaner sources of power, the capabilities facilitated by pulse modulation will come to play ever more irresistibly in promoting functionality and integration of renewable technology in today's energy infrastructure (Buswig et al., 2011).

4.2.1. Digital Communication Systems

The evolutionary changes in technology of digital communication systems have revolutionized the face of data transmission, particularly by incorporating pulse modulation techniques. The systems incorporate discrete signals, which provide effective communication across a broad band of channels, including optical and radio frequency media. In this case, the pulse modulation techniques such as pulse code modulation (PCM) and pulse position modulation (PPM) possess certain advantages such as noise immunity and ability to carry high data rates.

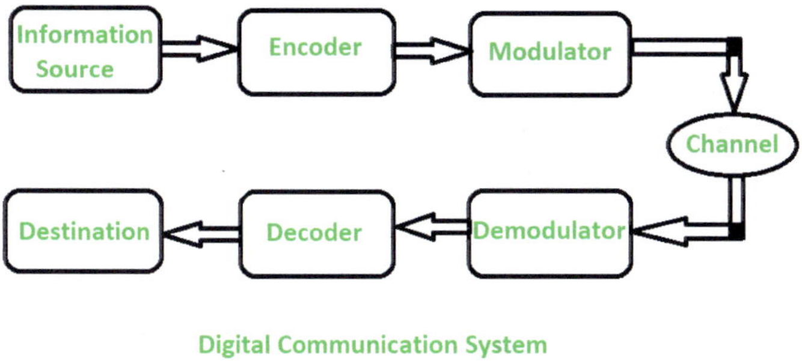

Figure 4.7. *Digital communication systems.*

Source: https://media.geeksforgeeks.org/wp-content/uploads/20200824101738/digital.png.

These techniques convert analog signals into digital signals in a way that information is properly indicated and transmitted. The adaptability of pulse modulation offers the basis for creating more evolved systems that enhance communications networks' efficiency and reliability, which are crucial in meeting the requirements of advanced digital communication needs. In addition, pulse modulation techniques are important in minimizing the degradations of communications which commonly happen in digital transmission systems. Problems such as multipath induced intersymbol interference (ISI) and fluorescent light interference (FLI) can compromise performance and necessitate advanced solutions for maximizing system efficiency. Current studies show the use of discrete wavelet transform (DWT) along with artificial neural networks (ANN) as a powerful solution to manage these issues, particularly in indoor optical wireless communication systems. Applying these emerging techniques, the DWT-ANN receiver is found to be superior to traditional filters. These advancements further sustain the incorporation of modern DSP techniques to reverse signal degradation and digital communication system integrity as a whole, ultimately having pulse modulation techniques deliver maximum performance in different applications (Angelova et al., 2011). Within optical communication systems, pulse modulation techniques have been central to enhancing spectral efficiency and combating signal distortions. The introduction of coherent detection systems in wavelength-division multiplexing (WDM) high-capacity networks revolutionized this sector. The systems utilize advanced digital signal processing (DSP) algorithms to counter various kinds of impairments including linear and nonlinear effects due to the Kerr effect. With increasing data transmission speeds to 400 Gbps and beyond, fighting fiber nonlinearity becomes a key area of concern. By means of the utilization of many methods for nonlinearity compensation, some of which is the inter-subcarrier nonlinear interference canceler technique, transmission performance can be greatly enhanced along transmission networks. The marriage of pulse modulation techniques and advanced DSP solutions, thus, is at the heart not just of maintaining signal integrity but also of pushing the frontiers of digital communication systems (Amari et al., 2017) (Figure 4.7).

4.2.2. Audio and Video Signal Processing

The discipline of audio and video signal processing is greatly influenced by pulse modulation methods, which are the foundation in the transmission of multimedia data. Pulse modulation is the process of transforming an information signal into a sequence of pulses which enables effective data transmission over various media. This is marked by the need to provide reliability, particularly in high-definition audio and video systems where reception quality is greatly sought after. Through the use of techniques such as Pulse Code Modulation (PCM) and Delta Modulation (DM), signal integrity is achieved despite interference and noise, which are inevitable in the transmission environment. The fidelity that these techniques can provide to audio and video is of great significance in order to guarantee that the final user has the uncompromised experience, irrespective of the signal deterioration limits in transmission. For Fiber-to-the-Home (FTTH) networks, pulse modulation techniques have been further developed to cater to the rising demands for high-definition content. The application of ultra-wideband (UWB) radio techniques is just one illustration of how pulse modulation may be employed to enhance transmission capacity. Experimental results demonstrate that standards-compliant UWB signals, particularly those employing orthogonal frequency-division multiplexing, are increasingly capable of transmitting high-definition audio and video signals for considerable distances in FTTH configurations. These advancements lead towards the conclusion that pulse modulation is not a theory anymore but a reality with the potential to achieve both high-quality and high-speed data transfer through existing infrastructure. The conclusion of such observations brings to mind the call for further research in modulation strategies to meet the evolving demands of multimedia delivery systems (Alves et al., 2008). Further, the use of error resilience in video coding also accentuates the significance of high-error resilience signal processing techniques in pulse modulation systems. As has been depicted by recent studies, traditional techniques of error correction, for example, Automatic Repeat reQuest (ARQ), are not always effective, particularly for real-time streams where latency is unacceptable. One alternative method that can be used is to re-arrange the receiver to recover and interpret erroneous packets without the need for a retransmission request, thereby enhancing video streaming's tolerance to real-world degradation. This groundbreaking method emphasizes the adaptability of pulse modulation techniques in optimizing communication standards for facilitating the uninterrupted transmission of audiovisual media, bridging the gap between abstract concepts and practical use in diverse environments (Bull et al., 2000) (Figure 4.8).

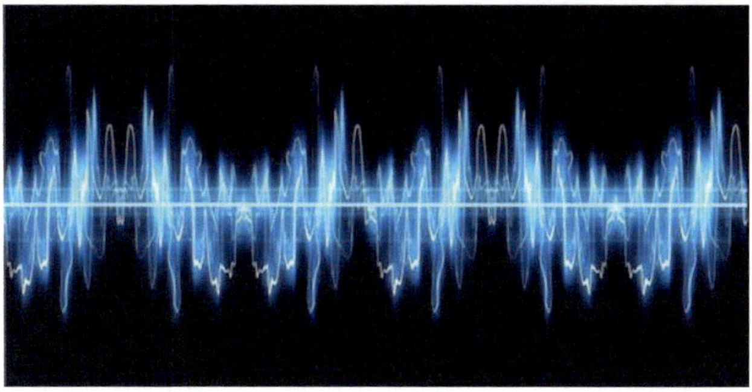

Figure 4.8. *Audio and video signal processing.*

Source: https://www.icdrex.com/wp-content/uploads/2023/03/1678784946.jpg.

4.2.3. Control Systems in Electronics

Electronic control systems are necessary to the control of electronic device and system behavior, particularly as related to pulse modulation techniques. They are employed to maintain desired output levels by dynamically modifying inputs as a function of feedback, hence enhancing performance and efficiency. For pulse modulation techniques, an effective control system is essential since they tend to adopt the variation of signals in order to capture data in an effective way. An expert control system can ensure accurate amplitude and timing of pulse signals, which can enhance the efficiency of communication in electronic devices. Integrated intelligent control algorithms with modulation techniques further elevate this effectiveness and leads to improved resource optimization as well as saving energy. Among the various pulse modulation techniques, the multilevel converters are prominent particularly due to their ability to handle high power and provide improved power quality. The operation modes of the converters and other power circuit topologies emphasize the significance of control systems for their functioning. Attention is laid on established technologies in numerous industries, emphasizing proper control systems integration to facilitate safe operation. García et al. (2009) describes how the systems not only cancel out harmonics and contribute to increased overall conversion efficiency but also offer the facility for improved performance in difficult applications. By optimizing multilevel converter modulation methods, control system designers can create control systems that meet special functional requirements, further establishing their flexibility among industries. Coupled with multilevel converters, advanced techniques like multicarrier phase-shifted sinusoidal pulse-width modulation (MPS-SPWM) and selective harmonic elimination PWM (MSHE-PWM) are the innovative control system-pulse modulation interplays. The MPS-SPWM technique, in particular, boasts greater overall bandwidth that is achieved due to the effective management of several carrier signals. This technique emphasizes the need for stringent control mechanisms as the location of dominant harmonics can be controlled with high accuracy in an attempt to optimize system performance. The MSHE-PWM approach's proposal that tries to control non-triplen harmonics using the same switching transition as the initial strategy serves to highlight the advantages of employing advanced control systems. Agelidis et al. (2005) shows how by expanding modulation operating ranges, the versatility of such techniques is significantly enhanced, offering robust performance in a broad spectrum of applications based on pulse modulation for effective signal processing.

4.3. ADVANTAGES OF PULSE MODULATION

The advantages of pulse modulation techniques are varied, offering significant advancements over traditional analog techniques of modulation. A significant advantage is the enhanced immunity to noise provided by pulse modulation. Conversion of analogue signals into discretized pulses renders the systems used for communication less susceptible to distortion and interference, a critical factor in maintaining signal quality over long communication. This is particularly relevant to systems operating in noisy environments or that require high data fidelity. Also, pulse modulation forms allow bandwidth to be efficiently used, which is critical in today's communications climate where spectrum availability constantly decreases. As pulse modulation allows data to be packed into short time intervals, it has the capability of transferring more data within a given frequency range, enhancing communication efficiency and quality. Another advantage of pulse modulation is its versatility to be used in various applications. The versatility of the modulating method can enable it to be used on several platforms including digital data communication, audio signals, and optical communication. For instance, with free-space optics (FSO), pulse modulation techniques can enhance signal transmission with some modulation and coding techniques to facilitate improved performance under varying atmospheric conditions such as fog or turbulence. These upgrades become indispensable for applications requiring strong connections of communications, especially in satellite and aerospace systems. The inherent flexibility of pulse modulation enables it to be employed to apply sophisticated control systems and methods, for example, those employing neural networks for DC-AC converters in renewable power systems. This flexibility again renders pulse modulation a prized gem in modern communication technology (Buswig et al., 2011; Awan et al., 2009). Lastly, the simplicity of use in utilizing pulse modulation versus analog options adds to its overall appeal. Improvements in digital signal processing technologies have made it easier to design and utilize pulse modulation systems, enabling easier error detection and correction. Such features are particularly beneficial while creating systems that require real-time data transmission and processing. Moreover, pulse modulation techniques can easily be integrated into existing digital and wireless technology, making it possible to upgrade to new systems with ease without significant resource overhauls. This versatility is critical in a more dynamic technological environment where the ability to expand systems with minimal interference is crucial. Thus, advantages of pulse modulation in respect of noise immunity, flexibility of usage, and simplicity of implementation establish it as a foundation of contemporary means of communication (Buswig et al., 2011; Awan et al., 2009) (Figure 4.9).

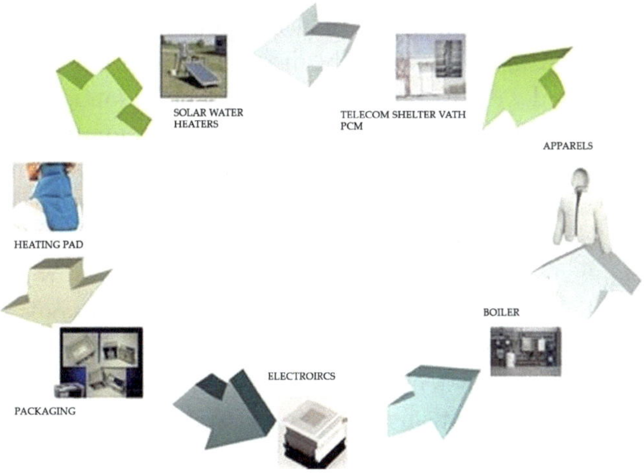

Figure 4.9. *Advantages of pulse modulation.*

Source: https://www.polytechnichub.com/wp-content/uploads/2017/05/pcm-application.jpg.

4.3.1. Improved Noise Immunity

In the context of pulse modulation techniques, improved noise immunity is a vital consideration which significantly optimizes communication system efficiency and reliability. As the possibility of noise significantly degrading data signal fidelity during transmission, sophisticated modulation techniques are designed to mitigate its impact. Sophisticated techniques such as Pulse Position Modulation (PPM) and On-Off Keying (OOK) possess features that facilitate discrimination against noise. By embedding information tactically in the timing of pulses, such techniques inherently have greater noise-immunity resistance, thereby guaranteeing data quality during transmission. Greater noise immunity ultimately provides for transmission of cleaner signals over a variety of environments, making such modulation techniques important for modern-day digital communications. Use of artificial intelligence, in the shape of Artificial Neural Networks (ANNs), is increasingly recognized as a valid technique for enhancing noise immunity in pulse modulation systems. For instance, research to DC-AC converters for wind turbine drives demonstrates how ANN control systems can stabilize output voltages against varied noise conditions. This adaptability allows for real-time adjustments that greatly minimize the adverse effect of noise on signal quality (Buswig et al., 2011). Through enhanced control strategies in modulation techniques, ANNs not only optimize performance but also ensure the integrity of the signals being sent is preserved, thereby widening the field of application of pulse modulation across different industries. Additionally, discrete wavelet transforms (DWT) is another sophisticated way of achieving improved noise immunity. DWT finds particularly good application in indoor optical wireless communication channels because artificial light interference significantly creates power penalties and thereby makes standard communication approaches worthless (Angelova et al., 2010). DWT is superior to classical high-pass filtering approaches in terms of increased complexity and performance in eliminating noise from transmitted signals. This ability to manage interference effectively enables pulse modulation techniques to maintain their signal integrity, and hence new possibilities for secure communication even in challenging environments. Such an investigation of noise immunity mechanisms like DWT and ANN implementation is, therefore, an indication of the ongoing development and advancement of pulse modulation techniques in telecommunications.

4.3.2. Efficient Bandwidth Utilization

Efficient use of bandwidth is very important in the realm of pulse modulation techniques, especially with the demand for data transmission continually increasing. Bandwidth, or the maximum rate of data transfer in a network, has its influence on the performance and efficiency of communication systems. With the application of pulse modulation techniques, such as Pulse Amplitude Modulation (PAM) or Pulse Width Modulation (PWM), bandwidth can be optimized. These approaches encode information in discrete pulses to enable several signals to share a single bandwidth. The best compromise between noise resistance and maximum possible data transfer rate is the challenge, with a guarantee of enduring performance within real-world use. Improvements in modulation schemes constitute a significant contribution to such objectives, which necessitate continuous research and development within the sector to meet increasing global connectivity demands.

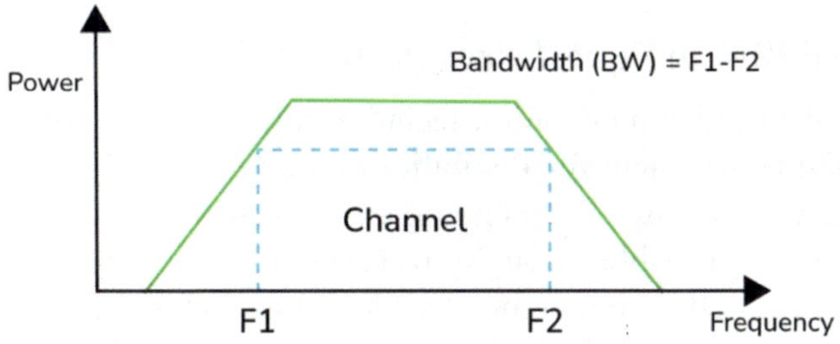

Figure 4.10. *Bandwidth efficiency.*

Source: https://media.geeksforgeeks.org/wp-content/uploads/20240702165801/Bandwidth-Efficiency.webp.

The possibility of combining artificial intelligence and machine learning with pulse modulation techniques is a new field of effective bandwidth utilization. By employing advanced algorithms, systems have the ability to change modulation parameters based on changing network conditions. Adaptive modulation techniques using neural networks, for instance, can modify pulse shapes and signal characteristics such that the available bandwidth is utilized to its optimal level. They not only improve data rates but also enable the networks to support quality of service in varied operational situations. For instance, control system developments for DC-AC converters show the potential of accuracy in devices like wind generation systems, which can facilitate the efficiency of energy transmission. The intersection of these technologies shows the important step towards optimizing bandwidth utilization, effectively serving an increasingly digitalized population (Buswig et al., 2011). Efficient use of bandwidth has implications extending beyond mere technological innovation; they intersect with broader sustainability goals, affecting environmental and societal aspects of network deployment. An inexpensive approach to bandwidth not only conserves the energy expended in data transmission but also caps the environmental footprint of communication infrastructure. By inextricably combining environmental imperatives with technical excellence in pulse modulation, stakeholders can maintain sustainability frameworks reconciling economic viability with ecological and social obligations (Hemdi et al., 2017). As bandwidth resources are further pushed to their limits, sustained innovation within modulation technologies will be the key to keeping pace with sustainable operational practices that align demand for connectivity with critical sustainability objectives. Where technological innovation, sustainability, and optimized resource exploitation overlap lies a thorough working structure that must propel telecommunications forward (Figure 4.10).

4.3.3. Flexibility in Signal Processing

The evolution of pulse modulation techniques has been facilitated primarily because of the requirement for flexibility in the process of signal processing. With the advancement of technology, modulation schemes have been formulated to be flexible enough to accommodate varying transmission environments as well as requirements. Flexibility holds the highest value in applications ranging from telecommunications to multimedia streaming, where data transfer efficiency and reliability are the priority. For instance, techniques such as Pulse Width Modulation (PWM) and Pulse Position Modulation (PPM) allow systems to experience controlled parameter variations in signals, which can be optimized as far as parameters like bandwidth and noise immunity are concerned. Utilization of these pulse modulation methods makes it simpler to design systems that are capable of altering parameters in real-time, thereby assuring stable performance across a broad range of communication scenarios. Within renewable energy utilization, flexibility in signal processing becomes even more highlighted. With wind energy projects, for example, their inverter topology makes use of pulse modulation schemes to convert DC to AC power in an efficient manner. A good example of such a work is the topology of a single-phase DC-AC converter used in wind generator systems where the output voltage has been stabilized by an Artificial Neural Network control strategy (Buswig et al., 2011). This project depicts how advanced modulation methods enhance the dynamic response features of signal processing systems in dynamic situations, making steady power supply possible despite any change in wind speed or direction. Thus, the capacity to process signals in a flexible way not only optimizes the delivery of energy but also helps in ensuring overall stability of the power grid. The application of digital signal processing (DSP) in advanced communications systems contributes further to the importance of the flexibility of signal modulation techniques. DSP allows the compensatory methods for fiber degradations, for example, the Kerr effect, to become stronger when data are transmitted at higher rates (Amari et al., 2017). As communications demands grow with larger needs—increasing more so with 400 Gbps and beyond—modulation schemes must also become more tolerant to accommodate nonlinear effects in light waves transmitted over fibers. Accommodation allows for the embedding of various nonlinear compensation techniques that improve spectral efficiency without impairing signal integrity. Consequently, the prime interaction between pulse modulation techniques and adaptive signal handling remains shaping the future boundary of energy and communication systems to ensure their efficiency and resilience to technology advancements.

4.4. LIMITATIONS OF PULSE MODULATION

Although the pulse modulation schemes offer many advantages of efficiency and quality of signal, there are some significant drawbacks that are evolved due to their employment. One such significant drawback is the increased system design complexity compared to traditional analog modulation schemes. The requirement of strong algorithms and signal processing renders the hardware requirement complex, and thus the design becomes costly and time-consuming. Moreover, often pulse modulation requires more electronics with high speed that can handle sustained high-rate signal state changes, and these may not always be readily available, resulting in potential delays and inefficiencies in implementation. Such challenges may deter smaller businesses or low-budget projects from embracing pulse modulation, thus making it less usable overall across fields. Another major limitation of pulse modulation is its susceptibility to interference and noise. These modulation techniques are extremely sensitive to timing errors as well, and errors may occur due to external causes or limitations in the hardware. Timing errors lead to signal integrity degradation, impacting the performance of the system negatively. The pulse-shaped signals also result in spectral spreading, producing unnecessary harmonics that have the potential to interfere with adjacent channels. When working in high-energy environments that have been discovered in particle physics experiments, the addition of pulse modulation is made increasingly challenging because of these issues of noise. For instance, as at CERN, the requirement of increasing the rates of transmitting data in environments like the Super LHC dictates that the susceptibility to interference of both modulations and the increasing physical boundary on hardware integration must be addressed (Amaral et al., 2010). Lastly, pulse modulation techniques may entail bandwidth restriction in contrast to other modulation schemes. Although they do provide certain advantages towards preventing fading from affecting amplitude modulation systems, successful transmission of data relies on accurately timed transmission and reception of pulses. If these pulses don't precisely match the characteristics of the transmission medium, bandwidth utilization can be greatly affected, causing underperformance and reduced data rates. This inefficiency can compound complexity in high-speed applications, particularly where reliable connectivity is a concern, such as in renewable energy applications like wind generation. The complexities involved in the delivery of a constant output voltage in pulse modulation technique-based systems can result in higher implementation complexity, with additional layers of monitoring and control required (Buswig et al., 2011). Thus, though pulse modulation has found its application in various industries, the outlined disadvantages necessitate careful use as a panacea across all sectors (Figure 4.11).

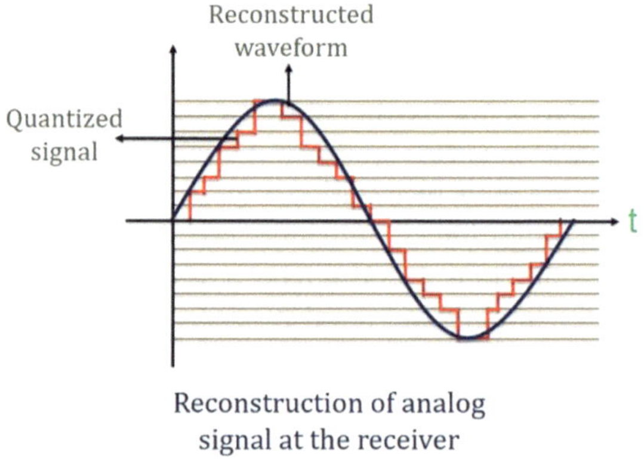

Figure 4.11. *Disadvantages of pulse modulation.*

Source: https://external-content.duckduckgo.com/iu/?u=https%3A%2F%2Felectronicscoach.com%2Fwp-content%2Fuploads%2F2018%2F10%2FReconstruction-of-analog-signal-at-PCM-receiver.jpg&f=1&nofb=1&ipt=912bd8093d9c838a2fde4b5fd923fe451825ae4b75196b8552a427b636ab0237.

4.4.1. Complexity in Implementation

The difficulty in applying pulse modulation techniques is primarily due to the need to control different system components, each for processing specific aspects of the signal. The key challenge is how to integrate disparate modulation schemes with digital signal processing techniques, particularly coherent detection receivers. Organized structures such as employed in the latest 100 Gbps wavelength-division multiplexing technologies need to be calibrated with specific parameters to realize maximum spectral efficiency and reduction of linear distortions within fibers. With a reliance on multi-level modulation schemes, more than one complexity layer is added since these schemes entail higher processing for the achievement of error correction as well as reconstruction of signals with precision. Without effective implementation strategies that account for these complexities, the performance of such systems would significantly degrade, making it essential to have a thorough understanding of both algorithmic and hardware design facets in contemporary communication networks (Amari et al., 2017). There is another dimension of complexity involved in handling non-linear impairments typical of high-speed communication systems as well. As data rates rise to 400 Gbps and beyond, the Kerr effect becomes increasingly impairing, leading to creative solutions for fiber nonlinearity compensation (NLC). Successful NLC techniques present challenging design criteria in terms of ease of deployment, typically requiring copious computational resources to maintain performance levels amid nonlinear distortion threats. Further, the coordination of these compensation techniques and system structure strongly dictates the design process because engineers have to balance trade-offs between performance gains and added complexity. It is this coordination between selection of modulation technique and NLC integration that decides not only the design selection but also scalability and cost efficiency of future systems of communication (Amari et al., 2017). For application in low data-rate communications application spaces, such as ad hoc wireless sensor networks, the modulation scheme can have a deep impact on implementation complexity. Non-coherent transmitting reference (TR) receivers are used in such cases due to their lower intrinsic complexity relative to their coherent counterparts. The reduced structure of the non-coherent receiver does not, however, solve the problems associated with interoperability and performance optimization. An efficient design must conquer several issues, such as signal fidelity and noise resilience, without breaking technology limits on the integrated circuit. This requires careful inspection of the trade-off between working efficiency and the strengths of the receivers architecture, thus stressing simplicity of design in terms of maintaining good performance parameters.

4.4.2. Sensitivity to Timing Errors

For all pulse modulation methods, sensitivity to timing error has significant consequences on the performance and reliability of the system. Timing error produces misaligned pulse signals, which result in interference and degradation of the transmitted information overall. For high-accuracy synchronization applications such as optical communications and synchronized laser-microwave networks, small differences in timing have significant effects. For instance, equipment that can function with attosecond-level synchronization in environments like very-long-baseline interferometry require extremely precise timing, whose deviation would cause less accurate outputs or even the failure to capture data. According to recent studies, developing methods to limit timing errors is crucial to ensure the validity of advanced photon science facilities tasked with studying atomic processes under varying conditions (Kaertner et al., 2016). Also, optical communication system behavior is dependent greatly on their ability to control environment parameters, particularly intersatellite links. The timing error sensitivity can lead to enormous power penalties, impacting the link budget as a whole.

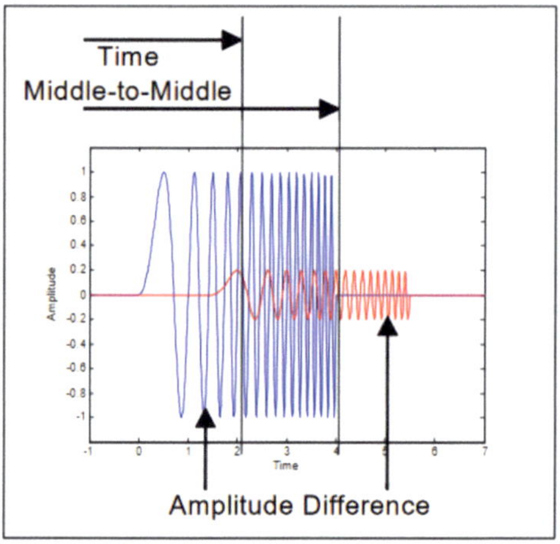

Figure 4.12. *Sensitivity to timing errors of pulse modulation.*

Source: https://www.tek.com/-/media/marketing-docs/a/amplitude-correction-im-pulse-response-measurement-radar-pulses/fig-2.png.

It has been established through research that antenna pointing and tracking accuracy are critical to optimum communication performance; suboptimum configuration will be to increase the effect of timing errors. More broadly, a test case suggests that while the direct detection systems show increased tolerance to spatial tracking errors, heterodyne systems are more degraded by synchronization errors and carrier phase noise (Chen et al., 2025). This is an important understanding that points to the importance of using strong timing error correction techniques to enhance the reliability of pulse modulation techniques in diverse applications. Lastly, addressing sensitivity to timing errors is crucial in order to spur innovation in pulse modulation techniques across numerous various applications. As technology advances further, demands for greater precision in data transmission and computation will continue to rise. However, only through ongoing innovation in research practice and methodology by engineers and researchers can such a performance be achieved. Wide experimental work and theoretical research, coupled with real-world applications of timing stabilization systems, are needed to break through the limitations imposed by timing errors. With advancements in synchronization techniques and error-correction mechanisms, developments in fields such as structural biology and material science become increasingly feasible. The convergence of improved technology and increased awareness of timing-related issues offers the potential for revolutionary progress in pulse modulation and beyond (Figure 4.12).

4.4.3. Limited Range in Some Applications

Even though pulse modulation techniques have numerous advantages for most applications, they come with limitations in their applicability. This weakness is particularly evident in high-speed optical imaging, where coherent Raman scattering techniques have emerged at the front due to their capability to produce good spatial resolution as well as three-dimensional sectioning capability. However, the compromise is between the speed of imaging and the identification of numerous compounds, which is more likely to require more time measurements to cover extensive spectral ranges. In order to bypass this, new approaches such as coherent anti-Stokes Raman scattering spectro-imaging have been developed, in which one can scan rapidly elaborate spectral information. Yet there are some occasions on which the limited range of these approaches precludes their application to differing situations, and their general applicability and value in resolving complex imaging challenges to the physical and life sciences (Bernhardt et al., 2013). Even the issue of limited range is not solely an imaging technique problem but extends into signal processing as well, where nonlinear effects impose enormous constraints. For example, variations in optical power of incoming signals onto a photodiode induce phase noise, which detriments the stability of microwave signals generated by photodetection. This phenomenon is more so felt with high-speed photodiodes, whose interaction of amplitude and phase noise makes it difficult to generate low phase noise microwaves. To be specific, maximum performance is achieved through accurate calibration of various parameters, like laser amplitude noise and photodiode structure. These are issues that reveal a fundamental limitation in the application of some pulse modulation techniques, such that meeting low phase noise conditions has a tendency to make them work best in very specialized regimes of operation (Datta et al., 2011). Lastly, the issue of constrained range shown by pulse modulation techniques reflects the importance of ongoing research and development. As programs demand increased precision and velocity, it is the task of the researchers to break beyond the constraints presented by current methodologies. Merging new technological advancements, including developing new photodiode structures or other signal processing techniques, can potentially offer solutions to some of these underlying limitations. By innovation in this field, it is possible to enhance the utility of pulse modulation techniques in various applications, ultimately broadening their range of use. The search for solutions that overcome present limitations not only facilitates improved performance but also enhances the prospect of breakthroughs of a dramatic nature in technologies based on advanced pulse modulation technologies (Bernhardt et al., 2013; Datta et al., 2011).

4.5. COMPARISON WITH OTHER MODULATION TECHNIQUES

Among modulation techniques, pulse modulation stands out as effective and versatile, especially in contrast with other techniques such as amplitude modulation (AM) and frequency modulation (FM). Pulse modulation divides signals into separate time intervals and allows for precise control of what information gets transmitted, which is incredibly beneficial to digital communication systems. On the other hand, AM varies the carrier wave's amplitude, thus more susceptible to interference and noise. FM, though stronger against such kinds of distortion due to its frequency variation techniques, can be less bandwidth-effective than pulse-based methods. Therefore, pulse modulation techniques like Pulse Width Modulation (PWM) and Pulse Position Modulation (PPM) gained prominence in high-fidelity and robustness applications, highlighting their excellence over traditional modulation techniques. The efficacy of pulse modulation techniques can be critiqued by comparing their modulation depth and signal-to-noise ratio against advanced techniques like laser-induced magnetic dipole spectroscopy (LaserIMD). Existing studies have shown the limitations of techniques such as light-induced triplet-nitroxide DEER (LiDEER) manipulation in applying rectangular pump pulses, which fall short of LaserIMD's modulation depth (Bieber et al., 2018). Latest developments of LiDEER, such as the integration of chirp inversion pulses, have shown possible improvements in modulation depth, pointing to a possible avenue of research that could further optimize pulse modulation techniques. These advances imply that while traditional pulse modulation methods are effective, design and implementation of them have to keep changing to compete with more specialized methods. Furthermore, hybridization of modulation methods can be done to create even better performance in some applications, particularly in optical communications. Techniques like electro-optical intensity modulation take advantage of interaction between amplitude and phase modulation and achieve synergy that optimizes overall system performance. By way of heterodyne measurements used to find phase changes in fiber-optic systems, researchers have created parameters such as the chirp parameter that directly affect modulation quality

(Auracher et al., 2010). This connection indicates the merits of comparing and integrating different modulation methods, each with its own capabilities that can be used in sum. Lastly, an understanding of pulse modulation techniques by contrast with these other techniques can facilitate breakthroughs for redefining the effectiveness of communication in increasingly more sophisticated transmission environments.

4.5.1. Amplitude Modulation (AM)

Amplitude Modulation (AM) is a basic pulse modulation method that facilitates information communication through a variety of communication channels. The modulation method alters the amplitude of the carrier signal based on the information to be conveyed, like audio, video, or data. One of the salient aspects of AM is how easy it is to implement since it is widely applied in broadcasting-based applications, such as AM radio. While less efficient than more modern techniques, the ability to reconstruct the original signal fairly accurately from amplitude fluctuations remains applicable. Additionally, the inherent characteristics of AM make it an easy demodulator, which simplifies receiver design and integration into numerous communications systems, therefore keeping its position in modern telecommunications. However, consideration should also be given to the drawbacks of Amplitude Modulation, namely that it is susceptible to noise and interference. The varying amplitudes make AM signals vulnerable to distortion from the outside world, which can seriously degrade signal quality. High ambient noise levels can suppress the amplitude variations, leading to less stable communication. Different methods and equipment have been developed to counteract these issues; for instance, use of high-quality photodiodes can increase the stability and signal integrity of the transmitted signals. Research has indicated that the synergy between low laser amplitude noise and adequate photodiode design is highly beneficial in enhancing system performance, achieving phase noise levels below –100 dBc/Hz offset at 1 Hz, thereby boosting overall communication fidelity (Datta et al., 2011). In contemporary use, AM is on equal footing with newer technologies such as photonic transmitters that utilize pulse modulation at high rates, as it shows convergence of methods supporting performance at reduced operation complexities. Integrated photonic transmitters operating at rates up to 100 MHz, for instance, illustrates the manner in which amplitude control can be effectively controlled for implementation in quantum communication and beyond. The use of optoelectronic devices reduces power consumption and miniaturization, which makes such systems suitable for today's needs, like satellite communication (Anzolin et al., 2010). Thus, while AM is a very old technique, flexibility in its application and ongoing research to enhance its utilization guarantees its position among the emerging trends of pulse modulation techniques (Figure 4.13).

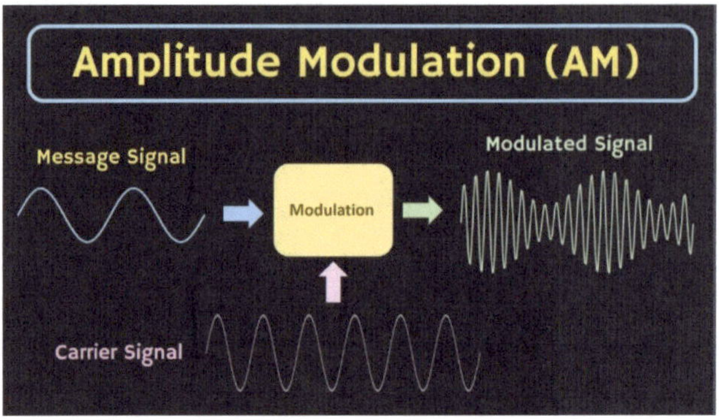

Figure 4.13. *Amplitude modulation (AM).*

Source: https://www.flyeye.io/wp-content/uploads/2024/10/Amplitude-Modulation.jpg.

4.5.2. Frequency Modulation (FM)

Frequency Modulation (FM) is a significant advancement in pulse modulation techniques, which provides a handy method of encoding data on to a carrier wave by varying the frequency of the carrier wave in relation to the amplitude of the input signal. Unlike the case of amplitude modulation (AM), where the amplitudes of the signals are altered, FMs unique method of operation reduces the susceptibility of signals to interference and noise, particularly in communication channels. This is most beneficially critical where applications are such as radio broadcasting, where sound clarity and purity take precedence. The presence of inherent resistance to signal fading in the case of FM also ensures that transmission gets stronger and more dependable, a clear testimony to the superiority of the techniques over traditional approaches. These characteristics render FM a first-choice option in many applications of technology, thus firmly establishing it in modern communication systems and optimizing overall efficiency of pulse modulation techniques. As the demand for reliable wireless communication grows, understanding the inherent dynamics of FM becomes increasingly essential. Frequency modulation as a process operates by translating instantaneous changes in the amplitude of the signal to changes in frequency, hence allowing for the transmission of high-fidelity audio and data over long distances across. Advances in optical and digital communications in the recent past have highlighted the importance of FM, particularly with improvements such as heterodyne-based phase noise reduction techniques as in the case of Aflatouni et al. (2018). Not only do such techniques ease signal processing, but also enable the use of mode-locked lasers in areas demanding high modulation formats. Thus, FM is constantly evolving, staying in line with the present needs and incorporating advanced technologies that enhance its utilitarian performance in many forms of communication media. Besides that, FM technology is also being applied beyond the usual audio broadcasting into the realm of digital communications as a fundamental building block in several modern modulation schemes. For instance, its backing for next-generation modulation formats is crucial in that it offers avenues of effective transfer of higher-order signals without compromising signal integrity. FM research has also ushered in understanding in the realm of phase noise and its impact on signal quality, which is furthermore related to findings like those gained through the analysis of the jitter of mode-locked lasers as discussed in Aflatouni et al. (2018). As research unrolls, situations of FMs potential not only enrich theoretical processes but also create gateways for technological innovation in telecommunication. This interaction of theory and practice consolidates the role of FMs among the range of pulse modulation techniques in being a part of evolving communication technology.

4.5.3. Phase Modulation (PM)

Phase modulation, or PM, is a key technique in the discipline of pulse modulation and holds significant value in enhancing the reliability and efficiency of digital communication. Unlike in amplitude modulation, where information is conveyed through variations in the signal amplitude, phase modulation varies the phase of the carrier wave in accordance with the input signal. This capability makes PM greatly resilient to noise since it enables robust signal integrity in harsh transmission conditions. The operation of PM is best seen in fiber-optic communication environments, where distortions resulting from phase shifts compromise signal quality. It is thus important to understand these distortions and how they influence transmission while designing PM systems for optimization. In order to appreciate the nuances of PM, one should examine the underlying mechanisms that accumulate to phase changes in optical signals. Novel techniques have been employed in latest research to quantify how residual phase modulation affects performance parameters in all types of devices, particularly fiber-based modulators. With the use of a heterodyne method to find the phase shift due to intensity variation, researchers have created a feasible system to analyze chirp parameters, which describe how quickly a signal's frequency changes with time. Testing reveals that such chirp parameters can easily be compensated with other phase modulators, giving a better overall quality of the signal and minimizing distortion (Auracher et al., 2010). In addition, the interaction between amplitude fluctuation and phase noise in photodetection systems is another demonstration of the complexities of PM in actual real-world applications. Power fluctuations of optical beams onto photodiodes can unintentionally induce phase fluctuations to impact overall timing stability required for microwave signal generation at high frequencies. With the study of high-speed InGaAs P-I-N photodiodes, researchers have outlined amplitude noise-to-phase noise conversion, reflecting the delicate balance that subsists between laser amplitude noise and optimizing photodiodes for utmost efficiency. The study identifies that low phase noise, particularly at high frequencies, calls for a harmoniously coordinated approach to the control of several parameters such as average photocurrent and photodiode fabrication (Datta et al., 2011). This marriage of phase modulation with pulse modulation techniques reveal not only the significance of surmounting signal transmission problems but also the pivotal position of phase modulation in the evolution of current telecommunication (Figure 4.14).

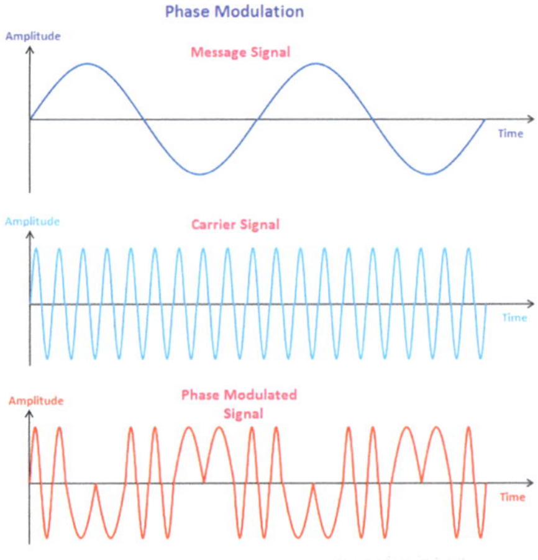

Figure 4.14. *Phase modulation (PM).*

Source: https://www.physics-and-radio-electronics.com/blog/wp-content/uploads/2018/04/phasemodulation.png.

4.6. FUTURE TRENDS IN PULSE MODULATION

As the need for high-speed communication continues to escalate, pulse modulation techniques stand to undergo revolutionary advances. Future communication systems, particularly with the introduction of fifth-generation

(5G) networks, will most likely employ these techniques to achieve maximum bandwidth usage and enhance data rates. With the inadequacy of the available radio frequency (RF) spectrum to accommodate increasing high-data-rate multimedia services, new modulation schemes, such as orthogonal frequency-division multiplexing (OFDM) and pulse width modulation, will be increasingly important. Such systems, not only will they lead to improved spectral efficiency but also have robust performance in dense networks, providing support for multiple users concurrently without sacrificing signal integrity and quality of service. Furthermore, the implementation of optical wireless communication (OWC) provides an encouraging trend for the future in pulse modulation. As new research finds, OWC operates on a non-regulated frequency spectrum, avoiding the constraints of conventional RF communications. The throughput speeds of more than 100 Gb/s illustrate the potential in OWC technologies, such as visible light communication, free-space optical communication, and so forth. They will not just enhance indoor and outdoor connectivity but also transfer heavy data traffic away from RF networks, keeping efficient usage of the spectrum. The move towards optical systems is a paradigm change in pulse modulation utilization with higher speed and reliability and addressing the increasing demand for wireless capacity (Chowdhury et al., 2018). With the advancements in the modulation method, overlaying cooperative techniques such as fractional frequency reuse and distributed antenna systems (DAS) will be critical to enhancing cellular communication. They improve cell-edge rates without negatively impacting the performance at the cell center, thus attaining even resource distribution. The cooperative nature of the systems will enhance more effective handling of users and better delivery of service in multiuser systems. With the use of sophisticated modulation methods and infrastructural development, future trends are positioning pulse modulation as an inherent form of emerging networks that promises to be more efficient and pleasing to the user in more complex communications settings (Ghafoor et al., 2011).

4.6.1. Advancements in Digital Signal Processing

The domain of digital signal processing (DSP) has undergone revolutionary changes, particularly in pulse modulation techniques. The developments that have been observed in DSP technology have supported the development of more efficient schemes for modulation that enable bandwidth optimization and maximum signal integrity. Multilevel pulse amplitude modulation (PAM) is a case in point that has enabled data rate enhancement without tagging bandwidth proportionally. Through sophisticated algorithms, DSP offers error-correction methods that reverse transmission errors and ensure system dependability. In addition, advancements in electrical and optical devices enable real-time signal processing, which is key to adaptive modulation methods that react differently to various environmental situations and user demand. This breakthrough results from the intricate interconnectedness between DSP advancements and advancements in pulse modulation theory, giving way to future communications systems. Further, fractional frequency reuse techniques are a significant advancement in cooperative communication environments, particularly in multi-cell systems. As described in recent work, these techniques, when applied with distributed antenna systems (DAS), have demonstrated improved throughput, primarily for users on the cell-edge. In such topologies, cooperative linear transmit processing can be employed to significantly enhance the efficiency of multi-user interactions, allowing for overall optimization of transmit power across the network. The benefit of achieving higher cell-edge throughput without loss to the central part of the cell is an inherent improvement in digital signal processing technology. This merging of developments in DSP with pulse modulation methods leads to an even more dynamic and scalable communication system that can accommodate the growing demands of next-generation networks (Ghafoor et al., 2011). Further, studies on next-generation satellite communication systems provide yet another side of the evolutionary process in digital signal processing. Current research supported by organizations such as NASA-Lewis focuses on enhancing onboard processing as well as technology development for various communication networks. VSAT network installation in meshed form utilizing frequency division multiple access (FDMA) for uplinks and time division multiplexed (TDMA) downlinks is a major step towards optimizing satellite communications. Among the pillars of such technologies is the multichannel demultiplexer/demodulator (MCDD) technology, which employs a synergy of optical and digital technologies for its signal enhancement. This innovation is an indication of the increasing influence of DSP in optimizing the utilization of pulse modulation techniques, in the long term enhancing satellite communication with regards to data output and sustainability (Abramovitz et al., 2025).

4.6.2. Integration with Emerging Technologies

The integration of new technologies into methods of pulse modulation is becoming all the more vital with the increasing need for data transmission. The onset of fifth-generation (5G) networks and the subsequent boom of multimedia services has made the conventional radio frequency (RF) spectrum inadequate. New ideas such as Optical Wireless Communication (OWC) are emerging to address these challenges effectively. OWC's ability to employ an unlicensed frequency band is a future alternative to RF networks with the ability to transfer large data. For instance, OWC has the ability to deliver data rates of up to 100

Gb/s, thereby unloading large traffic from traditional communication channels. This innovation not only enhances existing communication technologies but also opens up avenues for new technologies to be created in the area of data transmission, making it an essential aspect of modern networking practices (Chowdhury et al., 2018). Further, the development of pulse modulation techniques together with optical technologies indicates a general trend towards multidimensional design in communication systems. By systematically analyzing the systems, individual component requirements can be defined that will be compatible and run at an optimum level. Enhanced signal processing techniques embedded into the system allow effective control over transmission conditions in such a way that the system is made flexible enough to operate under any given situation. Input filtering techniques, frequency and spatial domain, are supported by well-designed antennas that co-operate with modulation techniques. This comprehensive system design highlights the innovative ability of integrating future technologies into traditional systems in order to enhance data transmission as well as broaden the scope of application in communication (Gotthans et al., 2012). Furthermore, the fusion of pulse modulation techniques with new technologies boosts innovation in communication systems such that high-performance and effective networks are made possible. With techniques such as visible light communication and free space optical communication, the limitations to RF systems can be addressed. The ability to ascertain properties of existing frequency spectrum allows designers to allocate resources in an efficient manner across available channels so that maximum transmission outcomes are achieved. This is particularly significant as communication requirements change, requiring systems that are flexible enough to accommodate sudden changes in data traffic. The interaction between modulating techniques and new technologies creates a vibrant platform for research and development that ensures communication infrastructures stay robust and able to meet future needs. In response to the issues created by the information deluge, incorporation of these technologies is key to communication efficiency and reliability enhancement (Chowdhury et al., 2018; Gotthans et al., 2012).

4.6.3. Potential for Enhanced Data Transmission

The evolution of pulse modulation techniques has brought significant opportunities for enhanced data transmission, particularly with the introduction of metamaterials into telecommunications. Metamaterials possess characteristics that allow electromagnetic functionalities beyond those of natural materials, enabling achievement in wave control and data rates. Using these materials, scientists are able to design systems that improve signal quality and reduce latency in data communication systems. Such precise control of the amplitude and phase of signals being propagated is essential to high-capacity communication systems. This is particularly true in environments involving low loss and high fidelity over long distances, where traditional methods are apt to fail. The meeting of pulse modulation techniques and metamaterials thus provides a viable platform for the realization of the next generation of high-rate data transmission networks, facilitating improved communication infrastructures that are crucial in today's data-oriented age. With further development of pulse modulation techniques, the role of terahertz (THz) technology in the technology cannot be overlooked. The unique characteristics of THz waves, coupled with the tunability afforded by metamaterials, are a configuration in which increased capacity in data transmission is facilitated. The materials allow for the generation of nonlinear responses, which can be utilized to optimize signal processing in scenarios that need high bandwidth and high data rate transfer. For instance, InAs plasmonic disk arrays, as recently discussed in research, have high nonlinear responses that are utilized for designing devices like nonlinear absorbers. These devices can be applied as optical limiters, making the system more resistant to interference and thus overall data transmission quality. By merging THz and pulse modulation, the potential of building robust channels of communication able to process immense amounts of information is significantly enhanced, showcasing the synergy between these cutting-edge technologies. Finally, the implications of such advances extend beyond theoretical inquiry and go on to real-world applications, particularly in satellite systems and mobile communications. The dynamic characteristics of metamaterials make it possible to create devices that can adapt based on varying environmental conditions, a key attribute in data integrity and data rate consistency. By having the capability to create tunable and adaptive materials, engineers can devise sophisticated signal processing mechanisms that react in real-time to fluctuations in signal strength and quality. This capacity for flexibility would have a deep influence on user experiences for telecommunications, where ambient connectivity is now simply assumed. Apart from that, with the increased need for high-speed networks, pulse modulation method optimization through the use of metamaterials is not just advantageous but imperative, opening doors to future improvement in data transmission techniques which can keep pace with rising user needs and technological capabilities (Averitt et al., 2015; Kodama et al., 2000).

SUMMARY

Pulse modulation techniques are crucial in modern communications to enable successful transmission of information over various mediums. By modulation of the carrier signal using the signal information, pulse modulation techniques offer a guarantee that data will be encoded and transmitted in a format that is noise-insensitive as well as robust. Pulse modulation analysis demonstrates that all methods, such as Pulse Position Modulation (PPM) and Pulse Width Modulation (PWM), each provide unique advantages and disadvantages in relation to application context. Each method's success relies on its ability to maintain signal integrity and minimize interference, which plays a significant role in high-scenario environments like satellite communication and telecommunication networks. Furthermore, the progress in pulse modulation, particularly when considering more sophisticated methodologies such as those discussed in Asgekar et al. (2001), has opened up new possibilities for study and application. An understanding of the processes governing the behavior of pulses, as described by pulsar science's stable system of sub beams near magnetic axes, can enhance more effective signal processing techniques. Such understanding enables engineers to optimize modulation schemes for specific applications, optimizing performance and reliability. The synergy of theoretical research and real-world implementation therefore shows the ability of pulse modulation techniques to grow in complexity and utility. Lastly, incorporation of cutting-edge technology, including new spin labels and spectroscopies, illustrates the dynamic nature of pulse modulation research and application. The findings presented in Bieber et al. (2018) about light-induced spectroscopy indicate ongoing efforts at enhancing measurement efficiency and accuracy, a milestone for the field. As scientists continue to explore and develop new techniques, opportunities for pulse modulation to transform communication systems will continue to grow. On the basis of both a legacy of success and new technological advancements, the future of pulse modulation techniques promises to redefine standards in several technological areas.

A complete understanding of the methods of pulse modulation introduces some basic considerations that are important for theoretical analysis and technical implementation. Pulse modulation is a basic technique in modern communication systems, and it makes it possible to transmit data efficiently with the use of various media. Pulse modulation converts analog signals into digital signals and enables information to be transmitted clearly, losing minimal amounts through distortion. The significance of methods such as Pulse Width Modulation (PWM) and Pulse Position Modulation (PPM) is evidenced by their versatility across different applications, from industrial control to communication. The capability of pulse attributes' control allows for better performance in power consumption and bandwidth efficiency, making these techniques core to technological progress. Furthermore, practical applications of pulse modulation persist to gain prominence in contemporary engineering applications. For instance, telemetry systems transmitting data from remote sources are enabled by the reliability and economy of pulse modulation techniques. A peculiarity is described in weather rocket payloads employing S- or L-band microwave links for data transmission.

These systems utilize low-cost oscillator-type transmitters to transfer data economically, an example of cost-saving and ease of use of pulse modulation devices. These applications demonstrate not only the cost benefits but also the capability of pulse modulation in supporting different types of data with promises of synchronization and accuracy in the transmit signals. Furthermore, breakthrough in complex systems has shown that there is no reliable solution to communications without pulse modulation. Through the utilization of time and frequency multiplexing techniques, data can be transmitted simultaneously via multiple channels, hence improving the efficiency of data relaying. As is seen in specific examples, mapping different channels of data with changes in pulse repetition rate and duty cycle proves the versatility of pulse modulation (Hartman et al., 2025). Real-time control under a dynamic environment is particularly valuable for mission-critical applications, such as communications in spacecraft, where sample configurations require predictable performance under varying conditions (Hartman et al., 2025). In the end, the continuing evolution of pulse modulation techniques demonstrates their significance in the area of communications as a basic element of the constant advancement of technology.

REFERENCES

1. Abramovitz, I., Budinger, J. M., Courtois, H. A., Ivancic, W., et al. (2025). Multichannel demultiplexer/demodulator technologies for future satellite communication systems. https://core.ac.uk/download/pdf/42814118.pdf (accessed on 18th September 2025).
2. Aflatouni, F., Arkhipov, I., Calò, C., Drzewietzki, L., Freude, W., Gioannini, M., Habruseva, T., et al. (2018). Numerical investigation of a feed-forward linewidth reduction scheme using a mode-locked laser model of reduced complexity. https://core.ac.uk/download/287724151.pdf (accessed on 18th September 2025).
3. Agelidis, V. G., Balouktsis, A., Balouktsis, I., Cossar, D., et al. (2005). Five-level selective harmonic elimination PWM strategies and multicarrier phase-shifted sinusoidal PWM: A comparison. https://core.ac.uk/download/11241418.pdf (accessed on 18th September 2025).
4. Alexander, T., Aprahamian, A., Back, H. O., Calaprice, F., et al. (2015). Scintillation efficiency measurement of Na recoils in NaI(Tl) below the DAMA/LIBRA energy threshold. http://arxiv.org/abs/1503.07212 (accessed on 18th September 2025).
5. Alves, T., Beltran, M., Cartaxo, A., Llorente, R., et al. (2008). Ultra-wideband radio signals distribution in FTTH networks. https://core.ac.uk/download/4148630.pdf (accessed on 18th September 2025).
6. Amaral, L., Dris, S., Gill, K., Grabit, C., et al. (2010). Feasibility of using bandwidth efficient modulation to upgrade the CMS tracker readout optical links. http://arxiv.org/abs/1010.1691 (accessed on 18th September 2025).
7. Amari, A., Ciblat, P., Dobre, O. A., Jaouën, Y., et al. (2017). A survey on fiber nonlinearity compensation for 400 Gbps and beyond optical communication systems. http://arxiv.org/abs/1708.06313 (accessed on 18th September 2025).
8. Aminzadeh-Gohari, D., Djahani, N., Gellar, M., Kahn, J. M., Kim, J., Lee, H., Lee, C., et al. (2007). Slot error rate performance of DH-PIM with symbol retransmission for optical wireless links. https://core.ac.uk/download/4147294.pdf (accessed on 18th September 2025).
9. Angelova, M., Ghassemlooy, Z., & Rajbhandari, S. (2010). A study of discrete wavelet transforms based denoising to reduce the effect of artificial light interferences for indoor optical wireless communication. https://core.ac.uk/download/4149283.pdf (accessed on 18th September 2025).

10. Angelova, M., Ghassemlooy, Z., & Rajbhandari, S. (2011). Wavelet—artificial neural network receiver for indoor optical wireless communications. https://core.ac.uk/download/4149364.pdf (accessed on 18th September 2025).

11. Anzolin, G., Gardelein, A., Jofre, M., Mitchell, M., et al. (2010). 100 MHz amplitude and polarization modulated optical source for free-space quantum key distribution at 850 nm. http://arxiv.org/abs/1009.1469 (accessed on 18th September 2025).

12. Asgekar, A., & Deshpande, A. A. (2001). Topology and polarization of subbeams associated with pulsar B0943+10's 'drifting'-sub pulse emission: II. Analysis of Gauribidanur 35-MHz observations. https://core.ac.uk/download/25321861.pdf (accessed on 18th September 2025).

13. Auracher, P., Bakos, G., Bakos, K., Rogers, C. E. III, Courjal, N., Devaux, F., Dorrer, C., et al. (2010). Characterization and compensation of the residual chirp in a Mach-Zehnder-type electro-optical intensity modulator. http://arxiv.org/abs/0910.5698 (accessed on 18th September 2025).

14. Averitt, R. D., Bank, S. R., Fan, K., Keiser, G. R., et al. (2015). Nonlinear terahertz devices utilizing semiconducting plasmonic metamaterials. http://arxiv.org/abs/1508.03183 (accessed on 18th September 2025).

15. Awan, M., Brandl, P., Capsoni, C., Ghassemlooy, Z., et al. (2009). Current optical technologies for wireless access. https://core.ac.uk/download/4149148.pdf (accessed on 18th September 2025).

16. Bernhardt, B., Guelachvili, G., Holzner, S., Hänsch, T. W., et al. (2013). Coherent Raman spectro-imaging with laser frequency combs. http://arxiv.org/abs/1302.2414 (accessed on 18th September 2025).

17. Bevilacqua, A., Blazquez, B., Carbonelli, C., Cassioli, D., Chen, S., et al. (2005). Implementation aspects of a transmitted-reference UWB receiver. https://core.ac.uk/download/pdf/11370300.pdf (accessed on 18th September 2025).

18. Bieber, A., Bücker, D., & Drescher, M. (2018). Light-induced dipolar spectroscopy–A quantitative comparison between LiDEER and LaserIMD. http://arxiv.org/abs/1806.04190 (accessed on 18th September 2025).

19. Brandt-Pearce, M., & Noshad, M. (2013). High-speed visible light indoor networks based on optical orthogonal codes and combinatorial designs. http://arxiv.org/abs/1308.0743 (accessed on 18th September 2025).

20. Bull, D. R., Canagarajah, C. N., Czerepinski, P. J., Nix, A., et al. (2000). Robust and scalable matching pursuits video transmission using the Bluetooth air interface standard. https://core.ac.uk/download/29025279.pdf (accessed on 18th September 2025).

21. Charbonnier, B., Gaudino, R., Greiss, I., Koonen, A., et al. (2011). Recent results from the EU POF-PLUS project: Multi-gigabit transmission over 1 mm core diameter plastic optical fibers. https://core.ac.uk/download/pdf/11424445.pdf (accessed on 18th September 2025).

22. Chen, C. C., & Gardner, C. S. (2025). Comparison of direct and heterodyne detection optical intersatellite communication links. https://core.ac.uk/download/pdf/42837883.pdf (accessed on 18th September 2025).

23. Chowdhury, M. Z., Hossan, M. T., Islam, A., Jang, Y. M., et al. (2018). A comparative survey of optical wireless technologies: Architectures and applications. http://arxiv.org/abs/1810.02594 (accessed on 18th September 2025).

24. Collado Escolano, A., Gálvez, E., Lumbreras Carrasco, D., Mon González, J., et al. (2019). Efficiency analysis of wide band-gap semiconductors for two-level and three-level power converters. https://core.ac.uk/download/286456302.pdf (accessed on 18th September 2025).

25. Datta, S., Diddams, S., Hati, A., Joshi, B., et al. (2011). Characterization of power-to-phase conversion in high-speed P-I-N photodiodes. http://arxiv.org/abs/1102.2540 (accessed on 18th September 2025).

26. García Franquelo, L., Kouro, S., León Galván, J. I., Martín Prats, M. Á., et al. (2009). Multilevel converters: An enabling technology for high-power applications. https://core.ac.uk/download/286564367.pdf (accessed on 18th September 2025).

27. Ghafoor, S., Hanzo, L., Xu, X., Zhang, Y., et al. (2011). Imperfect digital fiber optic link based cooperative distributed antennas with fractional frequency reuse in multicell multiuser networks.

28. Gotthans, T., Hrubos, Z., Kadlec, P., Kincl, J., et al. (2012). Communication subsystems for emerging wireless technologies. https://core.ac.uk/download/30311907.pdf (accessed on 18th September 2025).

29. Hartman, D. J., Kahnke, J., & Kirk, R. J. (2025). Conceptual mechanization studies for a horizon definition spacecraft communications and data handling subsystem. https://core.ac.uk/download/pdf/85247138.pdf (accessed on 18th September 2025).

30. Hemdi, A. R., Mahmood, S., Mat Saman, M. Z., Zakuan, N., et al. (2017). Potential impacts for monitoring sustainability: Case study of hollow fiber membrane. https://core.ac.uk/download/12008222.pdf (accessed on 18th September 2025).

31. Herceg, M., Matic, T., & Svedek, T. (2009). Performances of hybrid amplitude shape modulation for UWB communications systems over AWGN channel in a single and multi-user environment. https://core.ac.uk/download/44395598.pdf (accessed on 18th September 2025).

32. Kaertner, F. X., Kalaydzhyan, A., Muecke, O. D., Peng, M., et al. (2016). Attosecond precision multi-km laser-microwave network. http://arxiv.org/abs/1603.07963 (accessed on 18th September 2025).
33. Kodama, C. H., & Coutu, R. A. (2000). THz metamaterial characterization using THz-TDS. https://core.ac.uk/download/213088137.pdf (accessed on 18th September 2025).
34. Melrose, B. T. (2025). A unique modulation system for two channel data transmission. https://core.ac.uk/download/pdf/85232908.pdf (accessed on 18th September 2025).
35. Mohamad Nur, H. (2002). Persepsi pelajar sarjana muda kejuruteraan elektrik terhadap program latihan industri, Kolej Universiti Teknologi Tun Hussein Onn. https://core.ac.uk/download/12007367.pdf (accessed on 18th September 2025).
36. Oghorada, O. J. K., & Zhang, L. I. (2016). Performance evaluation of pulse width modulation techniques for losses reduction in modular multilevel flying capacitor converter. https://core.ac.uk/download/74233154.pdf (accessed on 18th September 2025).
37. Vaziri, A., & Weisenburger, S. (2018). A guide to emerging technologies for large-scale and whole brain optical imaging of neuronal activity. http://arxiv.org/abs/1805.03197 (accessed on 18th September 2025).
38. Yonis Buswig, Y. M. (2011). Development of a DC-AC power conditioner for wind generator by using neural network. https://core.ac.uk/download/12007862.pdf (accessed on 18th September 2025).

CHAPTER 5

DIGITAL MODULATION TECHNIQUES

Contents

INTRODUCTION

The landscape of digital communication has evolved at a rapid rate during the past few decades, necessitating the development of advanced modulation techniques to achieve the optimum efficiency of data transmission. Digital modulation relies upon variation of the parameter of a carrier wave to convey information and is indispensable to be employed in wireless communication to optical networks. With the growing use of digital media for communication by society, familiarity with such modulation techniques is required. The development of high-speed systems has been followed by a shift towards more sophisticated methods capable of supporting higher bandwidths with signal integrity maintained.

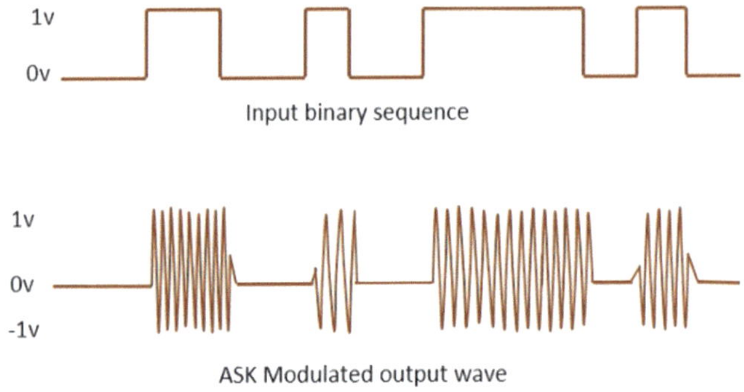

Figure 5.1. *Digital modulation techniques.*

Source: https://www.tutorialspoint.com/principles_of_communication/images/ask.jpg.

All these have fairly straightforward implications for the performance of the transmissions, particularly against the background of new technologies and the pressures placed upon network infrastructure. The introduction of coherent detection systems has heavily influenced fiber optic communication networks, which facilitate greater management of optical fiber impairments like nonlinearity and dispersion. As stated, these fiber impairments can be completely removed with digital signal processing (DSP) algorithms, which enhance the performance of high-speed systems based on wavelength-division multiplexing (WDM) standards. Through the application of multi-level modulation formats together with DSP techniques, data transmission spectral efficiency can be increased significantly. Additionally, at data rates of 400 Gbps and beyond, nonlinearity resulting from the Kerr effect becomes increasingly critical to manage. The reality that such issues can be addressed by digital modulation techniques continues to be a major sign of research significance here. In addition to improving the efficiency of optical systems, digital modulation techniques play a critical role in content protection and data integrity problems. The emergence of digital piracy has prompted the need to enforce rigorous protection against illegal copying and dissemination of digital content. As discussed, several solutions such as digital watermarking have been created to address these problems by embedding identification data within media content.

This development of methods illustrates the interaction between technological innovation and the measures required for safeguarding intellectual property in the digital world. The comprehension of digital modulation techniques hence involves the recognition of their role in communication efficiency along with the protection of content, which underlines their intrinsic significance in contemporary digital settings (Figure 5.1).

Definition of Digital Modulation

Digital modulation is one of the fundamental techniques in communication systems that converts digital information into an analog signal to be transmitted through various media. The conversion is necessary since the majority of the channels used for communication, such as radio waves or copper wire, are in analog form. Digital modulation employs a number of methods for expressing binary data in terms of variations in the carrier wave's amplitude, frequency, or phase. Consequently, it facilitates good transmission of data with less effect of interference and noise, thereby supporting better signal integrity. Digital modulation is also pertinent due to its flexibility, with different modulation methods being able to be tailored to specific applications and channel conditions, which means more bandwidth utilization and transmission reliability. Among the modulations that have been used digitally, including Phase Shift Keying (PSK), Frequency Shift Keying (FSK), and Amplitude Shift Keying (ASK), each carries different strengths and uses in different communications. PSK, for instance, codes information into the phase of the carrier wave, making it extremely resilient to noise and the interest of researchers looking to advance communication systems. Apart from this, hybrid techniques like Orthogonal Frequency-Division Multiplexing (OFDM) have gained popularity, especially in broadband networks of communication. These kinds of methods allow effective utilization of the spectrum and minimize inter-symbol interference and are specifically helpful in high-speed networks and wireless systems. Application of these methods proves that digital modulation is responsive to the demands of modern data transmission. Digital modulation technologies also come into play when combined with newer technologies for enhanced performance in specific environments. For example, pairing digital systems with Galois-Division Multiple Access (GDMA) is a case of technological innovation in making the most efficient use of bandwidth while serving multiple users by means of finite field transformations. These technological innovations pave the way for applications in homes and offices, where there can be high-speed data transmission over the use of fiber optics. In polymer optical fiber (POF) applications, techniques like eight-level pulse amplitude modulation (8-PAM) and OFDM have been explored in order to facilitate high capacity transmission over large distances. These developments highlight the evolution of digital modulation in line with technological needs, revealing its central position in meeting the contemporary requirements for communication and ongoing efforts aimed at improving and expanding its applications in different environments (de Oliveira et al., 2015; Abrate et al., 2010).

Importance in Modern Communication

The development in digital modulation techniques has played an important role in making modern communication systems efficient. The techniques enable data to be transmitted at a high rate of efficiency through various distances and mediums, thereby meeting growing bandwidth needs in the modern era. The example of modulation techniques such as Phase Shift Keying (PSK) and Quadrature Amplitude Modulation (QAM) enables data transmission at higher rates while being immune to interference as well as noise. As the users are becoming increasingly reliant on wireless technologies for communication, these developments in modulation methods ensure that data integrity and service quality are always prioritized. Such advancements are not only technical developments but also foundation enablers for future developments in internet connectivity and cellular communication and, therefore, emphasize their importance as drivers of the future of modern communication infrastructures. In addition, the achievement of digital modulation techniques to address electromagnetic compatibility (EMC) is central to their role in modern communications. With the future moving toward integrative interconnectedness, it is vital that various wireless systems be guaranteed they coexist without injurious interference. Coming out of these developments are statistical approaches based on amplitude probability distribution detection, which have proved to estimate disturbances more accurately than currently established quasi-peak detection compliance techniques. The development shows the need for new EMC standards and test methods (Azpúrua et al., 2019). By prioritizing such innovation, communication systems can be rendered more effective and more dependable, thereby ensuring a secure environment for the application of future-oriented apps such as web of things (IoT) and intelligent cities. Moreover, BbB-adaptive multicarrier systems are the ongoing evolution of digital modulation techniques that respond favorably to the dynamic behaviour of wireless communications environments. Utilizing adaptive coding and modulation methods, these types of systems can adapt to channel conditions in real time, maximizing throughput alongside imparting robustness against variable quality. Such timeliness is particularly crucial, as it maximizes the operation of modern communication networks so that information can be communicated even during unfavorable conditions with interference (Choi et al., 2007). Their inclusion in addition to diversity methods such as Rake receivers and space-time block coding aids these adaptive systems in turn to counter channel impairments. Thus, development and implementation of such advanced modulation schemes are a cornerstone to achieving robust communication in the very digital world of the present day, and their significance cannot be overstated in shaping the modern communications environments (Figure 5.2).

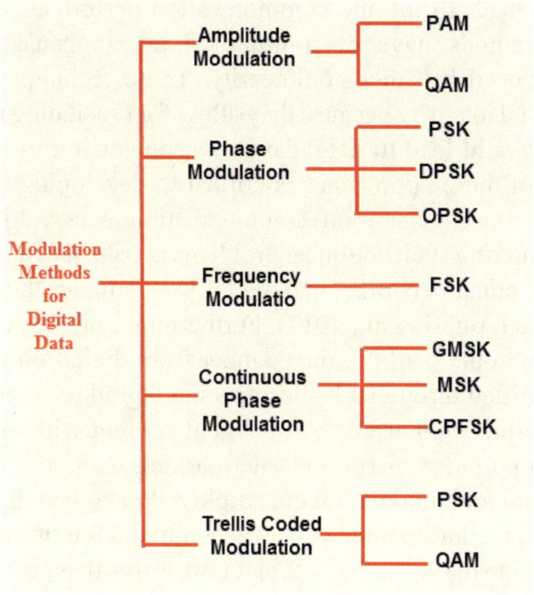

Figure 5.2. *Digital modulation techniques importance in modern communication.*

Source: https://www.watelectronics.com/wp-content/uploads/types-of-digital-modulation1.png.

Overview of Key Techniques

A detailed analysis of digital modulation techniques involves understanding coherent detection systems, which revolutionized optical communication. These systems utilize advanced digital signal processing (DSP) algorithms to counteract distortions in optical fibers, particularly in the context of 100 Gbps wavelength-division multiplexing (WDM) standard technologies. With multi-level modulation formats combined with DSP, coherent systems effectively counteract linear fiber optic distortions. With the rising needs of contemporary communication, awareness of these methods is crucial to enhancing spectral efficiency; thus, they are no less crucial in theory and practice in fiber technology.

Understanding the integration and performance of these methods informs current research and development, hence future communication planning to support higher data rates and improved transmission quality (Amari et al., 2017). For digital modulation, compensation methods for nonlinearity (NLC) are paramount, particularly with data rates approaching 400 Gbps and beyond. The Kerr effect, with growing significance at increased input power, presents formidable challenges to infinite communication performance. Consequently, various NLC methods have been proposed as compensation against the intricacies introduced by fiber nonlinearity. These techniques are crucial in guaranteeing signal integrity because they allow for precluding nonlinear effects that, otherwise, would lead to degradation. A comparison of the effectiveness and complexity of these approaches is critical to developing systems with the ability to handle next-generation communication needs. Finally, enhanced capability in countering such nonlinear problems is critical in the development of high-bandwidth optical networks, making NLC techniques the focus of today's research activities (Amari et al., 2017). Furthermore, digital watermarking is a critical technique in the battle against danger from digital piracy proliferation, a challenge magnified through advancements in digital technology. The ability to produce numerous similar copies of digital content without loss of quality highlights the importance of robust watermarking methods. Techniques such as spread-spectrum and modulation are employed to embed digital watermarks appropriately into media streams with an accommodation of trade-offs between invisibility and robustness. Use of wavelet transforms that possess significantly improved performance over previous methods further reflects the resilience of modern watermarking systems. In addition, the two-layer watermarking approach, employing the spatial domain and the wavelet domain techniques, illustrates a change in strategy towards guaranteeing watermark security against various forms of attacks. The multi-aspect approach points to the inclusion of advanced methods to make digital watermarking systems more robust and efficient, as it stresses their irreplaceable value in the protection of digital media as technology continues to advance.

5.1. FUNDAMENTALS OF DIGITAL MODULATION

Digital modulation is an elementary technique in modern communications systems, applied for the transmission of information through various communication channels. Analog modulation modulates the amplitude, frequency, or phase of the carrier signal directly, whereas digital modulation represents information as discrete states and hence offers noise and interference immunity. Some of the digital modulation methods, including Phase Shift Keying (PSK), Frequency Shift Keying (FSK), and Quadrature Amplitude Modulation (QAM), possess specific advantages depending on the requirements of the application. These schemes provide efficient use of the spectrum and straightforward digital implementation, which is critical for high-data-rate demands in contemporary applications such as wireless communication. Secondly, choice of the modulation technique drastically impacts performance metrics such as bit error rate (BER) and bandwidth efficiency. The greatest problem in digital communications is channel degradation like fading and interference control. Techniques such as adaptive modulation are employed to adaptively adjust the modulation scheme based on real-time analysis of the channel condition. For instance, Burts-by-burst (BbB) adaptive techniques enhance throughput through modulation type modification to counteract instantaneous channel condition in order to enhance robustness with respect to fixed modes (Choi et al., 2007). Such adaptability enables systems to maintain performance even under correlated fading or time-varying channels. An optimized digital modulation scheme can significantly offset the effect of channel degradation such that data communication is guaranteed. This adaptability is essential, especially in mobile phones, since the environmental conditions may vary irrationally. A good example of advanced digital modulation techniques is seen in high-energy physics research, for instance, in efforts like the Super LHC at CERN. As data volumes increase, efficient digital modulation techniques are inevitable to manage the enormous data streams from detectors. Research in applying QAM as a method has shown potential for achieving data rates of about 3–4 Gbit/s, highlighting the role of digital modulation in enabling future growth and technological advancements (Amaral et al., 2010). Under such a high-risk environment, the trade-off between digital modulation and hardware complexity must be precisely established so that performance would be complemented by feasibility to serve stringent operational requirements. Briefly, continuing innovation in the field of digital modulation techniques is a necessity to enable the unremitting growth of data-driven scientific inquiries.

5.1.1. Analog Versus Digital Signal

The distinction between digital and analog signals forms the fundamental basis for the examination of digital modulation techniques. Analog signals are characterized by their continuity, in which information is carried as varying electromagnetic waves. This makes analog signals capable of capturing slight differences in data, such as sound waves in music or changing light intensity. However, their continuous spectra make them susceptible to degradation and interference, particularly in longer-distance transmission. On the other hand, digital signals transfer information in discrete form, typically in the shape of binary code, and hence are resistant to noise and distortion. Both representations have their own advantages and disadvantages. These fundamental differences must be kept in mind while contrasting various digital modulation techniques as they are used to communicate and transfer data efficiently overcoming limitations present in each type of signal. Moving into modulation techniques further, Quadrature Amplitude Modulation (QAM) is a leading technology, positioned at the crossroads of analog and digital communications. Where it is the combination of amplitude and phase modulation, QAM allows for multi-bit per symbol transmission, increasing data rates significantly. This benefit is particularly applicable in high-trafficking environments such as data centers and telecommunications.

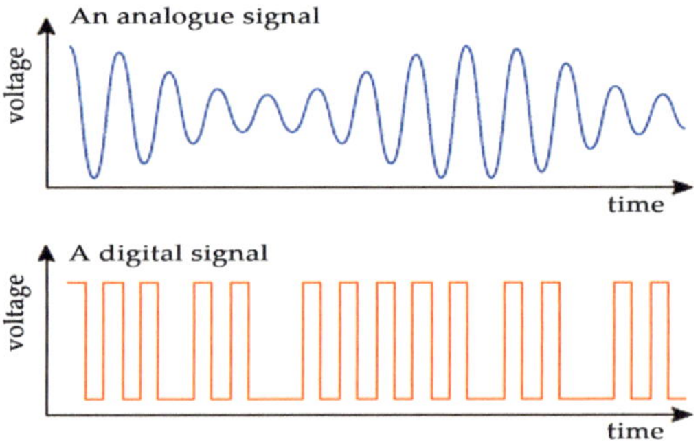

Figure 5.3. *The distinction between analog and digital signals.*

Source: https://vu2nsb.com/wp-content/uploads/2020/02/Analog-digital-wave-form.png.

A pertinent example is the Super LHC at CERN, where the requirement for faster data readout pushed the exploration of QAM in the upgrade of the CMS Tracker analog readout optical links. The need to achieve data transmission speeds of approximately 3-4 Gbit/s through the utilization of QAM indicates the technique's capacity to transcend the limitations of traditional analog systems and accommodate the demands of the modern times in cutting-edge technology (Amaral et al., 2010). Also, the application of digital signals alongside advanced modulation techniques like QAM as a coping mechanism to the challenges posed by analog systems in terms of information integrity and efficiency in the aspect of data transmission. Digital modulation not only optimizes against noise but also optimizes spectral efficiency, a key factor in bandwidth-limited scenarios. Laboratory settings, as quoted in the design of telecommunication courses in communications, experiential understanding of these techniques equips students to design creative solutions in complex communication settings (Albert et al., 1998). These teaching paradigms are essential for the acquisition of expertise in digital modulation so that practitioners can function in the areas of peak efficiency and optimal signal processing. It is evident today that the transition from analog to digital signals is not a mere technological advancement but an imperative to fulfill the high-performance requirement of contemporary communication systems (Figure 5.3).

5.1.2. Modulation Process Explained

The modulation method is a fundamental technique in digital communication, employed to represent the information onto the carrier signal for the purpose of improving transmission efficiency via various media. Modulation is a transformation of certain characteristics of the carrier signal, that is, amplitude, frequency, or phase, for conveying the digital information transmitted. There are many different modulation methods, including Amplitude Shift Keying (ASK), Frequency Shift Keying (FSK), and Phase Shift Keying (PSK), each of which has its own advantages and disadvantages that make one of them more appropriate than others to a particular situation. For instance, while ASK is simple to achieve, it is more susceptible to noise than the more powerful PSK. This inherent diversity in the performance of modulation techniques underlines the importance of selecting the appropriate method depending on the special requirement of the transmission channel and receiving end. Another key aspect of modern modulation techniques is digital signal processing (DSP), which has revolutionized the use of modulation in communications. DSP enhances the efficiency of signal transmission by supporting more sophisticated modulation formats that can incorporate methods like Orthogonal Frequency Division Multiplexing (OFDM). This method allows multiple subcarriers to transmit information at the same time and hence improves the utilization of bandwidth without compromising signal quality. Coherent detection systems employ these multi-level modulation formats, which can combat various linear fiber distortions. As discussed in current studies, coherent systems have made it possible to implement high-capacity transmission lines, such as those in current 100 Gbps wavelength-division multiplexing (WDM) norms, which effectively break the bandwidth limitation produced by fiber impairments by implementing sophisticated algorithms (Amari et al., 2017). Moreover, with the continued development of optical communication technologies, future systems will be able to support even greater data rates, for example, 400 Gbps and 1 Tbps. However, such systems are now more severely faced with nonlinear effects, for example, those caused by the Kerr effect in the high input power regime. In the interest of preserving transmission quality, the modulation process should incorporate countermeasures against such non-linearities using various compensation schemes. Effort has been put into enhancing fiber nonlinearity compensation

(NLC) methods and, as a result, the overall efficiency of modulated signals. Techniques correcting such impairments not only enhance the accuracy of data but also enhance the working ability of existing communication networks, unveiling the prospects for the next generation ultra-high-speed digital communication networks (Amari et al., 2017).

5.1.3. Significant Parameters of Modulated Signals

The key parameters of modulated signals are highly relevant in examining digital modulation techniques because they have a direct bearing on the performance and efficiency of communication systems. Signal amplitude, frequency, and phase are vital parameters that decide the overall portrayal of transmitted information. Amplitude modulation, for instance, alters the signal strength relative to a specific carrier wave, affecting the signal strength against interference and noise. Proper manipulation of these parameters allows for improved signal integrity and communication reliability. Moreover, by using enhanced modulation techniques, such as QAM, the ability to offer increased data throughput is provided, thereby satisfying the requirements of modern digital communication systems. While evaluating the performance of multifarious modulation techniques, one needs to consider various performance metrics like bandwidth efficiency, noise robustness, and bit error rate (BER). Bandwidth efficiency is the information conveyed through a specific bandwidth, which has a direct impact on the total data rate of a communications channel. Noise robustness describes the capability of a modulation technique to counteract interference and maintain signal integrity within different operating conditions. Bit error rate is a measure of incorrect received bits in relation to total transmitted bits and is one of the vital metrics of a modulation technique's reliability. With advances in systems, these parameters become increasingly significant, enhancing system designs that achieve the efficiency over robustness tradeoff, particularly in cases of large data transmission, such as in Buswig et al. (2011). Besides, the use of sophisticated technologies in signal processing provides an opportunity to optimize the parameters of modulated signals and promote the development of advanced communication systems. New technologies, such as optical signal processing and artificial neural networks, enable the possibility to optimize modulation and demodulation. For example, application of homodyne techniques to multiplexing can give rise to increased resolution in measurements in interferometric sensors, with signal processing at the center of obtaining accuracy in modulation. As seen in the achievements of Charrett et al. (2013), elements such as electronic bandwidth requirements and noise elements become important in ensuring effective digital modulation. In this way, through their understanding and utilization of significant parameters of modulated signals as well as the technologies that improve them, communication systems can achieve record performance, catalyzing progress in both industrial and academic environments (Figure 5.4).

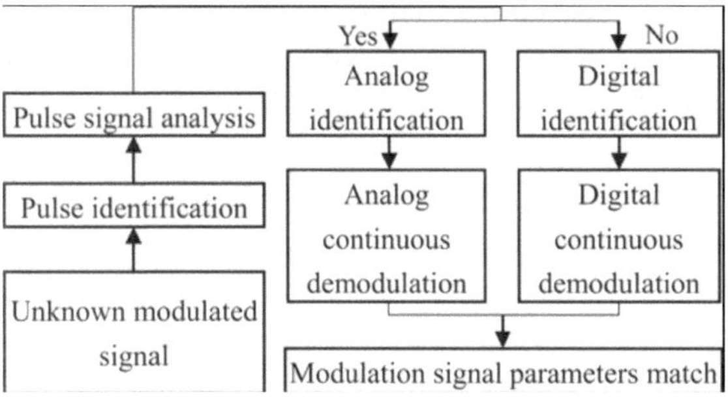

Figure 5.4. *Key parameters of modulated signals.*

Source: https://www.tandfonline.com/cms/asset/3a094d4d-43f9-4a1b-9ef8-d2d1eafb9bd3/tetn_a_2210297_f0002_b.gif.

5.2. TYPES OF DIGITAL MODULATION TECHNIQUES

Digital modulation techniques are the cornerstone of modern communication networks, permitting data to be transmitted effectively over media. The most widely used are the Amplitude Shift Keying (ASK), Frequency Shift Keying (FSK), and Phase Shift Keying (PSK). ASK varies the amplitude of the carrier wave to transmit digital information, and it is simple to employ but susceptible to interference. On the contrary, FSK uses distinct frequencies for data encoding, providing higher immunity to interference, more suitable for radio transmission. PSK changes the phase of the carrier wave, allowing for higher data rates and enhanced spectral efficiency. Both techniques possess their particular merit and demerit, respectively, and regulate their application in distinct situations, such as wireless communications, satellite communication, and data networking, thereby creating the modern scenario of digital communication systems. With the development in technology, newer and more advanced modulation techniques have emerged to meet higher demands on data capacity and transmission reliability. Quadrature Amplitude Modulation (QAM) is particularly significant as it is capable of incorporating both phase changes and amplitude changes, thus creating much higher data throughput. This technique is observed to be widely used in digital TV and broadband internet technologies as it delivers high efficiency in bandwidth usage. Furthermore, higher-order PSK, such as

Differential PSK (DPSK), promotes phase ambiguity resistance. Techniques of multilevel modulation are also gaining traction, allowing multiple bits to be sent per symbol, thus further maximizing bandwidth efficiency. These advancements are crucial because they are in line with growing requirements for fast internet and mobile communications and with filling an increasingly filled spectrum. The application of digital modulation techniques extends beyond traditional telecommunications to a great extent, extending to embedded systems and robotics. In both fields, accurate data exchange between sensors and processors is of vital importance. For instance, the employment of Phase Locked Loops (PLLs) within digital systems ensures clock stability, facilitating more secure data transmission in various applications, including navigation in robotics and sensor data processing (Burbidge et al., 2008). The performance of mobile robots' communication systems may also be enhanced through state-of-the-art digital modulation methods for providing real-time feedback in avoiding obstacles and mapping the environment to facilitate intelligent path planning (Carnegie et al., 2011). Through the advancement of these technologies, the continued advancement of digital modulation techniques will be essential for addressing the complex demands of modern and emerging communication systems (Figure 5.5).

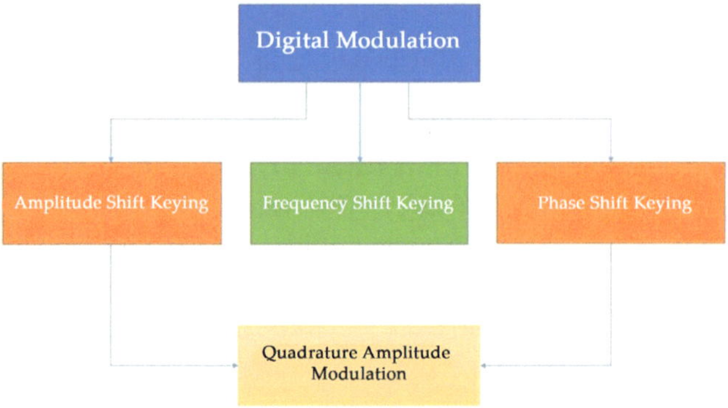

Figure 5.5. *Types of digital modulation techniques.*

Source: https://www.researchgate.net/profile/Diponkor-Bala/publication/348364945/figure/fig15/AS:978360746381315@1610270787739/Classification-of-Digital-Modulation.png.

5.2.1. Amplitude Shift Keying (ASK)

Amplitude Shift Keying (ASK) is one of the simple digital modulation techniques where information is transmitted by varying the amplitude of a carrier wave. The simplicity of ASK makes it very useful for simple applications, particularly in cases where the transmission channel is very noise-free. The technique operates by representing different amplitudes as different symbols or digital bits, which are commonly used to encode binary digits. For example, a higher amplitude may be used to represent a binary 1, and a lower amplitude to represent a binary 0. Although relatively simple to implement, ASK is also prone to noise and interference, and this can reduce the quality of the received signal by a considerable amount. This very susceptibility requires strong design techniques to ensure the integrity of the data being transmitted, particularly in real-world applications where environmental interference and other aspects may influence the quality of transmission.

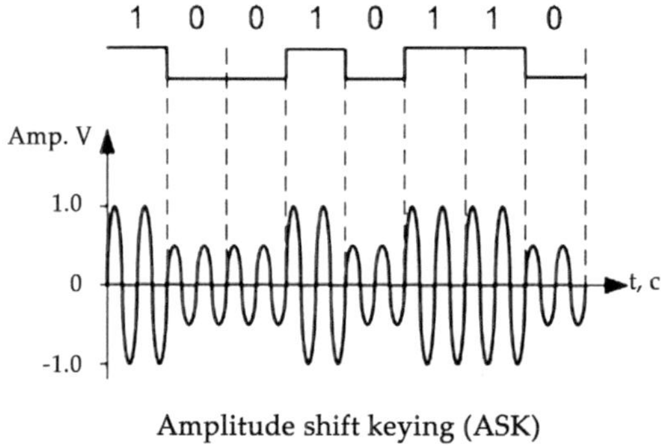

Figure 5.6. *Amplitude shift keying (ASK).*

Source: https://www.tmatlantic.com/upload/images/keying_ASK.jpg.

There is no use of ASK in today's communication networks without issues, but the latest technology is creating new opportunities for improvement. There is study into the use of ASK in conjunction with other modulation methods to leverage the ease of use and power of each. Techniques such as Amplitude Phase Shift Keying (APSK) present an exciting option through the combined use of amplitude and phase modulation, thereby enhancing the signal's spectral efficiency and immunity to noise. In particular, high-technology system advancements, such as those supported by integrated circuits, have facilitated the application of ASK practically in high-speed communication networks with improved bit rates and throughput (Kot et al., 2025). This underscores the very necessary need for sophisticated modulation techniques that will be capable of meeting the growing needs of data transfer in an increasingly networked world. Also, prospecting the future of digital communication, exploring high-dimensional modulation techniques, such as those like ASK, appears to be essential. Techniques like carrier less amplitude-phase modulation (CAP) have been put under the limelight for treating multiple dimensions of data, which is beneficial in optical communication systems (Daud et al., 2012). These advancements signal the shift towards more sophisticated modulation techniques that can potentially pose an upper hand over traditional approaches. By the understanding and enhancement of ASK, along with the implementation of new methods, the telecommunication industry can enhance the delivery of several services and obtain high-efficiency data transfer. As such, the study of ASK in the broader context of digital modulation methods remains crucial to the advancement of dependable and flexible communication frameworks required for today and tomorrow's applications (Figure 5.6).

5.2.2. Frequency Shift Keying (FSK)

Frequency Shift Keying (FSK) is a prominent digital modulation technique that conveys information by varying the frequency of a carrier wave. This technique employs different frequencies to represent binary digits, with each frequency corresponding to a bit value—typically a 0 or 1. FSK is particularly advantageous for its resilience against noise and interference, making it suitable for various applications, such as radio communications and data transmission. The inherent robustness of FSK stems from its ability to utilize multiple frequency shifts, which allows the receiving end to accurately detect the transmitted signal even in the presence of external perturbations. This characteristic of FSK underscores its importance among digital modulation techniques, especially when reliable communication is paramount in environments riddled with potential disruptions. An extension of FSK, termed Multi-Frequency Shift Keying (MFSK), enhances the data rate by employing multiple frequencies for encoding more bits simultaneously. In this approach, the space between frequencies is judiciously chosen to optimize spectral efficiency while minimizing the risk of adjacent channel interference. By increasing the number of frequency states, MFSK can transmit data at greater speeds compared to traditional binary FSK. However, this increase in speed often comes with a corresponding rise in system complexity and the need for sophisticated demodulation techniques to accurately interpret the combined frequency shifts. The balance between the benefits of higher data rates and the challenges presented by increased complexity exemplifies an ongoing area of research in digital modulation techniques. Furthermore, the versatility of FSK extends beyond traditional applications, as demonstrated in emerging technologies such as Visible Light Communication (VLC). As detailed in recent studies, FSK has been successfully implemented in VLC systems that transmit data through LEDs, showcasing its adaptability to various mediums. Such applications reveal the potential of FSK to facilitate high-speed data transfer in environments where traditional radio frequencies may be less effective. The integration of FSK with VLC also highlights the ongoing evolution of modulation techniques, as researchers explore methods to optimize data rates while tackling issues like ambient noise exposure. The exploration of FSKs applications in diverse contexts emphasizes its enduring relevance and adaptability in the constantly advancing field of digital communications, where understanding these modulation strategies remains critical for the development of robust communication systems (Harathi et al., 2018; Lee et al., 2011) (Figure 5.7).

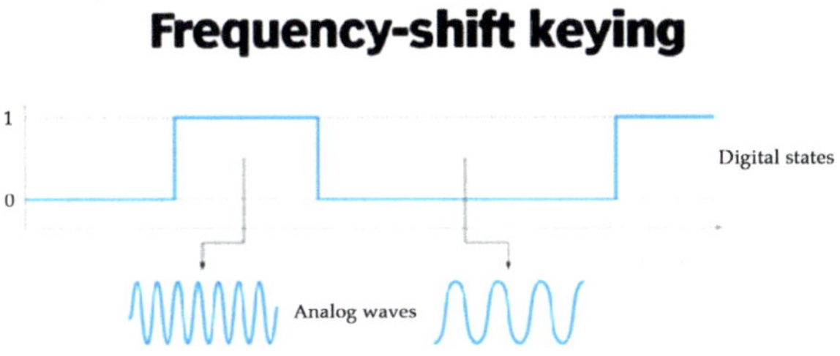

Figure 5.7. *Frequency shift keying (FSK).*

Source: https://www.techtarget.com/rms/onlineimages/frequency_shift_keying-f_mobile.png.

5.2.3. Phase Shift Keying (PSK)

Phase Shift Keying (PSK) is a basic digital modulation technique widely applied in telecommunications due to its robustness and efficiency in transmitting signals. PSK transforms data into discrete symbols corresponding to specific phase shifts by varying the phase of a carrier wave. PSK can transmit binary data successfully with two phases (binary PSK) or multiple phases for more data rate (M-ary PSK). Each stage has an associated unique bit pattern, which yields increased spectral efficiency. The modulation method is particularly useful where there is interference and noise in that data can be relied on to be transmitted via various channels. The ability of PSK to maintain signal quality and minimize error rates makes it a preferred method of usage when employed in applications like satellite communications to wireless networks. One of the main strengths of Phase Shift Keying is that it can achieve low bit error ratios (BER), especially in coherent detection. Coherent PSK systems use phase reference signals, which improves demodulation accuracy and makes them more resilient to channel distortions. Under turbulent atmosphere conditions such as free space optical communications, PSK has proved to perform better than other modulations such as Amplitude Shift Keying (ASK). The application of PSK in these situations has the tendency of delivering greater signal-to-noise ratio (SNR), which is critical to realizing high integrity of data over large distances. Analytic evaluations, as highlighted in more recent studies, suggest that PSK not only withstands physical layer issues but also achieves BER performance levels equivalent to more complex modulation techniques at the cost of lower power (Lee et al., 2011). Recent advancements in PSK also involve its integration with Multiple-Input Multiple-Output (MIMO) technology, enhancing its utilization in diversity techniques. MIMO-based Phase Shift Keying enables spatial signature exploitation and thus enhanced data throughput and integrity. The novel Space-Time Shift Keying (STSK) technique supports PSK principles, further enhancing signal performance in multipath situations. This hybrid methodology promises to operate better than traditional systems by the prevention of fading effects and expanded overall system capacity. The merging of PSK with MIMO illustrates the perpetual growth of digital modulation techniques and the commitment to continued research and development of communication systems. With companies requiring more data rates and robust communication channels, the adaptability of PSK in new forms makes its usability in the contemporary era (Chen et al., 2012) (Figure 5.8).

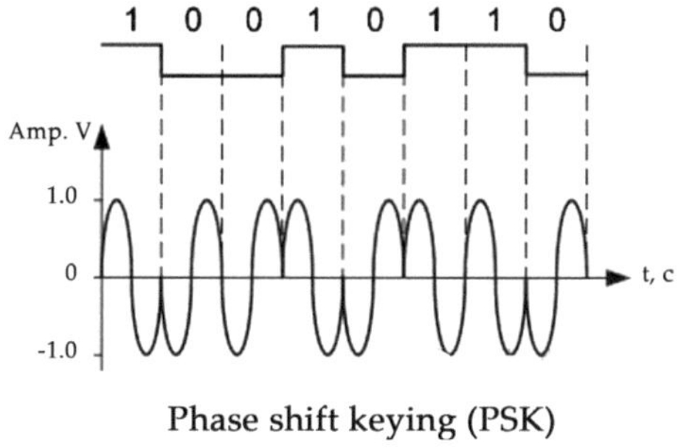

Figure 5.8. *Phase shift keying (PSK).*

Source: https://www.watelectronics.com/wp-content/uploads/keying_PSK-1-1.jpg.

5.3. ADVANCED DIGITAL MODULATION TECHNIQUES

More advanced digital modulation schemes have been crucial innovations in telecommunication, allowing more efficient data transmission across various media. Certain such schemes, such as Quadrature Amplitude Modulation (QAM) and Orthogonal Frequency-Division Multiplexing (OFDM), facilitate high data rate communication by more efficient spectrum utilization. QAM, for instance, employs both amplitude and phase modulation to transmit multiple bits per symbol, significantly increasing the data throughput. Moreover, OFDM also divides the bandwidth available into smaller sub-carriers, which can be modulated independently, thus creating improved resistance to multipath fading and interference. Thus, with such sophisticated modulation methods utilized, not only is data transfer improved in terms of efficiency and reliability but also the growing demand for bandwidth in modern communication systems is accommodated. The incorporation of sophisticated digital modulation methods is also a sign of a paradigm change toward supporting intermittent faults in connection systems. The past monitoring systems have been inadequate to support such faults because they are transient in nature, thus making it essential to develop efficient detection methodologies. A new method proposes the injection of a constant frequency sinusoidal signal to modulate the intermittent faults so that it becomes possible to detect them through demodulation and spectrum analysis. This innovation facilitates real-time monitoring of electronic interconnection systems, and it has been able to distinguish faults from ordinary operational noise successfully. As mentioned, the capability of monitoring low amplitude signal channels is highly beneficial in the detection of precursors to permanent failure, eventually leading to improved system reliability and lifetime (Ahmad et al., 2015). Secondly, the appeal of these advanced modulation techniques extends beyond fault detection into the realm of solid security features in digital communication. Digital watermarking, for example, has been a key technology in protecting intellectual property, particularly digital media. The method enables the insertion of a watermark into digital content—either in spatial or transform domain—to prevent unauthorized copying or piracy. The use of robust modulation techniques, such as spread-spectrum techniques, has significantly improved watermarked data resistance against various types of attacks, so that the integrity of the content is maintained even in extreme situations. The use of error correction and sophisticated embedding techniques also strengthens the resistance of these systems, hence making it a secure digital communication system for the modern world (Serdean et al., 2002).

5.3.1. Quadrature Amplitude Modulation (QAM)

Quadrature Amplitude Modulation (QAM) is a sophisticated digital modulation technique that allows the simultaneous transmission of multiple bits per symbol, making it extremely efficient in modern communication systems. QAM achieves a larger data rate than other simple modulation techniques like Binary Phase Shift Keying (BPSK) or Quadrature Phase Shift Keying (QPSK) by both altering the amplitude and phase of the carrier waves. Its modulation versatility allows QAM to accommodate various requirements in data transmission, and because of this, it is an integral part in digital TV broadcasting technology up to wireless data communication technology. This is why QAM becomes even more pertinent for applications with a high data throughput demand since it can pack data more densely without the requirement for more bandwidth. In the field of high-energy physics, the utility of QAM has been well demonstrated in the recent advancements at research institutes such as CERN.

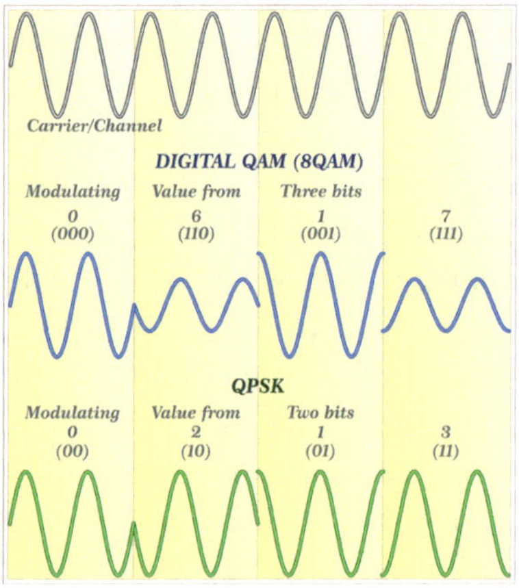

Figure 5.9. *Quadrature amplitude modulation (QAM).*

Source: https://phoenixnap.com/glossary/wp-content/uploads/2022/06/digital-qam.jpg.

Hence, the investigation of QAM in the refurbishment of the Large Hadron Colliders' data acquisition systems highlights its practical application in dealing with higher volumes of data. Present projections foresee a tenfold rise in luminosity, which, along with QAM's efficient data management, would allow for data rates of approximately 3–4 Gbit/s under optimal circumstances, with minimum error correction required (Amaral et al., 2010). The search for QAM as a modulation method proves not only its technical savvy, but also its value in providing the high level of performance that is needed in high-performance situations where timely delivery of data is the highest priority. Furthermore, the flexibility of QAM is underscored by the development of high-level programmable digital modems that are able to maximize many modulation techniques, including QAM. These modems are designed to operate under a variety of transmission rates and modulation orders, from 2-ary to 16-ary QAM (Poklemba et al., 2025). Because they have the ability to dynamically change parameters, they can provide best performance based on the transmission channel, and for this reason, they are particularly valuable in mobile and high-speed applications. Thereafter, the integration of QAM in such programable devices not only enhances data capacity but also supports ongoing efforts to minimize operational simplicity and power intake, hence resulting in the subsequent generation of telecommunication technology (Figure 5.9).

5.3.2. Differential Phase Shift Keying (DPSK)

Differential Phase Shift Keying (DPSK) is a popular digital modulation technique, where the phase of the carrier wave is modulated with respect to the previous symbol to modulate the data. It is particularly beneficial in situations where the transmission is plagued with phase ambiguity due to high noise or interference. As opposed to the traditional PSK that requires an absolute phase reference, DPSK employs difference encoding of the phase, thus offering more insensitive performance under non-coherent detection conditions. Such a property makes DPSK an appealing choice in wireless communication systems, especially those deployed within fading channels where it becomes difficult to maintain a phase reference.

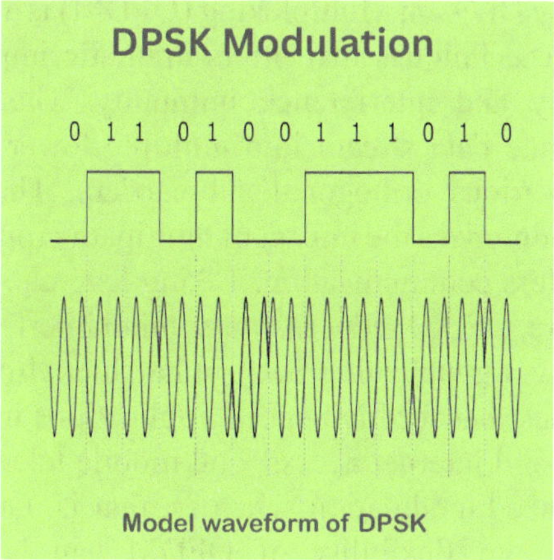

Figure 5.10. *Differential phase shift keying (DPSK).*

Source: https://cdn.fiberopticx.com/wp-content/uploads/2023/12/DPSK-Modulation.webp.

One major innovation in DPSK usage is embedding adaptive demodulation schemes, such as the rate-adaptive channel-state-information-free transmitter. Such systems as adaptive decision region boundaries enhance DPSK and M-ary differential phase shift keying and result in high reliability and throughput for a variety of channel conditions. For example, soft-decision type demodulation based methods have been shown to improve substantially in terms of performance for bit error analysis. These enhancements hold promise in using DPSK to more advanced, rate-adaptive arrangements without the introduction of additional overhead, demonstrated by the high-end designs using reliable bit detection even in unfavorable situations (Abouei et al., 2010). Also, the low-complexity non-coherent detection forms of Differential Amplitude and Phase Shift Keying (DAPSK) complement DPSK by giving higher coding gains and throughput characteristics for wireless use. Research evidence establishes that schemes like 64-ary DAPSK are capable of outperforming traditional 64-DPSK counterparts, albeit with significant coding gains and efficient throughput. The characteristic highlights the versatility of DPSK-like techniques, particularly as communication needs evolve. With the advancement of the digital communication environment, incorporating these new DPSK modifications can lead to more efficient and hardened systems that can handle the rising rates of contemporary networks (Hanzo et al., 2011) (Figure 5.10).

5.3.3. Orthogonal Frequency Division Multiplexing (OFDM)

Orthogonal Frequency Division Multiplexing (OFDM) is a precious technology in digital modulation techniques that offers dramatic improvements in terms of spectral efficiency and interference immunity. The method effectively decomposes a high-rate data stream into multiple lower-rate streams, which are transmitted on various orthogonal sub-carriers. The method facilitates bandwidth use, as it minimizes the impact of multipath propagation—a common phenomenon in wireless communication. Taking leverage of the orthogonality among the sub-carriers, OFDM reduces intersymbol interference to a minimum, thus enabling robust communication even in harsh environments. This aspect renders OFDM the best bet for high-speed applications in today's world such as digital TV, broadband internet access, and mobile telephony systems. With the addition of advanced modulation schemes, that is, Quadrature Amplitude Modulation (QAM), the flexibility of OFDM can be further increased. The combination of QAM and OFDM enhances the average data rate while preserving the advantages that flow to OFDMs immunity to multipath. Recent developments, for instance, to counter the prospect of applying QAM-OFDM in optical links, demonstrate the method's capability to offer higher spectral efficiency compared to traditional methods (Axer et al., 2005). By altering the paradigm of data transmission, such integration offers new avenues for increasing the capacity of existing systems and thus satisfying growing need for data-driven applications in contemporary communication networks. On top of that, the feasibility of OFDM has been confirmed by both simulation and experimental methods. For instance, single-phase DC-AC converter development is the concern of projects that put stable output voltage in control systems into the limelight (Buswig et al., 2011). Such advancements show the practicability of OFDM and its variations not merely in theory but also practically when it comes to real-time usage. Ongoing research on optimizing OFDM techniques even more highlights its pivotal role in shaping the future dynamics of digital communication. With the passage of time and developments in technology, OFDM will be at the forefront, bridging the gap between theoretical feasibility and practical application in an increasingly interconnected world (Figure 5.11).

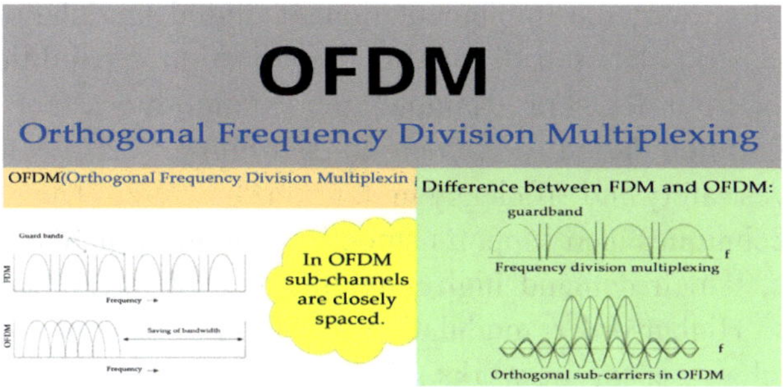

Figure 5.11. *Orthogonal frequency division multiplexing (OFDM).*

Source: https://www.flyeye.io/wp-content/uploads/2024/10/OFDM.jpg.

5.4. APPLICATIONS OF DIGITAL MODULATION

Digital modulation techniques are important in contemporary telecommunication systems for the successful transmission of information via various media. It provides means for the embedding of digital information on carrier waves, with higher signal integrity and strength upon transmission. It is utilized in various fields, including mobile communications, satellite broadcasting, and wireless networking. By incorporating techniques such as Phase Shift Keying (PSK) and Quadrature Amplitude Modulation (QAM), systems can achieve higher data rates with low noise, interference, and fading effects. The implementation of complex modulation techniques is highlighted in projects such as the Adaptive Wireless Networking (AWGN) project, where the development of reconfigurable mobile terminals for future communications becomes an important agenda (Hoeksema et al., 2003). This adaptability is required to meet the increasing demands of users and the evolving world of digital communications. In the use of broadband internet access, digital modulation techniques further enhance the performance of broadband systems. HSDPA technology, for example, exploits advanced modulation techniques that are adaptive to changing channel conditions in order to achieve optimal data throughput. The review of burst-by-burst adaptive multicarrier systems focuses on their capacity to dynamically adjust modulation parameters based on real-time channel assessments (Choi et al., 2007). Flexibility enhances the user experience by providing assured and high-speed connectivity needed for video streaming and online gaming applications. Moreover, the adaptive systems can even cope with the deterioration of channel quality, thereby proving to be a stronger alternative for traditional fixed-mode systems, ultimately resulting in improved network performance and customer satisfaction. Moreover, the implementation of digital modulation techniques in various industries has transformed communication capabilities and given innovative opportunities. For example, the automotive sector increasingly relies on digital modulation for vehicle-to-everything (V2X) communication, which improves safety and efficiency in transport networks. The adaptability of modulation techniques also supports smooth integration of Internet of Things (IoT) devices, which demand high-reliability data delivery across different environments. Higher-order modulation techniques have the capability to optimize wireless sensor networks, facilitating the easy interworking and scalability of connected devices. With the development of digital modulation techniques, they will necessarily continue to expand in use, affecting many industries, and demanding even more sophisticated communication solutions.

5.4.1. Telecommunications

The rapid growth of telecommunications has encouraged new, high-performance digital modulation techniques required in efficient communication systems. Due to the requirements for higher data rates and reliable transmission, modulation techniques have emerged to achieve this. Specifically, downlink packet access (HSDPA) has gained heightened significance since it can adaptively regulate burst-by-burst transmission. This adaptability is significant because it diminishes channel quality variations in real time, thereby offering more throughput and reliability than traditional fixed-modulation approaches. Adaptive OFDM also serves as a robust choice for changing wireless conditions, establishing a balance between efficiency and reliability. These advancements are critical to next-generation telecommunications as they enable systems to be capable of adjusting to evolving consumer demand without sacrificing quality of service demands. In addition to adaptive techniques, channel parameter analysis reveals that telecommunication requires advanced diversity techniques. To mitigate the effects of fading—a most serious issue for wireless communication—some methods have been developed, including Rake receivers and space-time block coding (STBC). These processes enhance the performance by reducing the bit error ratio (BER), especially in the case when the individual antenna elements experience independent fading. However, the performance of STBC is compromised with the effect of correlated fading environments, such as shadowing. This realization highlights the importance of selecting appropriate modulation and coding methodologies at the system design stage since the interaction between these approaches can contribute to a direct impact on communication dependability. The employment of adaptive modulation strategies based on channel states is thus important for achieving maximum performance in adverse conditions. Moreover, the latest studies in dual-hop wireless communication systems have also added richness to the understanding of end-to-end performance in telecommunication. Studies on fixed-gain relaying in Generalized-Gamma fading channels introduce some significant implications on modulation selection. Closed-form expressions for moments of end-to-end SNR obtained from the research provide the engineers significant information on the behavior of the system to estimate average bit error probabilities over coherent and non-coherent modulation schemes. These analyses not only provide a stringent evaluation of outage probabilities but also justify the requirement for mathematical modeling in the estimation of system performance under varying conditions. The interlinking of theoretical models with the output of simulation highlights the importance of digital modulation techniques in enhancing telecommunications, bridging the gap between theoretical performance and real achievement (Choi et al., 2007; Mansour et al., 2010) (Figure 5.12).

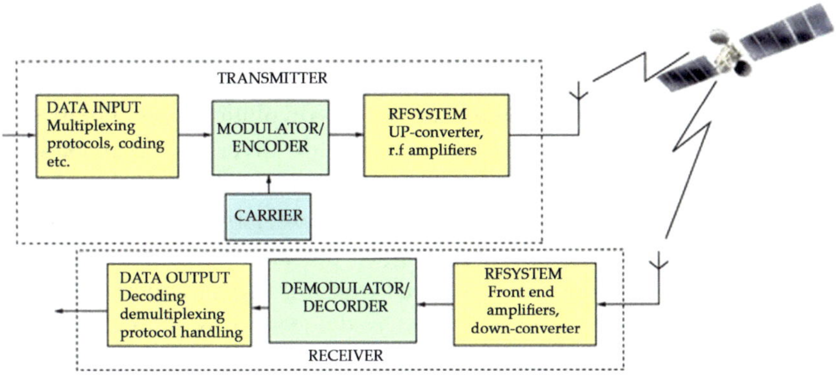

Figure 5.12. *Telecommunications in digital modulation techniques.*

Source: https://idstch.com/wp-content/uploads/2020/12/digitalmodem.gif.

5.4.2. Broadcasting

The broadcasting environment has radically changed, especially following the advent of digital modulation techniques, which have enhanced signal quality and transmission efficiency. Broadcasting is the dissemination of audio or video content to a dispersed audience through various media channels. Digital broadcasting offers enormous advantages over traditional analog technology in the form of enhanced sound and video quality, greater immunity against interference, and multi-channel broadcasting in a single bandwidth. The introduction of standards like DVB-H and DVB-SH indicates technological improvements in mobile TV, where signal reception uses advanced channel models characteristic of fading and multipath propagation. These models together with digital modulation enable maximum signal processing that is specially designed for the unique challenges of mobile environments, ultimately delivering improved user experience and access to multiple media forms in different locations (Polák et al., 2013). Moreover, the application of diversity techniques has been a key contributory factor to broadcast systems robustness and efficiency enhancement. These techniques exploit the many paths of signal transmission and utilize the statistical character of multipath-induced fading. With transmitter and receiver diversity, broadcasting systems can reduce the incidence of deep fading, thereby increasing the Signal-to-Noise Ratio (SNR) of the received signal. These enhancements are particularly important to digital broadcast networks that use Orthogonal Frequency Division Multiplexing (OFDM), as these methods enable the reliable transmission of continuous content across varying conditions. Not only does this approach minimize complexity, but it also means fewer transmitter sites required, which lessen cost and environmental impact in broadcast infrastructures (Bard et al., 2006). Lastly, practical implementation of these digital modulation techniques and diversity schemes demonstrates their implication in revolutionizing broadcasting paradigms. In practical application, the use of diversity schemes in the DVB-H system presents a compelling case for enhanced mobile broadcasting experience. System simulations have realized appreciable improvement in Bit Error Rate (BER) performance, testifying to the efficacy of these measures. By overcoming the challenges of various transmission environments, digital modulation techniques have offered increased access to multimedia material, thereby fulfilling the growing need for quality broadcast services. As technology improves, the identification and incorporation of new modulation techniques will remain an important factor in determining the future of digital broadcasting (Bard et al., 2006).

5.4.3. Data Transmission

In digital data transmission, digital modulation techniques are the cornerstone for efficient communications schemes. Digital modulation techniques allow for the representation of digital signals in analog waveforms so that data can be transported over various media like terrestrial and satellite channels. The evolution of developments in digital modulation has been necessitated to a large degree by the requirements for more bandwidth and better spectral efficiency. With a growing degree of communication networks and increasing data rate, the introduction of sophisticated modulation technology, such as Quadrature Amplitude Modulation (QAM) and Phase Shift Keying (PSK), has become necessary to enhance the quality of transmission as well as support real-time operations such as video streaming of high-definition content. Through this manner, the inherent linkage between modulation technology and effective data transmission emerges in the context of modern telecommunications. The use of digital modulation techniques can be further examined with their application in advanced communication systems, as exemplified by digital video broadcasting.

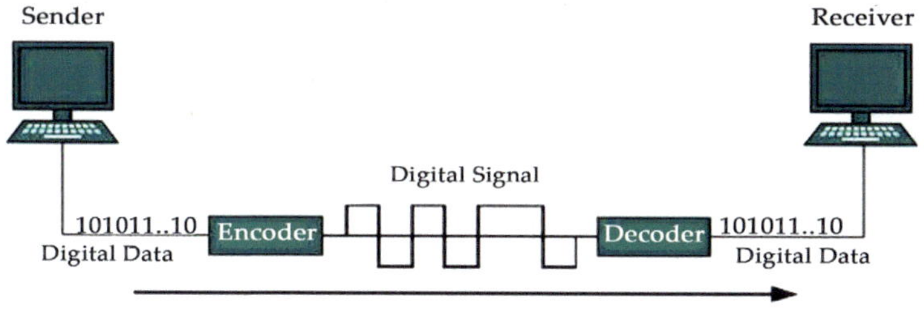

Figure 5.13. *Data transmission.*

Source: https://www.tutorialspoint.com/data_communication_computer_network/ images/line_coding.jpg.

Advances in bandwidth-efficient schemes are transforming the way video signals are processed and sent, particularly in achieving broadcast-quality standards. For instance, a hardware implementation of a Differential Pulse Code Modulation (DPCM) algorithm has been shown to be able to compress NTSC composite color TV signals successfully with an excellent average of 1.8 bits per pixel. This outstanding effectiveness reflects the role played by digital modulation to enhance data transmission through bandwidth usage control without diminishing signal quality. These trends are indicative not only of technological progress but also of the development of the television broadcasting industry in the field of transmission and data processing technologies (Shalkhauser et al., 2025). Furthermore, optical communication has revolutionized data transmission using coherent detection systems. These systems operate through DSP algorithms to correct for fiber impairments, which have become increasingly prevalent with advances in wavelength-division multiplexing technology. Optical communication systems utilize multi-level modulation formats that, when combined with DSP techniques, enhance spectral efficiency and enhance transmission resilience. With growing data rates in networks, fiber nonlinearities such as the Kerr effect-induced ones represent challenges that need to be efficiently compensated. Research on various nonlinear compensation schemes goes on and is an important factor in achieving optimal performance in upcoming WDM systems with high capacity, and underscores the interdependency between the modulation schemes, the transmission quality, and the cutting-edge DSP applications in contemporary data transmission techniques (Amari et al., 2017) (Figure 5.13).

5.5. ADVANTAGES OF DIGITAL MODULATION

Digital modulation techniques offer numerous advantages which enhance communication system performance and reliability. One significant advantage is that they are interference and noise-resistant, which is the greatest consideration in today's saturated electromagnetic spectrum. Analog modulation degrades with noise exposure, yet digital signals maintain their integrity to a greater extent, allowing cleaner transmission over greater distances. This power is particularly crucial when applied to industrial application where output reliability is needed. This power is what enables digital modulation techniques to perform higher in terms of bit error rates than their analog counterparts. By ensuring that signals are always recoverable even in less-than-ideal conditions, digital modulation is essential for contemporary communication infrastructure, further justifying its application in numerous different technological domains. Furthermore, digital modulation can efficiently use bandwidth, a critical concern in the now overcrowded wireless spectrum. Techniques such as Quadrature Amplitude Modulation (QAM) and Phase Shift Keying (PSK) enable multiple bits of information to be transmitted per symbol. This significantly increases the data rate without the need for additional bandwidth, enabling more information to be transmitted faster and more efficiently than older modulation methods. Furthermore, advancements in digital processing methods, such as Artificial Neural Networks, have optimized these methods even more by improving synchronization and reducing signal transmission errors (Buswig et al., 2011). As a result, the utilization of digital modulation not only optimizes bandwidth but also facilitates applications where high data rates are demanded, thereby optimizing the overall performance of communication networks. Aside from these technical advantages, digital modulation techniques also offer great flexibility and scalability advantages. Digital systems are inherently simpler to update and alter than analog ones, with new modulation techniques and protocols being able to be added without drastic hardware updates. This flexibility is critical for the quick development of communication technologies, particularly in situations where standards are routinely updated. For instance, advancements in synchronization methods, such as those explored in contemporary research on digital frequency discrimination, demonstrate how digital modulation can be tailored to meet specific requirements but still achieve high efficiency levels (Bobula et al., 2008). Such adaptability ensures digital modulation is consistently upgraded and effective in a wide area of use, including industrial communications and consumer electronics. Lastly, the advantages of digital modulation techniques position them as fundamental building blocks in modern communication systems (Figure 5.14).

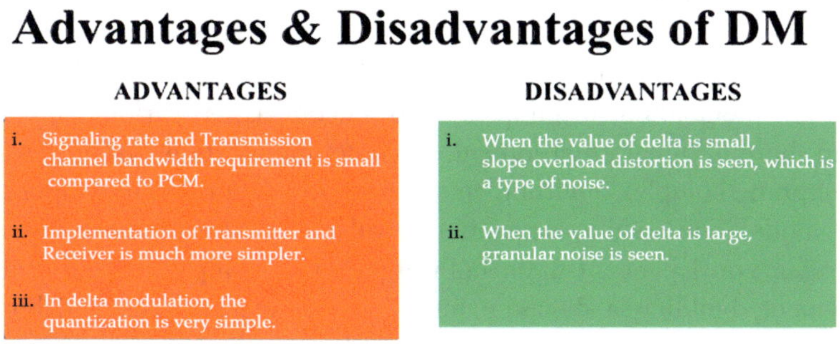

Figure 5.14. *Advantages of digital modulation.*

Source: https://cdn.numerade.com/ask_images/45c9e851584546f7a1f8a885d88d5451.jpg.

5.5.1. Improved Noise Immunity

Maximum noise immunity is a vital issue in the design of digital modulation techniques, particularly as communications equipment deals with increasingly prevalent interference. The need is exemplified in the case of wind power generation, for example, where DC-AC converters must provide output stability under fluctuating circumstances. Through mathematical modeling and incorporation of high-end control techniques like Artificial Neural Networks, systems can make high levels of resistance to noise a reality. In particular, voltage output stability requirements are highlighted in the design, which is of central importance in ensuring secure communication. In addition, simulation as well as experimental evidence justifies the effectiveness of such control techniques, demonstrating that higher noise immunity is associated with higher tolerance of performance for a vast range of operating conditions, ultimately resulting in more secure communication outcomes (Buswig et al., 2011). In addition, advances in denoising techniques, such as the discrete wavelet transform (DWT), have revolutionized optical wireless communication by surmounting the challenges posed by artificial light interference (ALI). In indoor communication environments, where sunlight has the potential to deprioritize signal quality, effective noise suppression is critical. The DWT stands out from the rest in offering an evolved solution that surpasses the performance of traditional high pass filtering methods in both efficiency and complexity. By utilizing DWT-based denoising algorithms, researchers have showcased dramatic reductions in optical power penalties caused by ALIs, thereby maintaining the fidelity of various modulation schemes like on-off keying (OOK) and pulse position modulation (PPM) (Angelova et al., 2010). This advancement illustrates how increased noise immunity does not only make signal quality but also makes complex communication systems feasible. Pressure for greater noise immunity in digital modulation techniques also encompasses various applications beyond the traditional environment. With communication systems extending to cover more complex environments and greater data rates, the importance of robust modulation techniques cannot be ruled out. Methods that successfully fight noise are dependent upon both algorithmic complexity and real-world deployment, building a dynamic interplay that increases dependability. No matter whether through sophisticated encoding methods or adaptive filtering processes, the unifying aim is the same: improving system performance under an array of potential interference conditions. As testified by the progress in control systems and filtering techniques, these technologies as a whole serve to enhance not only the user interface but also the veracity of the underlying communications infrastructure, demonstrating a long history of commitment to developing strong digital transmission solutions.

5.5.2. Efficient Bandwidth Utilization

Maximum bandwidth efficiency is a cornerstone of modern digital communications for achieving the highest data rates with the least amount of interference to the signals. Novel modulation techniques such as Orthogonal Frequency-Division Multiplexing (OFDM) are one of the most effective means to achieve such an optimization. The technique allows us to transmit a high-speed information stream by breaking it down into numerous low-speed sub-channels that overlap frequency bands. By taking advantage of the properties of OFDM, systems can be of high spectrum efficiency, a feature that is particularly valuable in frequency dense environments. Its efficiency allows for better management of the bandwidth, causing tremendous boosts in transmission capacity and quality, enabling it to accommodate the increased demand for data-intensive applications in the current digital age (Leenheer et al., 2013). Besides, the flexibility of OFDMs is essential in ensuring efficient bandwidth utilization under various channel conditions. The modulation scheme inherently possesses resistance to inter-carrier interference and inter-symbol interference, which can be employed in environments where signal attenuation is prevalent. With evolving communication systems towards higher data rates and higher connectivity demands, OFDM-based technologies are being explored for employment in optical networks as well, extending their significance beyond the classical wireless employment. For example, optical OFDM (O-OFDM) transmission technologies have been of considerable interest due to the potential they present for ultra-high-speed optical communication and are expected to be a good candidate for future networks. This flexibility ensures that signals are passed on accurately and, thus, maximizes the utilization of the allocated bandwidth as well as ensures effective communication under different circumstances (Leenheer et al., 2013). Finally, the pursuit of efficient bandwidth utilization also intersects with applications of renewable energy, wherein digital modulation techniques have the potential to render control systems optimal. A good example is the development of a single-phase DC-AC converter for wind generator application with the purpose of enhancing energy conversion efficiency. The use of advanced control schemes, such as Artificial Neural Networks, to provide output voltage stabilization in situations of variability suggests the application of modern modulation techniques to renewable energy sources. This confluence of digital modulation and energy systems is an expanding area of bandwidth efficiency, in which both communication and energy transport leverage the same fundamental principles to achieve overall system optimization. As research on these fronts progresses, collaboration between solutions that are efficiency-focused intensifies and shapes the future of digital communication and energy generation.

5.5.3. Enhanced Security Features

The contribution of advanced security mechanisms to digital modulation techniques cannot be overemphasized, particularly in an era where digital attacks are becoming increasingly common. The highly sophisticated nature of modern-day communication systems requires robust security to prevent unauthorized entry and safety of sensitive data. Techniques such as chaotic masking modulation come in handy in this regard as they utilize chaotic systems to mask transmitted signals, making them inaccessible to possible eavesdroppers. The architecture of robust adaptive observers, particularly for chaotic synchronization, significantly improves security by opening the way to resilient systems that can counteract attack and uncertainty. Adding variations to the chaotic dynamics system enhances security further, demonstrating the efficacy of applying sliding mode observer designs to uncertain nonlinear dynamics (Raoufi et al., 2006). This dynamic answer serves to identify how innovative modulation schemes can radically reshape the security landscape of wireless communications. The incorporation of artificial intelligence (AI) into digital modulation techniques contributes to security attributes, particularly in heterogeneous and complex network environments. Machine learning algorithms are key components in modern wireless communication systems, as they enable real-time analysis and decision-making capabilities. Through adaptive learning, these algorithms are able to identify and respond to possible threats in real time, thus security protocols remain a notch ahead of invaders. The integration of AI has proved to be effective in application such as cognitive radios and machine-to-machine networks, where the traditional ways may be inadequate. By using supervised and unsupervised learning mechanisms, the potential for security features is enhanced significantly, making these systems more secure and robust (Chen et al., 2019). The complementary aspect of modulation techniques and intelligent systems indicates their cooperation in enhancing communication security. Apart from technological advancement, the integration of innovative security features also needs acute understanding of the underlying network structures. As wireless networks evolve, particularly in military and civilian use, their structure becomes increasingly complex, with an attendant need for increasingly complex ways of securing them. Interoperability between devices and networks also underscores the necessity to employ adaptive techniques that can handle emergent vulnerabilities. The future of wireless networks is based on their ability to deliver high-rate, low-latency, and reliable information services along with robust security features simultaneously. The scholarly focus on the history of machine learning and how it will affect the dynamics of the network can provide valuable insights to integrate these emerging security features appropriately (Chen et al., 2019). Lastly, with the advancement of digital modulation technologies, the balance between security and communications efficiency will be crucial in addressing the challenges of the future in the digital age.

5.6. CHALLENGES IN DIGITAL MODULATION

Digital modulation is an important part of current communication systems, and it has a number of advantages over analog techniques, including better spectral efficiency and immunity to noise. But applying digital modulation techniques involves many challenges that can greatly affect signal quality and reliability in transmission. One of the main challenges is preserving the integrity of transmitted signals in the presence of various impairments in the channel, such as noise, interference, and fading. Since the demand for these high data rates keeps on growing, particularly in wireless communications and in fiber optics, the susceptibility of digital modulation techniques to such interference is the concern of the day. Additionally, with the utilization of multilevel modulated formats, which are used for enhanced spectral efficiency, the effects of linear and nonlinear distortion effects are also increased, calling for sophisticated algorithms for effective compensation (Amari et al., 2017). Another major challenge in digital modulation is efficient EMC management, particularly in wireless systems where various devices co-exist. The growing density of devices and access points leads to increased electromagnetic interference, which deteriorates the signal quality and can result in poor system performance. Up-to-date EMC standards tend to be insufficient, and it is required to adopt new methods, which can be more precise in quantifying the impact of these interferences. Recent advancements recommend applying statistical methods in a bid to give a better realistic estimate of the degradation towards EMI, hence directing optimal modulation strategies (Azpúrua et al., 2019). Revising these standards and compliance requirements is important in keeping the reliable coexistence of various wireless systems intact, especially as the movement toward a networked society remains strong. Lastly, the continued evolution of fiber optic communication systems poses a unique set of problems with digital modulation. The introduction of coherent detection systems, which use digital signal processing (DSP) techniques to reverse the impact of fiber impairment, has revolutionized the transmission capacity. But as the data rates increase close to 400 Gbps and beyond, the problems associated with nonlinearities, particularly those induced by the Kerr effect in optical fibers, become increasingly significant. The necessity for compensation of such nonlinearities exists to enhance transmission performance and maintain signal integrity within high-speed systems (Amari et al., 2017). New compensation techniques and policies must be tried extensively on their performance and complexity to determine their viability for overall application. Despite such difficulties, innovation in digital modulation techniques will persist to shape the future evolution of communication technologies.

5.6.1. Interference and Signal Degradation

Interference is a critical phenomenon in the effectiveness of digital modulation techniques, particularly in wireless communications systems. Interference can be from various sources, including adjacent channels, manmade electronic equipment, as well as environmental factors such as weather conditions. The interference typically distorts the received signal, resulting in signal degradation that can compromise information integrity transmission. Hence, the performance of digital modulation techniques such as Quadrature Amplitude Modulation (QAM) or Phase Shift Keying (PSK) can be negatively affected, leading to increased bit error rates (BER) and degraded communication. To prevent such interference, researchers are interested in fault-tolerant modulation techniques that incorporate error correction techniques and adaptive filtering algorithms to enhance signal immunity against interference so that the quality of the communication link is ensured. In addition, the advancements in the field of electromagnetic compatibility (EMC) knowledge spurred the development of more effective ways to measure the interference effect on signal integrity. Traditional compliance methods utilized quasi-peak detector levels and simple pass/fail decisions, which do not capture the nuances of real-world environments affecting digital communications. Recent research indicates that applying new statistical techniques, such as amplitude probability distribution detectors in combination with time-domain measurements, yield more information about the extent of degradation caused by electromagnetic interference. These new measurements highlight the necessity for international bodies to redefine EMC standards to keep up with existing challenges in digital communication systems to yield a more assurance framework for interference as well as for signal quality (Azpúrua et al., 2019). No exaggeration can be used to describe how important signal simulation is in providing a defense against interference. Communication system-based projects such as Space Station Communications and Tracking System Modeling highlight the utmost need for overall-system simulation environments. By developing frequency division multiple access (FDMA)logy" system simulations and RF link analysis models, scientists are able to study and predict how interference can hamper communications performance under a broad array of circumstances. These simulation networks, in addition to enabling testing of modulation schemes under various interference conditions, provide some system design and tuning insights. This strategic focus on simulation with careful empirical study results in a clearer understanding of the dynamics of interference and its impact on signal integrity (Chie et al., 2025). Lastly, they are the key to enhancing digital modulation techniques to enable resolute, reliable communication in increasingly complex environments.

5.6.2. Complexity of Implementation

For the case of digital modulation schemes, implementation complexity emerges as the most important factor that has an effect on performance as well as realizability. New-generation optical communication networks, particularly the ones over 400 Gbps per wavelength, require sophisticated modulation methods to realize high spectral efficiency without inducing linear and nonlinear effects. Application of sophisticated digital signal processing (DSP) is needed to address these problems. However, the inclusion of these advanced methods can significantly enhance the degree of system design intricacy, causing massive loads on computational facilities in addition to increasing overall system costs. As a result, awareness of the trade-offs between the improvements in performance offered by these complex techniques and the increase in implementation complexity becomes more critical to engineers and designers in the field. With the introduction of coherent detection systems, the nature of digital modulation has been revolutionized with the capability of enhancing spectral efficiency through multi-level modulation formats. Nevertheless, with the trend of systems shifting towards high data rates, the implementation complexity comes to the forefront. In particular, the requirement of effective fiber nonlinearity compensation (NLC) is a particularly demanding task, which necessitates effective algorithms to counteract the harmful impacts endured due to nonlinear optical effects, for example, caused by the Kerr effect. High-performance methods must also contend with higher computational complexity, which can create signal processing bottlenecks. A comprehensive analysis of such methods and complexity of deployment highlights the underlying difficulties that engineers face in optimizing communication systems for performance as well as efficiency in high-capacity networks, as realized through various studies (Amari et al., 2017). In achieving such complexities, a balanced evaluation of various equalization methods and techniques reveals stark variations in computation requirements. For instance, a comparative analysis of conventional orthogonal frequency-division multiplexing (OFDM) coherent receivers and filter-bank-based OFDM receivers illustrates the latter offers major complexity savings without compromising important performance parameters. These studies highlight the importance of modulating techniques being optimized both in terms of performance as well as implementation practicability. Additionally, the compromise between conventional equalization techniques, such as digital back-propagation, and computationally lighter counterparts enhances the debate on complexity in implementation. The aspect of both performance enhancement and computational expense comes into play, enabling stakeholders to make better-informed choices about digital modulation methods to use in upcoming communication systems (Bouziane et al., 2016).

5.6.3. Limitations in Bandwidth

The bandwidth constraints are a significant issue in the domain of digital modulation techniques. Since the communications technology is constantly changing, efficient bandwidth usage becomes highly crucial. Bandwidth constraints can restrict transmission data rates, which can subsequently impact the overall speed and capacity of digital communication systems. The constraints normally result from the regulatory frameworks that dictate spectrum assignment and physical constraints such as signal propagation and interference. Therefore, modulation techniques with the potential to efficiently leverage available bandwidth are essential for enhancing communication performance. Techniques such as Galois-Division Multiple Access (GDMA) emerge as effective solutions, with little bandwidth requirement and improved performance in heavy traffic scenarios (de Oliveira et al., 2015). Further, incorporating modern encoding and modulation schemes can potentially offset the impact of bandwidth limitation. Introduction of techniques like balanced encoding, which creates nulls in the transmitted signal, allows effective use of the available bandwidth. Addition of a calibration tone to the signal, which is both perfectly phased and frequency-matched to the carrier, allows coherent demodulation at the receiving end. Such calibration proves effective in mitigating issues such as multipath fading, therefore improving signal quality and reliability in bandwidth-constrained environments (Davarian et al., 1987). These developments reflect the importance of adaptive modulation methods that respond instantaneously to bandwidth constraints and still ensure data transmission quality. Further research into novel digital transforms and coding methods is also crucial in mitigating bandwidth constraints. Techniques such as the Finite Field Hartley Transform make encoding more effective because information can be encoded within p-ary systems rather than traditional binary systems. This not only maximizes the use of usable bandwidth but also makes the system more immune to obstructions from the environment. Adding such multidimensional modulation techniques improves more effective bandwidth utilization, which ultimately advances the boundaries of digital communication technologies. Since the demand for expanding bandwidth continues in numerous applications, ongoing research and development of modulation techniques continue to be essential in overcoming these inherent limitations.

SUMMARY

The evolution of digital modulation techniques has immensely accelerated the field of communications, moving it towards higher efficiency, improved utilization of the spectral resource, and improved noise immunity. In the essay, various techniques such as Pulse Width Modulation (PWM), Phase Shift Keying (PSK), and Quadrature Amplitude Modulation (QAM) were discussed, mentioning their individual merits and applications. Specifically, the shift from analog to digital modulation marks a watershed, as it allows systems to maintain higher data rates while preserving signal integrity. The integration of these techniques forms the core of modern technologies and underscores the importance of continued research and advancement. Additionally, field application of digital modulation techniques illustrates the relevance of these across many industries, including telecommunications and renewable energy systems. For instance, the design of a one-phase DC-AC converter for application in wind generators offers proof of feasibility in practical applications. As mentioned in recent studies, the controller implemented to regulate output voltage through mathematical models demonstrates the impact of advanced digital modulation on energy efficiency and power system reliability (Buswig et al., 2011). This link between theory and practice supports the argument to place even greater emphasis on research in digital modulation. Lastly, post-analysis of various modulation methods centers not only on technicality but also on the need for innovation in system design. Techniques like the three-phase three-level PWM inverter have been analyzed for their efficiency and output current harmonics under various modulation techniques, demonstrating their applicability (Dahono et al., 2010). Lastly, the efficient use and optimization of such digital modulation techniques pave the way for enhanced performance for communication systems, establishing the necessity of synergizing traditional and novel techniques in order to meet the demands of an evolving technology landscape.

Within the digital modulation techniques, salient concepts note the evolution of communication technology, particularly through the use of methods like coherent optical orthogonal frequency division multiplexing (CO-OFDM). The technique has gained ample attention due to its capacity to enhance spectral efficiency and simplify the digital signal processing (DSP) architecture. CO-OFDM inherently supports high data rate, and complementarily, it is insensitive to a broad class of linear degrading effects and hence a strong candidate for next-generation optical fiber communications. Nevertheless, its inherent large peak-to-average power ratio makes it susceptible to fiber-induced nonlinear effects and calls for advanced compensating techniques. Recent research indicates that the use of machine learning-based DSP is able to effectively combat these nonlinear effects without significantly contributing to computational burdens, demonstrating an evolution within the subject matter integrating traditional signal processing and contemporary machine learning methodologies (Aldaya et al., 2018). Another important facet of digital modulation includes phase-locked loops (PLLs) utilized for synchronization and stabilization purposes of on-chip clock operation. PLLs have many forms, including fully analog, fully digital, and semi-digital forms, the most widely used one being the charge-pump (CP) semi-digital PLL in system-on-chip (SoC) applications. The architectures have critical applications such as converting low-frequency inputs to clocks, thereby making effective data transfer feasible in high-frequency systems. Their reliability is of utmost importance because other on-chip components—for example, analog-to-digital converters (ADCs), digital-to-analog converters (DACs), and user-specified logic—depend on functioning PLLs. Thus, ensuring the integrity of the PLL design and test procedures is of paramount importance not only for device performance but also for the overall efficiency of digital communication systems (Burbidge et al., 2008). By reading the key highlights of digital modulation techniques, it is clear that the merging of traditional methods with emerging technologies such as machine learning can be termed as a significant advancement. The use of CO-OFDM brings to the forefront the importance of compromising on optimal performance with the realization of using compensation methods to combat the challenges of nonlinearities. Likewise, the application of solid PLL architectures indicates the necessity of maintaining synchronization to enable overall system reliability. Each of these addresses an ongoing progression in the field of digital modulation, pointing to the intersection of new technology and established methods, as it preludes future innovations in the field of communications systems. With further development and research, the potential for greater efficiency, reliability, and responsiveness in online communications will certainly develop, affecting various applications across the communications sector.

REFERENCES

1. Abouei, J., Brown, J. D., Pasupathy, S., Plataniotis, K., et al. (2010). Adaptive demodulation in differentially coherent phase systems: Design and performance analysis. http://arxiv.org/abs/1006.0379 (accessed on 18th September 2025).
2. Abrate, S., Gaudino, R., Nespola, A., Offenbeck, T., et al. (2010). High-speed communications over polymer optical fibers for in-building cabling and home networking. https://core.ac.uk/download/pdf/11420470.pdf (accessed on 18th September 2025).
3. Ahmad, S. W., Jennions, I. K., Perinpanayagam, S., Samie, F., et al. (2015). A carrier signal approach for intermittent fault detection and health monitoring for electronics interconnections system. https://core.ac.uk/download/42144211.pdf (accessed on 18th September 2025).
4. Aldaya, I., Barry, L., Giacoumidis, E., Tsokanos, A., et al. (2018). Harnessing machine learning for fiber-induced nonlinearity mitigation in long-haul coherent optical OFDM. https://core.ac.uk/download/287582088.pdf (accessed on 18th September 2025).
5. Amaral, L., Dris, S., Gill, K., Grabit, J., et al. (2010). Feasibility of using bandwidth efficient modulation to upgrade the CMS tracker readout optical links. http://arxiv.org/abs/1010.1691 (accessed on 18th September 2025).
6. Amari, A., Ciblat, P., Dobre, O. A., Jaouën, Y., et al. (2017). A survey on fiber nonlinearity compensation for 400 Gbps and beyond optical communication systems. http://arxiv.org/abs/1708.06313 (accessed on 18th September 2025).
7. Angelova, M., Ghassemlooy, Z., & Rajbhandari, S. (2010). A study of discrete wavelet transforms based denoising to reduce the effect of artificial light interferences for indoor optical wireless communication. https://core.ac.uk/download/4149283.pdf (accessed on 18th September 2025).
8. Axer, M., Dris, S., Foudas, C., Gill, K., et al. (2005). Maximizing the bandwidth efficiency of the CMS tracker analog optical links. http://arxiv.org/abs/physics/0511013 (accessed on 18th September 2025).
9. Azpúrua Auyanet, M. A., Pous Solà, M., & Silva Martínez, F. (2019). Improved electromagnetic compatibility standards for the interconnected wireless world. https://core.ac.uk/download/288625020.pdf (accessed on 18th September 2025).
10. Bard, M., Cosmas, J., Song, Y. H., Zhang, Y., et al. (2006). Diversity gain for DVB-H by using transmitter/receiver cyclic delay diversity. https://core.ac.uk/download/334219.pdf (accessed on 18th September 2025).

11. Bertran Albertí, E., Montoro López, G., & Tarrés Ruiz, F. (1998). An experimental course on digital communications. https://core.ac.uk/download/81582988.pdf (accessed on 18th September 2025).
12. Bobula, M., Danek, K., & Prokes, A. (2008). Simplified frame and symbol synchronization for 4–CPFSK with h=0.25. https://core.ac.uk/download/44395813.pdf (accessed on 18th September 2025).
13. Bouziane, R., Erkilinc, M. S., Killey, R., Nazarathy, M., et al. (2016). Comparison of linear and nonlinear equalization for ultra-high capacity spectral superchannels. https://core.ac.uk/download/79549921.pdf (accessed on 18th September 2025).
14. Burbidge, M., & Richardson, A. (2008). Phase locked loop test methodology. https://core.ac.uk/download/1558600.pdf (accessed on 18th September 2025).
15. Carnegie, D. A., Dorrington, A. A., Drayton, B., Jongenelen, A., et al. (2011). Design and construction of a configurable full-field range imaging system for mobile robotic applications. https://core.ac.uk/download/29198923.pdf (accessed on 18th September 2025).
16. Charrett, T. O. H., Kissinger, T., & Tatam, R. P. (2013). Fiber segment interferometry using code-division multiplexed optical signal processing for strain sensing applications. https://core.ac.uk/download/29409542.pdf (accessed on 18th September 2025).
17. Chen, K. C., Hanzo, L., Jiang, C., Ren, S., et al. (2019). Thirty years of machine learning: The road to Pareto-optimal wireless networks. http://arxiv.org/abs/1902.01946 (accessed on 18th September 2025).
18. Chen, S., Hanzo, L., & Sugiura, S. (2012). A universal space-time architecture for multiple-antenna aided systems.
19. Chie, C. M., Lindsey, W. C., & Tsang, C. S. (2025). Space station communications and tracking systems modeling and RF link simulation. https://core.ac.uk/download/pdf/42810779.pdf (accessed on 18th September 2025).
20. Choi, B. J., & Hanzo, L. (2007). Near-instantaneously adaptive HSDPA-style OFDM versus MC-CDMA transceivers for WIFI, WiMAX, and next-generation cellular systems.
21. Dahono, P. A. (2010). Design and analysis of three-phase three-level PWM inverter. https://core.ac.uk/download/12342426.pdf (accessed on 18th September 2025).
22. Daud, M. E., Ibrahim, A., Ibrahim, Z., Kamarudin, M. N., et al. (2012). High dimensional modulation and MIMO techniques for access networks. https://core.ac.uk/download/13799320.pdf (accessed on 18th September 2025).

23. Davarian, F. (1987). Antimultipath communication by injecting tone into null in signal spectrum. https://core.ac.uk/download/pdf/10396169.pdf (accessed on 18th September 2025).

24. De Leenheer, M., Morea, A., Mukherjee, B., Zhang, S., et al. (2013). A survey on OFDM-based elastic core optical networking. https://core.ac.uk/download/55762862.pdf (accessed on 18th September 2025).

25. de Oliveira, H. M., & Miranda, J. P. C. L. (2015). On Galois-Division multiple access systems: Figures of merit and performance evaluation. http://arxiv.org/abs/1502.03698 (accessed on 18th September 2025).

26. Hanzo, L., Liang, D., & Ng, S. (2011). Near-capacity turbo coded soft-decision aided DAPSK/Star-QAM.

27. Harathi, P. (2018). Building a bidirectional visible light communication link: Challenges and contributions. https://core.ac.uk/download/211332420.pdf (accessed on 18th September 2025).

28. Hoeksema, F., Potman, J., Rauwerda, G., Smit, G., et al. (2003). Adaptive wireless networking. https://core.ac.uk/download/pdf/11458471.pdf (accessed on 18th September 2025).

29. Kot, R. C., Oliver, J. D., Jr., Richards, C. R., et al. (2025). Advanced digital modulation: Communication techniques and monolithic GaAs technology. https://core.ac.uk/download/pdf/42851069.pdf (accessed on 18th September 2025).

30. Lee, C. G., Calvani, S., Rajbhandari, S., Popoola, W. O., Tang, X., & Ghassemlooy, Z. (2011). Coherent optical binary polarization shifts keying heterodyne system in the free-space optical turbulence channel. https://core.ac.uk/download/4149475.pdf (accessed on 18th September 2025).

31. Mansour, A., Peppas, K. P., & Tombras, G. S. (2010). Dual-hop transmissions with fixed-gain relays over Generalized-Gamma fading channels. https://core.ac.uk/download/pdf/107602.pdf (accessed on 18th September 2025).

32. Poklemba, J. J. (2025). Programmable digital modem. https://core.ac.uk/download/pdf/42814838.pdf (accessed on 18th September 2025).

33. Polák, L. (2013). Analysis and simulation of the signal's transmission in the DVB-H/SH standards. https://core.ac.uk/download/30311053.pdf (accessed on 18th September 2025).

34. Raoufi, R., & Zinober, A. S. I. (2006). Adaptive sliding mode observers in uncertain chaotic cryptosystems with a relaxed matching condition. https://core.ac.uk/download/49826.pdf (accessed on 18th September 2025).

35. Serdean, C. V. (2002). Spread spectrum-based video watermarking algorithms for copyright protection. https://core.ac.uk/download/29816683.pdf (accessed on 18th September 2025).

36. Shalkhauser, M. J. O., & Whyte, W. A. (2025). Digital CODEC for real-time processing of broadcast quality video signals at 1.8 bits/pixel. https://core.ac.uk/download/pdf/42826877.pdf (accessed on 18th September 2025).

37. Yonis Buswig, Y. M. (2011). Development of a DC-AC power conditioner for wind generator by using neural network. https://core.ac.uk/download/12007862.pdf (accessed on 18th September 2025).

CHAPTER 6

TRANSMISSION THROUGH COMMUNICATION CHANNELS

Contents

INTRODUCTION

The introduction of any research work is an essential column, providing the reader with an adequate background and roadmap of the purpose of the essays. In the case of Transmission Through Communication Channels, the pertinent research of traditional as well as virtual communication channels provides a basis for studying the inherent principles governing the pathway of information. As we navigate various channels, such as verbal, non-verbal, and electronic forms, it is discovered that effective communication is not merely the art of senders and receivers but also processes that facilitate or inhibit understanding. This preface aims to put into perspective how these channels work together in most settings, affecting human interaction while noting the importance of clarity and effectiveness in communication. Furthermore, the dynamic nature of communication channels has a vital impact on society's interactions. Each channel possesses distinctive attributes that influence the transmission process, impacting not only the transmitted message but also how it is perceived by its audience. For instance, technological innovations have redefined traditional communication models, creating new channels for information dissemination while presenting the challenge of maintaining message integrity in transmission. As previously suggested in research, effective communication strategies must adapt to such evolving channels to facilitate timely and unaltered delivery of pertinent information to the desired audience.

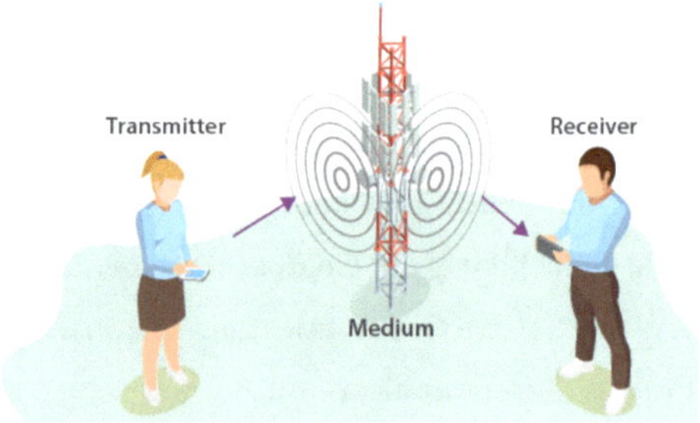

Figure 6.1. *Transmission through communication channels.*

Source: https://cdn1.byjus.com/wp-content/uploads/2021/04/Communication-Systems-01.png.

This essay seeks to investigate the complexities of these, examining how various modes of communication contribute to information clarity and the central position occupied by flexibility in modern communication. Further, researchers highlight the significance of understanding theoretical models of communication networks. For instance, gossip-based network models of communication explain the intricate patterns of information dissemination between nodes in a network, whereby each node serves as a sensor, a relay, and an information user simultaneously. The models allow us to comprehend the information age and how different transmission policies can affect the efficiency of communication. Drawing on these theoretical approaches, this essay aims to further an inclusive account of the various factors influencing information transmission, both in relation to human experience and to the structural aspects of communication channels. By means of this broad-ranging inquiry, the foundations laid out in this introduction will pave the way for a more in-depth consideration of the intricacies of information transfer and its implications for higher-level societal dynamics (Figure 6.1).

Definition of Communication Channels

Communication channel theory is the representation of the media through which messages are conveyed between a sender and a receiver. The model encompassed physical and abstract media such as telephone cables, radio frequencies, and computer networks. The definition stretches beyond hardware to incorporate processes and protocols through which message transmission is facilitated effectively. For instance, the evolution of mobile communication channels illustrates how technological advancements have transformed interaction paradigms. The transition from traditional telephone lines to advanced cellular networks is a revolutionary shift in the mode of information exchange, especially with the advent of mobile edge computing and the 5G ecosystem. These technologies enable efficient data transmission while addressing issues of latency and resource scarcity in communication systems, and this showcases the dynamism of communication channels in the presence of ever-evolving technological landscapes. Exploring further the nature of communication channels, it can be seen that they are not only characterized by their functionalities but also by their limitations. These limitations are such as limited bandwidth, interference, and requirements for synergy among disparate communication technologies. As in the scenario of transitioning to mobile edge computing (MEC), optimization of resource management is focused on in order to be in a position to support more advanced applications that demand high-speed connectivity. Integration of MEC into communications channels is one significant line of research that seeks to maximize radio and computation capacities. Placing these challenges into perspective requires concerted efforts by both scholars and practitioners in generating innovative solutions, thereby ensuring communication in an increasingly data-driven world. Addressing such elements of communication channels represent the nuanced ballet of technological potential versus practical use. The evolving image of communication channels is also reflected in the emergence of 5G technology, which promises a paradigm shift in mobile communication. Unlike its predecessors, 5G employs bigger carrier frequencies and giant bandwidths to significantly enhance the user experience. The transformation does not stop with fundamental infrastructure paradigms; it also involves harmonizing accessible technologies such as LTE and WiFi under one harmonized communication system. With scientists and experts identifying key issues concerning 5G, it is imperative to note the implications on communication channels as a whole.

The requirement for cost-effectiveness, intelligency, and flexibility is going to change not only how messages are transmitted but also how various technologies can blend together in a harmonious union to form an integrated society. This is further importance of knowing communication channels in the big picture of upcoming technologies like MEC so that transmission mechanisms continue to evolve with the needs of society.

Importance of Effective Communication

Communication is the cornerstone of effective relationships in every setting, most notably in companies. Through the establishment of an open exchange of information, people can establish relationships, reach agreements, and solve complicated issues. Perhaps the most significant component of effective communication is an awareness of the implications of different channels that are utilized for transmitting messages. Choosing a channel—verbal, written, or non-verbal—one affects the understanding and reception of messages. For instance, the process of transmitting its values by a non-profit organization requires it to note the way its stakeholders interact with various media. As described in a case study, the transmission model addresses the significance of multi-sensory channels that eventually mean more decoding of messages by stakeholders (Alhakami et al., 2014). An awareness of communication constructs can therefore enhance both clarity and effectiveness. In addition, the role of good communication is also to ensure security in communication systems, particularly in technological communications. In cognitive radio networks where efficient spectrum management has been envisioned, effective and secure communication becomes a top priority. In this scenario, the work is not just to communicate information effectively but also to secure the communication from threats. This environment raises new threats, including licensed user emulation attacks, which have an impact on communication integrity. According to Alhakami et al. (2014), common control channel security is crucial to maintain robust network operation. On this account, efficient communication in technology-based environments focuses on maintaining the clarity of message transmission and defense against emerging security threats. Apart from enhancing organizational effectiveness, the power of effective communication is also experienced in social settings. Effective value communication can have significant far-reaching social impacts, affecting community engagement and overall wellness. Research evidence from studies on nonprofit organizations underscores that organizations employing multiple channels of communication—traditional and innovative—can convey their values more effectively, engaging stakeholders on a deeper level (Wilks et al., 2016). This cascading behavior has the potential to develop social bonds, foster trust, and create participatory engagement in social endeavors. Therefore, the efficacy of communication processes extends beyond the simple exchange of data; it plays a key role in shaping societal attitudes as well as in instilling cooperative behavior. Communication is not just a process of operational necessity; it is a builder of society.

Overview of Transmission Methods

The evolution of transmission methods has been profoundly influenced by technological development and the needs of efficient communication. Modern communication media utilize a number of techniques, both wireless and wire-bound, to exchange information. Among these, the use of noise carriers has been an innovative approach to enhancing the security and reliability of communications. Researchers have extensively studied autocorrelation systems with noise carriers, and they are most effective in constructing confidential communication systems. The strategy is to provide top priority to making conveyed information secure from interception but keep the channel secure from unauthorized information injections. It therefore addresses major issues concerning sustainable communication, especially under adverse signal propagation conditions highlighted in literature (Didkowsky et al., 2013). Apart from carriers of noise, organizational mechanisms used in corporate networks also have an impact on influencing transmission methods. Corporate environments are often in need of overcoming the rising expenditure on leasing communication channels, particularly as the need for better quality and faster data mounts. In order to counter such issues, care should be taken to optimize the use of bandwidth effectively. Research indicates that the application of compression techniques for video and voice data and variable channel architectures can achieve significant cost reductions. The technologies facilitate optimum utilization of transmission channels, addressing bandwidth use issues without data integrity compromise. This channel design adaptability supports real-time response to variance in networking efficiencies and demands, which is crucial in today's business communication (Bondarenko et al., 2014). Furthermore, multi-speed channel integration is a central factor in the evolution of transmission technique. Through the establishment of multi-speed channels with different types of digital signals, communication systems can serve more varied demands from users. For instance, E1 signal-based channels can readily support various speeds using some encoding structures. This capacity not only maximizes bandwidth utilization but also ensures that various communication requirements are addressed. The mechanisms proposed for channel creation emphasize the importance of synchronization and form to allow for the co-existence of various digital signals. These innovations enhance the performance of overall communications systems by making them more versatile and more efficient, setting the stage eventually for other innovations in transmission technology (Bondarenko et al., 2014) (Figure 6.2).

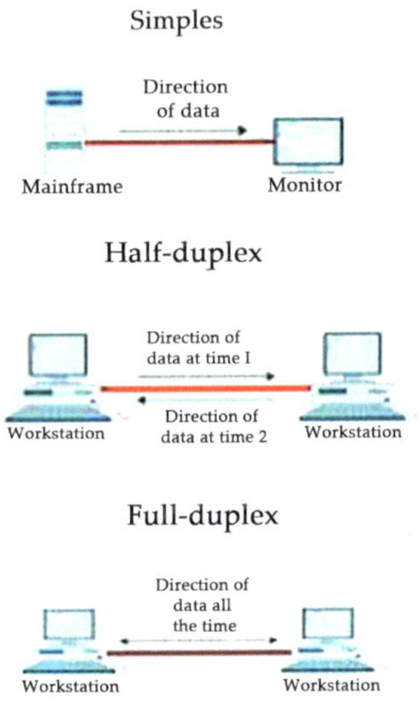

Figure 6.2. *Overview of transmission methods.*

Source: https://encrypted-tbn0.gstatic.com/images?q=tbn:ANd9GcTzszTyFVaPhz CkMVGol3MnxoSlHt5zyXAbaw&s.

6.1. TYPES OF COMMUNICATION CHANNELS

In communication channel discovery, various forms are identified, each serving different purposes and customers. Such a primary distinction exists between the traditional and digital channels. Traditional channels entail the utilization of face-to-face dialogue, print media, as well as telephone communication, all of which have, over time, facilitated individual and business interactions. Digital channels, on the other hand, have accompanied the advent of the Internet with an application involving the utilization of email, social media, and instant messaging services. These online media do not only facilitate connection but also make possible swift information transfer, serving real-time engagement. The performance of the channels usually hinges on the nature in which they are used, since every channel has unique strengths and weaknesses that determine the communication experience as a whole. Identifying these differences is crucial for individuals and organizations seeking to maximize their communications strategy in a more and more connected world. Also, channel classification involves the distinction between formal and informal communication. Formal channels typically are characterized by structured communication patterns in organizations, such as official emails, memos, and reports. Formal channels have established hierarchies and protocols, and information is shared accurately and efficiently. Informal communication channels, such as social gatherings and casual conversations, facilitate a free exchange of ideas and information.

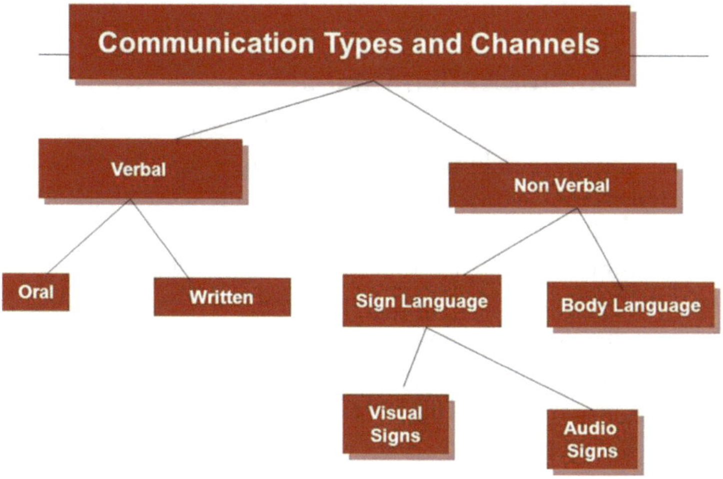

Figure 6.3. *Types of communication channels.*

Source: https://i.pinimg.com/736x/d8/70/4a/d8704ab065a90f0bb1fa3157dfae0f4d.jpg.

Although often overlooked, informal channels play a vital role in the development of relationships and cooperation. They permit the unobstructed flow of information, which fills gaps that formal channels cannot or do not fill. The relationship between formal and informal communication can impact directly on workplace dynamics and serve to emphasize the value of knowing the channels used in varying environments and their related role in effective communication. In mapping the channel landscape of communications, the arrival of new technologies such as Secure Device Pairing (SDP) schemes highlights the emergence of solutions tailor-made for certain technology settings. These schemes facilitate secure communication between devices—particularly in private mobile and Internet of Things (IoT) environments—where traditional security measures may no longer suffice. By ensuring secure data exchange under predetermined protocols, SDP schemes respond to the increasing digital sophistication of interactions. As noted, the analysis of SDP schemes reveals systemic security flaws that provide avenues for mitigation, particularly in integrating privacy requirements into their design (Fomichev et al., 2017). Furthermore, understanding the drivers of marketing practices on digital channels highlights flexibility and dynamism in the habits of communication, viral marketing being a prime example in the peer-to-peer dissemination of information potential (Bourlakis et al., 2008). More generally, these observations are used to construct an integrated analysis of the communication channels and their significance for marketing and security within present reality (Figure 6.3).

6.1.1. Verbal Communication Channels

Verbal communication channels are the most fundamental mode of human interaction, serving as conduits for the exchange of ideas, feelings, and information. Both verbal and written modes constitute these channels, and they have the onus of delivering links and disseminating information in business and interpersonal relationships. Clarity, tone, and background information typically regulate effectiveness in verbal communication. In many situations—be it in ordinary conversation, academic discourse, or corporate environments—the choice of words can have a significant impact on audience engagement and comprehension. As implied by research in multimodal communication, various modes, including verbal components, can work to enrich meaning when appropriately combined. The capability to convey thought through verbal mediums not only facilitates immediate interaction, but also sets the ground for enriching relations and collaborative activities among groups and individuals, establishing the formative power of words in society. Feedback mechanisms are an indispensable part of verbal communication since the speaker-listener process guarantees clarity and mutual understanding. Effective oral communication calls for responsiveness to the cues of the speakers, such as body language and tones, that convey more than explicit words. The interactive nature of communication also calls for the ability to practice "being flexible" in communication that enables the speaker to adapt his or her messages based on feedback acquired. For instance, in educational settings, lecturers always adapt their oral strategies based on students' levels of engagement, thus evolving a better learning environment. Additionally, the nature of verbal communication could impact significantly on relationships since research indicates that clarity and empathy improve relational dynamics. Through competent verbal discourse, one is able to engage in complexity management, rapport building, and ultimately construct more productive and meaningful interactions across contexts. Notwithstanding the prominence of verbal channels, there are challenges that can counteract effective communication. These can involve differences in languages, cultural differences, and cognitive styles of people which affect the reception and interpretation of messages. For instance, cognitive styles may have an effect on the individual's preference of the mode in which information is received, and this could result in miscommunication when verbal messages are conveyed without considering these differences. Research shows that users of various cognitive styles treat multimedia information in a unique manner, and therefore, communicators must be aware of the differences when crafting messages. Moreover, effective oral communication in an era dominated by technology needs to know how technology enables such interactions. Technologies such as video conferencing and instant messaging have transformed the traditional oral channels, necessitating the transformation of oral abilities to adapt to modern communication environments, a practice demanding frequent exercise and attention in a bid to thrive in diverse communicative contexts.

6.1.2. Non-Verbal Communication Channels

In looking at non-verbal communication channels, it is necessary to value the pervasive role they play towards building understanding that does not necessarily occur through verbal or written speech. Non-verbal messages, such as facial expressions, attitude, and body positioning, convey feeling and intent that has a tendency to go beyond words. The significance of these channels can be particularly highlighted in cross-cultural interactions, where discrepancies in non-verbal communication are likely to lead to misunderstandings or enhance rapport between individuals. For instance, a mere movement may have completely different meanings across cultures, emphasizing that one must be sensitive and mindful while communicating across cultures. It is an added advantage to the communication process to know how to interpret and respond to non-verbal signals, further substantiating the argument that non-verbal communication is also a necessary channel in communication that verbal communication alone cannot carry through. Furthermore, the integration of non-verbal communication channels with technological media also indicates the forward-looking path of interactions in contemporary society. Machinery and digital media become more likely to offer the interfaces through which non-verbal signals are being communicated. For example, emoticons and graphical symbols of computer-mediated communication mimic non-verbal cues for facilitating more emotional expression. According to multimodal communication studies, legal machines and computers can also communicate in other modalities, for example, visual messages or sound signals, and hence essentially have a conversation with users (Lachmayer et al., 2012). Such integration of human and machine communication raises new issues regarding the adequacy of these channels to replicate the nuanced dynamics of face-to-face communication. As technology continues to develop, understanding how such non-verbal channels of communication work in automated environments is vital to optimizing clarity and user engagement. Finally, the scope of non-verbal channels of communication also extends into fields such as information ethics and cognitive science. The philosophical examination of man and machine interaction takes into account the ethics of using non-verbal signals generated by technology. This cross-disciplinary practice calls into question authenticity, agency, and the possibility of misinterpretation in communicative systems, particularly in those requiring high-stakes decision-making. As exemplified in current works, the practice of studying non-verbal communication as part of cognitive studies contributes significantly to understanding broader contexts in which information travels (Slavkovsky et al., 2013). Through the lens of non-verbal communication channels, researchers attempt to uncover patterns of understanding and misunderstanding that emerge when individuals interact with machines and ultimately enrich our grasp of communication processes in an increasingly automated world (Figure 6.4).

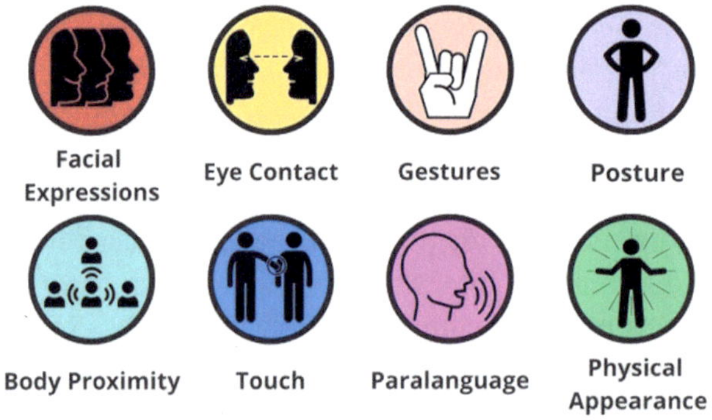

Figure 6.4. *Non-verbal communication channels.*

Source: https://i.ytimg.com/vi/5nLQl9e5NV8/hq720.jpg?sqp=-oaymwEhCK4FEI-IDSFryq4qpAxMIARUAAAAAGAElAADIQj0AgKJD&rs=AOn4CLCiEWS6sW1ji GMLdXAmCHsh9TMClA.

6.1.3. Digital Communication Channels

Digital communication channels are the backbone of modern telecommunication systems, enabling information to be communicated via various media. The channels can be optimized to carry a vast array of forms, for example, wireline and wireless systems, each characterized by their own attributes that influence their efficiency and performance. Bandwidth, error ratios, and modulation techniques are the principal parameters that control the effectiveness of these channels. For instance, fading-induced and interference-induced impairments in wireless communication require solutions that maximize the capacity and reliability of data transfer. The average bit error rate and ergodic capacity, both elementary performance measures, provide foundational insights into capacity limits and reliability of these channels, with far-reaching implications for their design and implementation (Alouini et al., 2011). It is important to comprehend these measures in a bid to further optimize communication systems towards fulfilling contemporary needs. The complexity of digital communication is also increased by the requirement for a low bit error rate (BER) and high data throughput. It is a problem, particularly for optical communication systems, where forward error correction (FEC) techniques are a necessary implementation. The traditional FEC bound model assumes that there exist decoding operations to reduce the BER to acceptable levels, making data transmission efficient. However, more recent studies questioned the applicability of this approach, citing that dependence on a channel-independent FEC limit may lead to underestimation of spectral efficiencies, particularly for more advanced modulation schemes (Agrell et al., 2015). With the continuing developments in communications channels, there is ever-growing need to revisit such fundamentals and use approaches that enhance BER knowledge under varying conditions. Besides, digital communication networks have to match the pace of technological progress and evolving user demands. The progress towards more sophisticated transmission methods brings with it the need for novel modulation schemes and error correction codes. Additional demands on data bring additional spectral efficiency, which typically requires a flawless balance between coding rate and the order of modulation. Implementation of advanced decoding techniques using generalized mutual information gives promising directions towards improved BER estimations and system performance (Agrell et al., 2015). Thus, analysis of digital communication channels not only reveals inherent problems but also indicates the requirement for new-era solutions to tackle the growing needs of global communications, delivering reliability and high rate in a world that is more interconnected than ever.

6.2. THE ROLE OF TECHNOLOGY IN COMMUNICATION

The advancement of technology has greatly impacted the landscape of communication, with the manner in which societies and people communicate and share information. Communication in the past relied on traditional methods such as face-to-face communication, letters, and publications. The use of digital technologies such as the internet and mobile phones has made communication a more real-time and globally accessible system. Such a transformation has not only facilitated more efficient exchanges but also given rise to new forms of expression, ranging from social network sites to communal online spaces. As such, the trends of interpersonal communication have been transformed, allowing for more enriching encounters that transcend geographical and cultural boundaries. Moreover, the integration of advanced technologies into practice of communication has enhanced how stories and cultural heritage are conveyed. In particular, artificial intelligence and interactive storytelling are revolutionizing the manner in which audiences engage with content.

Figure 6.5. *The role of technology in communication.*

Source: https://fastercapital.co/i/Lack-of-communication--Bridging-the-gap-for-seamless-collaboration--The-Role-of-Technology-in-Facilitating-Communication.webp.

For instance, museums that were previously perceived as being static warehouses of objects are becoming more dynamic spaces where virtual and real experiences meet. This phenomenon alludes to technology's capability of creating hybrid spaces where audiences can have a better connection with cultural stories. As they journey through these interactive museum experiences, they are not only consuming content, but the co-creation of knowledge upon which they can draw after the lifespan of their very first visit. These developments underscore how technology is essential in bringing about cultural heritage that is accessible and meaningful in today's globalized environment (Falco et al., 2017). Besides, the diffusion of intelligent devices and Internet of Things (IoT) technology has introduced new dimensions to safe communication protocols. With personal devices spreading everywhere, it is now more critical than ever to have secure security architectures that protect user data and privacy. Secure Device Pairing (SDP) protocols are typical examples of the technological breakthroughs that enable safe communication among interconnected devices. However, the heterogeneity of assumptions and models employed in SDP schemes hampers the evaluation of their effectiveness and suitability across contexts. With a system model and taxonomy of these schemes, researchers can identify common weaknesses and pave the way for privacy and security improvement. This emphasizes a key area of technology communication: the balance between cutting-edge interconnectedness and user data protection in an increasingly digitalized world (Fomichev et al., 2017) (Figure 6.5).

6.2.1. Impact of the Internet on Communication

The advent of the internet revolutionized communication, fundamentally altering the way individuals and businesses communicate. Traditional communication channels, including face-to-face communications and written communications, are increasingly complemented or replaced with digital media. Email, social media, and instant messaging apps have facilitated communication in real-time over distance, allowing ideas and information to be easily exchanged rapidly. This transformation has not only become more profound in terms of interpersonal relationships but has also widened the range of professional interactions. Consequently, organizations are experiencing changes in their communication tactics, opting for more flexible and adaptable approaches to keep up with such rapid evolution of technology. Since communication is digitizing, it is all the more necessary to understand the impact of this evolution on interpersonal as well as organizational communication. The convergence of the various modes of communication has ushered in the era of speed and convenience, changing the very essence of human interaction. Besides speed and convenience, the input of the internet to the process of communication is also the democratization of information. Without the internet, information was typically restricted to the elites who dominated dissemination networks. But the internet has killed these barriers, allowing individuals to share their opinions, views, and experiences freely. This closeness brings together a mosaic world of ideas and perspectives, which enriches public debate as well as enables informed decision-making. Strategic communication has also become more paramount as a key component in a number of areas, from politics to commerce, as a sign of greater awareness of its role in structuring perceptions and influencing behaviors. As developed in the foreign intelligence collection discourse, technology reconfigures the means and availability of exchange of information, therefore necessitating a rethinking of how communication frameworks operate within contemporary society (Donohue et al., 2014). The development mirrors the empowerment possibilities and threat of disinformation inherent in increased connectivity. In this context of such profound transformations, there is a need to consider the ethical aspects and complications that arise from the internets impact on communication. Issues of privacy, bullying, and disinformation present significant hurdles that must be overcome when society embraces digital communication mediums. The complexities of virtual interactions are prone to conflicts and misunderstandings, which put great emphasis on the use of robust communication strategies that incorporate responsibility as well as openness. Furthermore, the intersection of development and communication gains relevance, as communication strategies can contribute to robust societal transformation, a fact valued in contemporary times (Serrat et al., 2012). With these ethical dimensions, the multifaceted nature of internet-based communication and the heavy responsibilities that accompany its use in an interconnected world may be better comprehended. Ongoing discourse on these issues is needed for fostering a communication environment of integrity and for promoting good interaction.

6.2.2. Evolution of Communication Devices

Communication technology has been a major force of change in changing society throughout history. The first tools began with simple tools such as the messenger and the use of smoke signals to meet the needs of communication. The invention of the telegraph in the 19th century was a breakthrough where messages were transmitted across wide distances near real-time. This innovation led on to later innovations, opening the way to the telephone and radio. Each new machine made communication faster and more efficient, effectively revolutionizing human contact. These inventions suggested an increased dependence upon technology, which was a pointer to a sociological shift towards which information exchange was becoming more organized and swift. As communication equipment improved, the development of digital technology marked a new era in the conveyance of information. The introduction of the mobile phone in the latter part of the 20th century revolutionized communication to be not only portable but also available to the masses. The inclusion of the internet further elevated communications to a new level with users able to bridge geographical distances through instant messaging, video calls, and social networking websites. With flexibility and immediacy being the watchwords in this age of digital communication, new forms of social interactions were created and privacy, identity, and relationship were altered. Technologies such as smartphones integrated various functions into a single product, illustrating how various communication technologies merged and had profound effects on daily life. Emerging technologies such as low-power wide-area networks (LPWAN) are changing the face of communication devices in the future. The deployment of systems such as LoRaWAN is an emerging interest in adaptive communication technologies, particularly Internet of Things (IoT) applications. These networks are planned with global connectivity and flexible structures that support more than one device at a time. But, as the demand for connectivity is growing, misuse bottlenecks and technological barriers have to be planned carefully. Moreover, the role of artificial intelligence and machine learning in optimising communication networks will play a major role while moving toward 6G wireless communication systems. Not only will these technologies fulfill the increasing demand for secure connectivity, but they will also propel the incorporation of communication technologies into societal frameworks, deciding the trajectory of human communication and connectivity (Adelantado et al., 2017; Abbas et al., 2020) (Figure 6.6).

Figure 6.6. *Evolution of communication devices.*

Source: https://fastercapital.co/i/From-Microwaves-to-GaAs--Revolutionizing-Communication-Devices--The-Evolution-of-Communication-Devices.webp.

6.2.3. Social Media as a Communication Tool

The evolution of social media has revolutionized the landscape of communication as a vital tool for people and organizations. Capable of disseminating information at high velocity and interactively, social media websites enable users to share ideas, values, and experiences with global audiences. The age-old transmission model of communication based on linear channels must be modified in order to align with the evolving nature of social media communication. By engaging users with content through comments, shares, and likes, these websites put a lot of emphasis on feedback and two-way communication. Such interaction allows for community and collective identity building, which best works for non-profit agencies that are attempting to communicate their values and mission effectively. As revealed by empirical evidence in a recent research study, the importance of multi-sensory channels in communication highlights the compound way in which social media transmits beyond literal text messages that add richness to the user experience and strengthen stakeholder relationships (Wilks et al., 2016). Social media exists as a strategic communication tool in non-profit organizations whose purpose is to enhance stakeholder engagement and communicate organizational values' perceptions. By way of mixed methodological approaches, researchers have been able to observe the way that non-profits used these spaces to convey their missions and foster community involvement. The necessity of having an identifiable mechanism through which value is being conveyed gains a new level of importance, as social media not only enables organizations to broadcast but also to listen and respond to stakeholder feedback. This recursive procedure legitimates the transmission model by attributing importance to its situational relevance; stakeholders are having a conversation that evolves dynamically. The possibility of organizations changing their plans for communication in response to live interactions adds additional importance to the acceptance of social media as a polyvalent communication tool, one which is able to organize public opinion and engagement greatly in modern environments (Wilks et al., 2016). Lastly, the influence of social media as a communications tool is not merely information transmission; it's an open platform for discussion and cooperation with the ability to influence the course of society. Increased visibility of social media in today's communication highlights the changing relationship between technology and human interaction. As indicated in a study that looked at secure device pairing and communication protocols, emerging technologies have to incorporate privacy and security elements necessary to shield users' interactions. Thus, transmission model through communication channels needs to cope with the complexity introduced by social media. This involves regular revision of communication approaches that not only meet the current needs of organizations but also emerging needs introduced by evolving digital spaces. By realizing that social medias two functions are that of a channel and an enabler of communication, organizations stand a greater opportunity to leverage these media better in creating positive social change and building community relations.

6.3. BARRIERS TO EFFECTIVE COMMUNICATION

The effective communication challenges reveal themselves via numerous different mediums, influencing the transfer of ideas as well as the spreading of information. One of the most serious challenges is unavailability of research outputs and associated data. Availability issues such as material not being freely available are observed to hinder the utilization of valuable information among the academic community as well as elsewhere (Missingham et al., 2016). When research is locked behind paywalls or is not methodically classified to be retrieved, there is less opportunity for successful communication because stakeholders will not be able to retrieve or engage with the content. Furthermore, discoverability tends to be a hurdle, whereupon research that could facilitate successful discussion is not seen because it poorly ranks on search engines or lacks proper introductory context and therefore leads to decreased engagement by potential users who would be aided by such research. The other imperative barrier to communication is the prevailing technological environment that affects how information is shared. With continuous developments in digital means of communication, some practices do not keep up well, reducing user engagement and experience. A case in point is the rise in viral marketing, which has raised awareness of the need to synchronize campaigns with higher social contexts and consumer behavior. The complexity of viral marketing typologies emphasizes how a mismatch of message content and audience type can result in communication breakdown (Bourlakis et al., 2008). Customers may overlook or misunderstand messages if they are not related to their interests or the form of their communication channels. Thus, the task is to create meaningful messages that resonate within a technologically-driven, heterogeneous environment to achieve enhanced understanding. Lastly, interpersonal relationships add another degree of sophistication to overcoming communication obstacles. Both trust and perception are essential factors in determining how messages are perceived and understood. In contexts where communication is conducted via middlemen—for example, credible mediators or peer groups—the effectiveness of the messaging transmission may be impaired in the case that there is no credibility built or in the case that the middleman lacks extensive knowledge of the subject of communication (Missingham et al., 2016). Additionally, the relational context between the communicators might influence the clarity of the message and the communicative intentions. Consequently, the sender and the receiver are required to establish an environment of openness and honesty, which develops better exchanges of ideas and collaborations that ultimately make communication more effective across various fields (Figure 6.7).

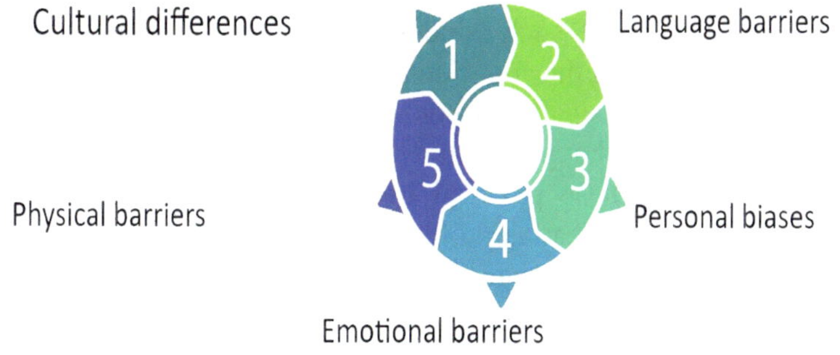

Figure 6.7. *Barriers to effective communication.*

Source: https://fastercapital.co/i/Communication-skills--Enhancing-Communication-Skills-through-Recoupling--Common-Barriers-to-Effective-Communication.webp.

6.3.1. Physical Barriers

Physical barriers can significantly impede the transmission of communication across various channels, affecting the clarity and effectiveness of interactions. These barriers manifest in diverse forms, including environmental obstacles such as noise, distance, and physical obstructions, which can disrupt the flow of information. For instance, in an organizational setting, a factory with heavy machinery may present auditory distractions, making verbal communication challenging among workers. Additionally, the spatial arrangement of workplaces can create physical separations that hinder collaboration and create silences that negatively impact teamwork. Understanding these barriers is crucial for effectively managing the communication process, as they must be identified and addressed to facilitate clearer exchanges. In the realm of digital communication, physical barriers are uniquely transformed, yet their implications remain pertinent. Connectivity issues, such as limited bandwidth or unreliable internet service, serve as modern physical barriers that can hinder the smooth transmission of information across communication channels. As highlighted in recent studies, the rapid expansion of internet services underscores the importance of overcoming these barriers to harness the full potential of digital communication (Coble-Neal et al., 2025). Organizations must consider technological infrastructure as a critical element in fostering effective communication. Failing to address these technological barriers not only impedes the transmission of messages but can also lead to significant economic consequences, as companies may struggle to keep pace with competitors who navigate these challenges more effectively. Finally, the influence of physical barriers on communication extends to the frameworks for innovation adoption within organizations. As identified in the literature, the channels through which innovation is communicated, such as cooperation and competition, are critical in facilitating or obstructing the adoption process (Autant-Bernard et al., 2025). When employees are physically disconnected or face barriers in accessing technological resources, the potential for collaboration diminishes, stifling innovation. Addressing these physical barriers is essential for fostering an environment conducive to creative problem-solving and the seamless exchange of ideas. By implementing strategic initiatives that enhance connectivity and reduce physical obstacles, organizations can enhance their communication channels and ultimately drive innovation within their teams.

6.3.2. Psychological Barriers

Psychological barriers play a significant role in the efficiency of information being communicated via various channels, particularly when it comes to health communication. HIV transmission is a prime example, where the complexity of human behavior coming into contact with the dissemination of information is apparent. Irrespective of scientific awareness regarding the transmission of HIV, psychological barriers such as fear, stigma, and misperception have a tendency to deter individuals from taking preventive measures. This case emphasizes the necessity of communication approaches that, in addition to presenting factual information, account for emotional and social factors that shape decision-making. If populations, especially vulnerable ones like women and adolescents in developing countries, are hindered from receiving and utilizing information about risks, any communication effort is less effective, calling for targeted communication interventions that account for such psychological factors (Imrie et al., 2003). Moreover, women's subservient position in certain societies contributes additional psychological barriers to negotiating safer practices such as condom use. This kind of dynamic is reinforced by societal pressure and family planning expectations, with the possibility of being reluctant to adopt safer sex behavior.

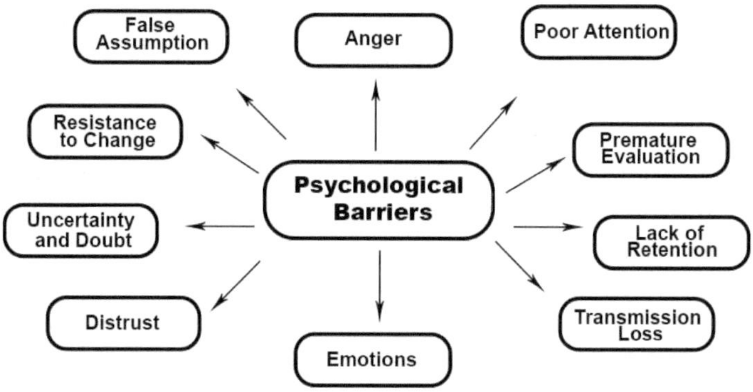

Figure 6.8. *Psychological barriers to effective communication.*

Source: https://parsadi.com/wp-content/uploads/2022/02/types-of-psychological-barriers.jpg.

Awareness of these psychological barriers is especially critical for public health programs aiming to reduce infection transmission rates of diseases such as HIV. Interventions that only aim at generating awareness without comprehending the underlying psychological context may be of limited success. Rather, having structural interventions coupled with competent communication may create a context where people feel empowered to adopt required preventive strategies. These interventions can also contribute to reducing social stigma, eventually increasing the overall effectiveness of health communication initiatives (Imrie et al., 2003). The effectiveness of communication channels in inducing behavioral change is also compromised by interaction between internal attitudes and external realities, such as urbanization, poverty, and conflict. These forces create a breeding ground for psychological blocks to communication, making high-minded attempts at communication far less effective. For example, those who reside in deprived environments might find that health messages are not applicable to their immediate survival needs, thus creating a gap between dissemination of information and adoption of behavior. Such disengagement can maintain loops of misinformation and unhealthy habits that ultimately counteract the general purpose of public health efforts. It stresses that one needs to adopt a holistic strategy not only prioritizing health communication but also integrating the entire psychosocial components in play, which ultimately allows more coordinated approaches in dealing with such public health issues as HIV transmission (Figure 6.8).

6.3.3. Cultural Barriers

Cultural barriers usually constitute significant impediments in the smooth flow of information through avenues of communication. Cultural barriers manifest themselves in many different ways, including variations in language, miscommunication in nonverbal cues, and societal norms that regulate effective modes of interaction. For instance, certain gestures or facial expressions may have varying meanings in different cultures and so are open to potential miscommunication. This holds particularly in collaborative settings, like university collaborations or international business ventures, where communication must be as clear as possible to be effective. Ignorance of these differences can slow knowledge sharing and sap the general efficacy of communication efforts. Their manufacture calls for ways of filling such gaps in cultures, calling for attention to training and sensitization for better knowledge among diverse groups engaged in information sharing. Moreover, availability is also essential in overcoming cultural blocks in communication channels. As Missingham et al. (2016) illustrates, availability and discoverability-related barriers greatly limit access by business organizations as well as researchers. The effectiveness of communication channels relies on the possibility that information not only apply but also be readily accessible to the target audience. Where the target users do not navigate easily or trust information sources, the potential for successful collaboration is reduced. Making access pathways and participations simpler, particularly in learning and research settings, translates to bringing initiatives into maximizing the dissemination of knowledge and optimizing it for discoverability. This involves investment in human capital and technology and in the type of settings within which there can be more than one voice feeding in and being enhanced by the diversity of information.

In global markets, cultural barriers are most severely felt in sectors such as audiovisual communications, where ownership and culture of content become highly sensitive factors. The delicate balance between trade liberalization and the protection of culture is an indication of the transnational complexity, as suggested in Mukherjee et al. (2025), where countries like India are unwilling to extend greater access to their audiovisual market due to concerns regarding cultural insensitivity. The stakes underlying such contexts are similar to broader negotiations made within institutions like the WTO, where countries are compelled to negotiate between their cultural agendas and globalizing pressures. Communications can create avenues to mutual benefit while maintaining cultural contexts through exchange and cooperative agreements, as in the proposed Indo-U.S. FTA. To overcome cultural barriers, then, involves not just technical solutions but also awareness of the complex landscapes shaping these communications.

6.4. THE PROCESS OF COMMUNICATION

Good communication is essentially founded upon the understanding of various communication channels and the process that regulates their transmission. The communication process at the center includes a sender, a message, a channel, a receiver, and feedback. The communication process is interactive; messages have to be encoded by the sender in the proper form for transmission by selected channels. The medium—whether verbal, written, or electronic—is a determining factor for the accuracy and efficacy of the received message. Context and what the audience sees also contribute to the effectiveness of communication. For instance, in mission-critical applications provided in the literature for ultra-reliable low latency communications (URLLC), the need for on-time and correct information exchange becomes paramount, the exploration of which necessitates knowledge of sophisticated network architectures (Arnau et al., 2017). Channel characteristics play a crucial role in defining communication reliability and latency. Various models demonstrate how selection and utilization of specific communication channels can facilitate the encoding and decoding process in a more simplified manner, with honesty and understanding of the message being conveyed. New-age communication technologies have rendered traditional paradigm obsolete, particularly mechanisms for feedback and error correction. In network design, the subtleties of multiple-input, single-output (MISO) systems highlight the critical need to envision possible delays and formulate strong countermeasures against them. This is particularly important in advanced communication systems where quick decision-making and response are required. The solution to these problems is integral to enhancing communication processes and sustaining the efficacy of information transmission across various platforms (Arnau et al., 2017). Furthermore, as the global village continues to shrink, the communication process is also vulnerable to interference by ill-fated players who use these channels for malicious intents. Network steganography presents an insight into the seedy underbelly of communication processes, illustrating how hidden information is implanted in innocent-looking transmissions. It is necessary to study these facets so as to make a move toward devising protective measures against potential breaches and uninhibited access to confidential information. With each passing day, research is progressing, and it is becoming increasingly important to overcome challenges brought about by such malicious attacks. With the strategy of communication security and efficiency, technology can be utilized to create stronger communication systems without sacrificing the processes so that they are still effective and reliable (Figure 6.9).

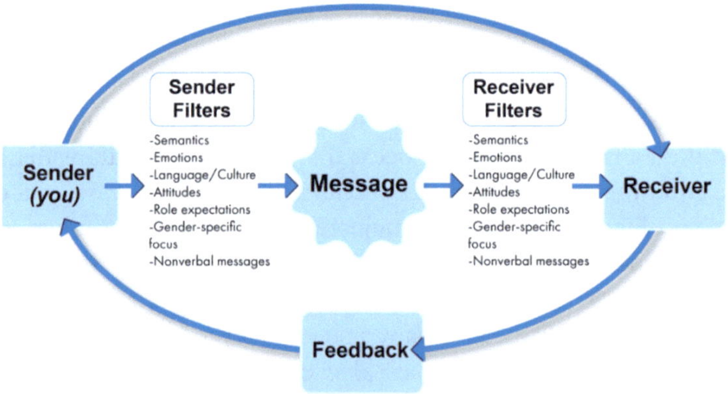

Figure 6.9. *The process of communication.*

Source: https://wtcs.pressbooks.pub/app/uploads/sites/38/2021/12/Communication-Process-by-Meredith-Pomietlo-1024x1024-1.png.

6.4.1. Sender and Receiver Roles

In communication theory, sender and receiver roles are the foundation of knowledge about information transmission dynamics. The sender, or the source or encoder as commonly referred to, has the responsibility of crafting a message that captures the intended information. This is not just a choice of words or symbols but also a regard for the receivers context so that the message successfully sends and receives through the target audience. The intricacies of the senders role continue in cases where multiple channels of communication are employed because different media can affect the presentation and meaning of the message. Through understanding the role of the senders, one realizes that how they act contributes to the conciseness and correctness of the communication in specific contexts, which in turn contribute to the effectiveness of the information transmission. On the other hand, the receiver has a pivotal role in the process of communication as they are the ones who decode and understand the message sent by the sender. This process entails not just passive reception but active engagement with the information, cognitive processing, and situational meaning-making in order to achieve correct meaning. A receiver's history, experience, and inclinations can significantly impact their interpretation, leading to likely variance of comprehension despite an effective sender message. Furthermore, the channel of communication employed can further make the process complex, as the workings of the different media may facilitate or hinder understanding. Having an understanding of the complexity inherent in the receiver's role is very important, as effective communication depends on both sides' ability to manage their respective roles effectively, illustrating the sender and receiver interdependency. The sender-receiver relationship is even more complicated by the existence of external factors that may affect communication dynamics. For instance, in quantum communication theory, the existence of a kindly third party may alter the character of the sender-receiver relationship because it introduces other factors that affect the transmission of the message. Such occurrences require a reinterpretation of the traditional models of communication, illustrating how external factors, such as helpers or environmental factors, can redefine the way in which information is sent and received. Outside of quantum environments, wireless sensor networks illustrate similar principles where sender and receiver functions must change with varying traffic patterns and communication situations. In such circumstances, the need for energy-saving protocols, as seen in present MAC schemes, illustrates that sender and receiver dynamics may be influenced by technological constraints, evidencing the need for adaptability in ensuring effective transmission across varied communications channels.

6.4.2. Encoding and Decoding Messages

Encoding and decoding messages are the essence of effective communication over multiple channels. Encoding maps information into a transmittable form with little distortion to the original message. This initial conversion can involve multiple forms, from simple text encoding to advanced digital signal processing. In digital communication, encoding can be data formatting into binary or the application of complex encryption methods for security reasons. The intricacy of encoding methods, such as fountain coding, facilitates stronger transmission. The adaptability of such methods, highlighted in studies of decentralized communication networks, highlights how good encoding can aid the transmission and integrity of messages within varying environments. Decoding, however, is the important process whereby the decoder reads the encoded message. This process must possess information about the encoding method applied during transmission. Effective decoding procedures must be able to incorporate potential distortions or imperfections inserted along the communication path. For instance, the incorporation of feedback channels can contribute significantly towards the effectiveness of decoding, as established in studies emphasizing the importance of dynamic symbol selection in feedback-based systems (Cassuto et al., 2015). These techniques not only ensure the maximum rate of information recovery but also enable the removal of errors that could be induced by noise or interference, so that the communication remains clear and meaningful. Moreover, the interaction between encoding and decoding cannot be overstressed, particularly where high security and reliability are required. Techniques such as clandestine communication demonstrate the sophisticated methods used to maintain message secrecy without losing their successful transmission through complicated channels. For the meta-analyses of research, covert processes demonstrate that secure message transmission is possible even in environments highly prone to eavesdropping, as demonstrated by recent experimental evidences (Arrazola et al., 2017). This sophisticated balance between encoding correctness and decoding effectiveness accentuates the necessity of merging cutting-edge tactics to maximize overall communication systems efficiency and security. The constant evolution of these methods demonstrates their central position in ongoing advances within the field of communication technologies (Figure 6.10).

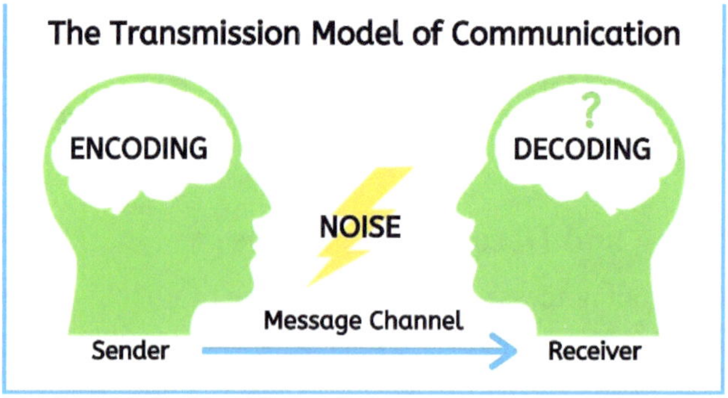

Figure 6.10. *Encoding and decoding messages.*

Source: https://openbooks.macewan.ca/professionalcommunication/wp-content/uploads/sites/21/2022/11/Transmission-Model-of-Communication.jpg.

6.4.3. Feedback Mechanisms

The concept of feedback mechanisms is significant in the optimization of communication effectiveness, particularly in wireless networks where the reliability of transmission is typically not guaranteed. Feedback mechanisms are effective as real-time dynamic tuning to regulate communication protocol according to real-time data to enable systems to learn to respond to different factors such as latency and signal strength. Giving due consideration to secure communication channels is crucial, particularly in situations where elements of the environment make signal transmission unfavorable.

Feedback loops can be utilized to allow systems to adjust parameters such as sampling rates adaptively in an attempt to ensure that data is transmitted effectively to maintain system stability and performance under uncertain environments. This adaptability not only enhances the user's quality of experience but also the overall robustness of wireless communication systems. In wireless control systems, feedback is a crucial means of adapting to the uncertainties introduced by unpredictable communication resources. Cross-layer adaptive feedback scheduling (CLAFS) is an example of one such novel solution for managing these uncertainties. By combining design for control systems with information about current channel states, CLAFS allows real-time adaptation for optimal overall system responsiveness. It does this by dynamically adjusting sampling periods according to parameters like deadline miss ratios and transmission rates, thus optimizing the performance of control.

Adaptive feedback scheduling has been shown, through simulation results, to not only compensate channel capacity variations but significantly enhance communication reliability in wireless local area networks (WLAN) (Andersson et al., 2008). With regard to future advances in communication technology, the integration of feedback mechanisms into 5G system design offers intriguing prospects. Feedback structures will be essential in meeting the increasing complexity of mobile communication environments.

Given the momentum behind more efficient systems, exploring the idea of signal sparsity is beneficial. Sparse signal processing enables communication systems to optimally exploit the inherent properties of data being communicated, facilitating more effective resource management and coding methods. By using such sparsity-based methods, particularly in situations such as massive MIMO and cloud radio access networks, engineers can tackle some of the greatest challenges 5G systems bring. Lastly, these new feedback channels could provide the means for increased efficiency and performance in future wireless communications (Boche et al., 2015).

6.5. THE IMPORTANCE OF CONTEXT IN COMMUNICATION

Effective communication also relies on context, which determines the reception and interpretation of messages. Context not only refers to the immediate environment in which communication occurs but also to the cultural, social, and personal factors determining both the receiver's and sender's perception. The communication pattern of a nonprofit arts foundation, for instance, may be very different based on the prior knowledge, values, and expectations of stakeholders. By taking context into consideration, organizations are more able to frame their messages, and as a result, they are rendered more evocative and clearer.

As shown in Wilks et al. (2016), the extension of the traditional transmission models to provide for multi-sensory channels shows how context does not simply have an ancillary role but is an instrumental variable that impacts directly on the successful decoding of communicated values. This nuanced approach acknowledges the complexity of human communication, eventually allowing organizations to create more meaningful interactions with their publics. Moreover, the advancement in communication technologies, such as fifth-generation (5G) networks, further supports the significance of context towards effective reception of messages. With its physical-layer security and physical-layer service integration (PHY-SI) features, the nuances of the communication channel must be considered in order to effectively transmit information. The context under which a message is sent may influence which specific technologies and how they are applied. For example, a high-transmission-rate, secure environment may necessitate specific integration techniques, as described in Chen et al. (2018), to satisfy the unique demands of different services.

By acknowledging and responding to these context variables, communicators will be able to maximize their effectiveness, making it crucial that the significance of context be underscored when formulating communication plans. Finally, consequences of context extend beyond influencing the immediate interaction between sender and receiver; they may also influence society's outcomes. When businesses emphasize context in their communications, they are not just enhancing their own engagement but also contributing to broader social dynamics. For instance, a successful nonprofit arts organization that can articulate its values will be able to influence cultural attitudes and civic involvement, which will add depth to the social tapestry. As highlighted by Wilks et al. (2016), maintaining shared values can have a positive effect on quality of life, and therefore effective communication on the basis of respect for context has deep rewards. Therefore, it is essential to recognize the multi-faceted nature of context both for effective communication within a single encounter and in supporting the social constructions which are the basis of collective knowledge and activity.

6.5.1. Situational Context

Awareness of the situational context in communication channels goes beyond technical specifications; it involves setting and conditions under which messages are received and transmitted. The context dictates the interactive dynamics between machine and human and the effectiveness of the communication. In legal information dissemination, for instance, machines are increasingly being utilized to communicate complex legal messages through forms that integrate multimodal inputs. As the recent studies have exhibited, not only can machines facilitate the transfer of information to be simpler, but even create institutional facts like human interaction and consequently obscure the line between machine and human agency (Lachmayer et al., 2012). This scenario raises pertinent issues with regard to the impact of machine-mediated communication in institutions and the manner by which these systems learn to deal with diverse situational contexts to perform their desired action. In emergency response, situational awareness plays the central role of facilitating effective members' communication. A good example is their application in incorporating wireless wearable sensor networks into firefighters' equipment, which facilitates real-time situational awareness when conducting operations. The networks use complex algorithms to perceive conditions and transmit data effectively to central command posts, illustrating how situational context drives the process of communication (Knockaert et al., 2014). With the application of this technology, responders can acquire critical information on their environment to facilitate decision-making in threat environments. The dialogical relationship between the physical and communicative spaces of the sensor networks shows how situational context coordinates both the uptake of information and the meaning of the information to favor mission success. In addition, the effectiveness of channels relies on the ability of participants to read and respond to situational cues, therefore further enhancing contextual comprehension. Use of multimodal communication strategies improves richness of interaction, allowing for more detailed transmission of information that reflects the subtleties of the environment. For example, as machines begin performing central tasks typically linked with human tasks, the output must be directed towards enabling varied situational contexts, leading to variant legal consequences and operating efficiencies (Lachmayer et al., 2012). Hence, attending to situational context not only provides richness to the process of transmission but also configures perceptions and conduct among actors. This points towards the sheer importance of situational awareness in human as much as machine-based communication, providing the ground for seamless transmission via multiple channels (Figure 6.11).

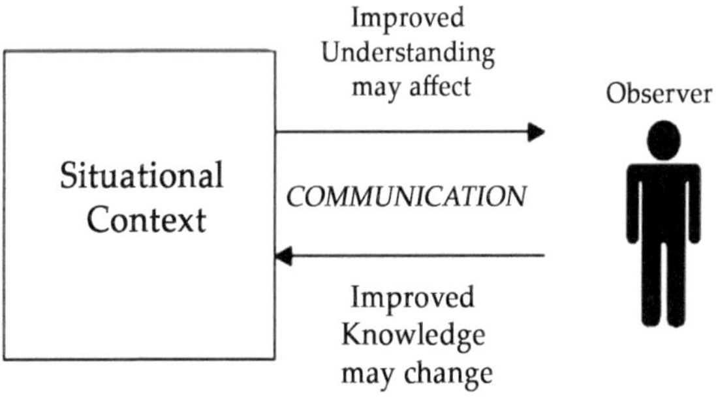

Figure 6.11. *Situational context in communication.*

Source: https://ebrary.net/htm/img/29/434/91.png.

6.5.2. Cultural Context

Cultural context is a critical lens to see communication channels, particularly in terms of thinking about how messages are delivered and received. Looking at nonprofit organizations, for instance, the cultural landscape they operate within is central to deciding on both the messages they transmit and methods they select to deliver. The transmission model of communication is often adapted to the specific cultural context because it must adapt to various perceptions and interpretations of stakeholders. A study of a newly-formed nonprofit arts foundation showed that incorporating multi-sensory channels and valuing the importance of contextual subtleties actually enhanced the effectiveness of communication. From this explanation, it is evident that the cultural context will dictate the effectiveness of value transmission, ensuring that the stakeholders indeed decode and connect with the purpose articulations within the organization (Wilks et al., 2016). Also, the interaction between culture and communication channels is evident in how various types, either traditional or unorthodox, affect the flow of information. For example, in the legal profession, machines serve as intermediaries that provide multimodal communication of rich content, crossing institutional frameworks and human understanding. The ability of machines to communicate content using diverse sensory modalities—visual or audio—to underscore the importance of grasping cultural differences in the manner in which individuals interpret these transmissions. Because machines move beyond the status of tools and become participants active in communication, their function must be examined on the basis of the cultural underpinnings inherent in the exchanges. Comparing machines to quasi-legal actors describes the manner in which cultural narratives and structures are able to recontextualize communication dynamics, thus negating traditional assumptions about interpretation of messages (Lachmayer et al., 2012). Lastly, the cultural background speaks beyond literal meaning, influencing the entire implications of communication in societal structures. Nonprofit organizations, as an instance, have a significant role to play in creating a sense of community by aligning their values with those of the stakeholders that they support. The cultural values by such organizations not only influence their internal communication practices but also resonate with the general external society, hence, giving rise to social cohesion. When organizations communicate their values in an effective manner, they have a ripple effect that enhances the quality of life in society and increases a shared sense of purpose and identity. This complex interaction explains that communication efficiency extends beyond the transmission model, requiring a nuanced appreciation of cultural variables that condition reception and interpretation of messages conveyed. As organizations operate in these cultural contexts, identification of stakeholder perspectives becomes critical to achieving effective communication (Wilks et al., 2016) (Figure 6.12).

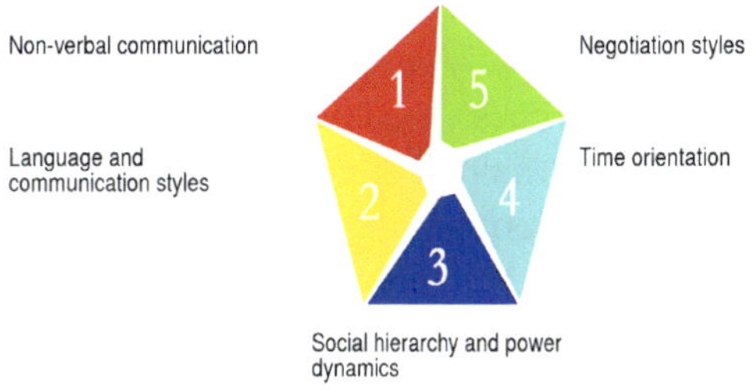

Figure 6.12. *Cultural context in communication.*

Source: https://fastercapital.co/i/Cross-cultural-communication-and-negotiation-Navigating-Cultural-Differences-in-International-Business-Ventures--Understanding-Cultural-Context.webp.

6.5.3. Historical Context

The social and political environment of communication channels creates a context necessary to explain the evolution of transmission modes and what they have represented. The history of communication technologies has been a response to the social-political and technological contexts of its era. From the development of the printing press, on to the emergence of electronic communication, each advance has been an accommodation to society's needs and challenges at the time. For instance, the transition from analog to digital communication not only improved efficiency but also increased accessibility, which impacted connectivity around the world dramatically. Besides, as communication channels evolved, they became increasingly affected by legal and regulatory bodies that sought solutions for the complexities of spreading information and rights implicated, like in the argument for machines as legal agents in the modern age (Lachmayer et al., 2012). When considering the historical context, one also must consider the influences from other disciplines that have propelled the evolution of communication technology. Computer science, engineering, and law teach together an integrated picture of how communication systems work and how legal principles are adapted to technological growth. Analysis of systems like OFDM highlights the intricacies of modern digital communication with their roots in previous research and development. For example, the evolution of MIMO systems illustrates how engineering methods have been crafted to address challenges such as rank-deficient conditions, a general ailment in real-world applications (Hanzo et al., 2007). This integration of the past evolution with existing issues attests to the dynamism of communication technology. Furthermore, the past lays bare the ongoing struggle between control, technological advancements, and popular demands. As machines progressively assume tasks classically performed by humans in the communicative process, the lawfulness of their activities becomes progressively more important. Analogical dependence on past events, such as how machines are perceived in terms of status-functions, makes it possible to gain better insight into how legal principles develop in relation to technology. The idea that machines, like people, can convey legal content between varying modalities shows the paradigm-changing potential of the communication channel. The research into legal machine architectures offers a critical lens to the consideration of the historical record and future development of the systems of communication, emphasizing respect for careful consideration of their implications for society (Lachmayer et al., 2012).

6.6. STRATEGIES FOR IMPROVING COMMUNICATION

Communication techniques with high efficiency are vital to improving information transmission over varied channels, especially in complicated systems such as wireless networks. Cooperative relaying techniques are a significant method, for example, that leverages the overall power of many single-antenna relays. With these techniques, networks can achieve performance improvement using space diversity and spatial multiplexing with or without the limitations typically experienced by traditional multi-antenna systems. This approach not only corrects the inherent flaws of wireless channels but also optimizes the overall data rate and communication reliability, which are crucial in achieving successful transmission. As there is an increasing demand for effective communication systems, the study of novel relaying mechanisms, such as those based on symbol-level modulation, becomes increasingly relevant to solving current wireless communication issues (Quan et al., 2010). In considering security as well as communication effectiveness, information-theoretic security solutions incorporated into it form a robust paradigm for protecting data while in transit. By utilizing the randomness of propagation channels, these approaches strengthen the security of wireless networks while enabling effective communication. The inherent communications graph (iS-graph) is an effective tool to visualize secure links of communication in a network. It becomes feasible to choose channels having the capability to support high levels of security, which is crucial in reducing the hazards in unauthorized access. Sectorized transmission and eavesdropper neutralization usage increases the connectivity and resilience of the iS-graph even further in establishing secure communication in adverse conditions. This three-pronged focus on communication security is essential in today's data-focused world, where protection of information is paramount (Barros et al., 2010). Finally, clear and effective communication demands the ever-adapting to technology and environmental changes. Methods such as the use of advanced algorithms for compressing data and error correction will be instrumental in enhancing communication quality. These methods are employed to minimize loss of information and improve the precision of messages being sent, resulting in a more efficient exchange. Moreover, ongoing evaluation of future technologies, such as artificial intelligence and machine learning, can provide insights into how communication systems can be optimized. By applying such innovations, stakeholders can make potential issues simpler to foretell and make communication channels more flexible. Such a forward-looking approach is crucial in mitigating the threats imposed by evolving communication needs to ensure that transmission systems remain effective, secure, and reliable for future challenges ahead.

6.6.1. Active Listening Techniques

Active listening skills are foundational tools for effective communication via various mediums, developing better understanding and enhancing interpersonal relationships. Active listening is more than just listening to words; it involves processing and interpreting the message of the speaker and providing feedback which facilitates further discussion. Paraphrasing, summarizing, and open-ended questioning are some of the skills not only demonstrating attentiveness but also verification of the feelings of the speaker, which allows a helping atmosphere. In addition, active listening is extremely important in advanced communication environments, such as in describing technical problems related to SDP schemes, where accuracy and comprehensiveness are of utmost importance to ensure effective cooperation (Fomichev et al., 2017). As a result of this reflective practice, not only are participants not passively engaged but they are actively participating in information transmission and receipt, leading to more effective exchanges. In the Internet of Things (IoT) context, active listening skills can significantly avoid misunderstandings arising as a result of complicated technical language.

Figure 6.13. *Active listening techniques.*

Source: https://fastercapital.co/i/Communication-and-Interpersonal-Skills-Mastering-Effective-Communication--Essential-Skills-for-Success--Active-Listening-Techniques.webp.

As there are more networks, the complication in communication increases, and thus a sense of mindful and concise listening behavior among the participants is needed. Active listening skills can help individuals handle complicated topics like medium access control (MAC) protocols more efficiently. This is particularly relevant because technical vocabulary can exclude some members of the audience. By heightening comprehension and encouraging discussion, these techniques can span gaps between different levels of expertise and enable problem-solving in collaboration (Alonso et al., 2016). Therefore, establishing a culture that upholds active listening not only improves one-to-one relationships but also encourages communication effectiveness overall within technical fields. Apart from this, the use of active listening is not restricted to interpersonal communication but extends to organizational communication practices in the current scenario. Organizational training of teams through active listening practices allows organizations to improve total performance, particularly where collaboration is called for to achieve innovation and success. In scenarios where SDP and IoT solutions are developed, active listening promotes a culture of open feedback and continuous improvement. Active listening teams are in a better position to counter deficiencies in the systems ascertained in security schemes or communication pitfalls ascertained in MAC layers. Teamwork driven by active listening promotes collective understanding of objectives and priorities, thus facilitating the transmission of critical information through communication channels. As technology evolves, the integration of active listening into organizational culture makes groups become responsive and adaptive to rise-up issues and opportunities (Figure 6.13).

6.6.2. Clarity and Conciseness in Messaging

Clear and concise communication is fundamental in effective transmission across various channels of communication. The simplicity of messages ensures that ambiguities that could lead to miscommunication are avoided and thus facilitates an effective flow of information. This is highly significant in learning institutions where the ability to pass complex information directly can lead to extremely enhanced understanding among students. In online media environments, where information is likely to be quickly consumed, the short presentation of messages is equally crucial. When teachers utilize clear and concise messages, they provide their students with frameworks through which they can build critical capabilities, which are vital to navigating the nuances of information evaluation and creation. Moreover, the quality of how content is written is crucial in generating interest and enabling the message intended to be delivered to stick with the audience (Lee et al., 2012). Clarity is also important in keeping the audience's attention in today's world that is dominated by speed. With the extensive spread of information, particularly through the digital platform, individuals tend to suffer from information overload. Clarity contributes to this challenge as incredibly complex and verbose communications can create disconnection or shallow understanding. The ability to convey ideas back to their essence—highlighting the most significant points without unnecessary information—ensures messaging is effective. For businesses, in advertising or training messages, employing clear and concise messaging not only makes the message more effective but also respects the time of the audience. As an example, good marketing is built on the rapid capture of the viewer's attention using straightforward and concise language, building ads that effectively present key messages in short, concise ways to build understanding and action (Trung et al., 2025). Further, the establishment of brevity and clarity is dependent on knowledge of context and audience demographics. Diverse audiences contain varying levels of background information and interpretive frameworks whereby a universal message can be ineffective. Understanding these nuances allows the communicator to craft messages in a way that is both familiar and comprehensible. Schools, for example, must use varying tactics when addressing different ages or skill levels. Clearness and concision become key weapons for bridging perception gaps, particularly in multicultural settings where population factors play a strong role in influencing interpretation. When communicators exert deliberate effort to construct messages that are uncomplicated as well as relevant, they ensure transmission is more likely to be effective, with higher levels of participation and greater meaningful interaction in every mode of communication (Lee et al., 2012).

6.6.3. Utilizing Appropriate Channels

The success of the process of information transfer mainly relies on the selection of communication channels. Various channels capture certain properties that influence the character of the communicated message in the context of its clarity and credibility. For instance, for the presence of stationary ergodic sources, channel selection becomes crucial, particularly when transmitting to non-ergodic composite channels. Implementing channel state information at the receiver (CSIR) allows for more knowing choices regarding coding methods in order to maximize performance. As described in current research, reformulating capacity analysis techniques—like outage and average capacity—delivers further understanding of the communication capacity, allowing for improved transmitting outcomes while adjusting for various signs problems (Effros et al., 2007). Furthermore, different methods of switching can significantly enhance the efficiency of communication channels by accommodating different kinds of services. As integrated services—voice, data, and video—are increasingly in demand, network design for integrated-service networks has emerged as a more prominent field. By implementing a single process with a single approach to switching throughout the bandwidth of the network, the systems offer cost-effective, application-independent service provision. This versatility is evident in methods such as burst switching and packet switching, both of which provide frameworks for improved management of bandwidth allocation (Bradley et al., 1989). These innovative strategies ensure that a channel's theoretical peak is assessed and harnessed, eventually enhancing the speed and reliability of data transfer. In addition to assigning utmost importance to channel choice and switching mechanisms, effective utilization of communication channels entails a clear understanding of the bottom-level performance measures. Employment of end-to-end distortion measures in combination with classical capacity definitions ensures system designers obtain optimal source and channel codes. This is founded on the interaction between source characteristics and channel conditions, with solutions optimized in accordance with specific communication requirements. In addition, making the smooth exchange of information between the source and the channel feasible through a proper interface is crucial for the sustenance of transmission efficiency. As indicated in recent research, this degree of synchronization helps in the separation of encoding and decoding operations, therefore, helping in the development of a precise and stable communications system (Effros et al., 2007).

SUMMARY

The movement of data through several channels of communication is a complicated phenomenon based on the effectiveness and reliability of a variety of factors. Through analysis of different paradigms of communication, it can be understood that the underlying technological framework, such as converters and control systems, plays a key role in enabling efficient data transmission. For instance, the design of a one-phase DC-AC converter for wind turbines illustrates how mathematical modeling and control techniques that stabilize output voltage are key to secure communication. Through integration of such systems into communication networks, it is possible effectively to enhance general performance and reliability. Thus, the blending of technology and communication design is an essential theme in maximizing transmission outcomes (Buswig et al., 2011). In addition, communication performance measurement involves important metrics such as ergodic capacity and mean bit error rates. Such measurements are used as a benchmark to assess system effectiveness and efficiency. There has recently been a suggestion of new correlations that link these measures of performance in wireless communication systems for which a meaningful comparison of their effectiveness within various environments can be made. The discovery that it is conceivable to interrelate these measures without direct complete statistical description of the communication channel is nothing but a breakthrough in optimization of communication techniques. Appreciation of these interrelations goes a long way in confirming the basic principle that communication systems must be designed precisely in order to possess high throughput and low error rates in practical applications (Alouini et al., 2011). Ultimately, the controversy regarding transmission through communication channels brings to light the importance of constant innovation and research in this field. The integration of advanced control schemes and performance metrics broadens the realm of how data integrity and reliability factors may be interpreted. The changing technology necessitates adaptable systems that can contend with varying conditions. Future endeavors need to focus on building these communication infrastructures still further, so they remain viable in the face of new challenges. In developing these advances, the broad goal remains simple: to optimize communication effectiveness on different platforms, and thereby help form a more integrated and efficient world society. The results of this research thus not only reflect current trends but also provide the foundation for further advances in the technology of communication.

In addressing the transmission of the communication channel, the significant position of legal documents like the International Health Regulations (IHR) (2005) must be highlighted. This is a binding international agreement for World Health Organization member states and therefore a significant instrument in the management of public health emergencies. Findings from studies conducted at various points of entry (POEs), such as the Julius Nyerere International Airport (JNIA), capture the challenges that have been encountered in implementing these regulations. While there are avenues of communication that offer effective risk communication, disjointed efforts in separate government departments complicate enforcement and coordination. Key identified barriers are the lack of established POEs for IHR (2005) implementation and ineffective infrastructure for emergency readiness, thus highlighting the need for policy harmonization and the deployment of resources to stimulate improved operational effectiveness at international travel hubs (Bakari et al., 2013). Transmission is also through another aspect of viral marketing, namely that communication strategies shift incredibly rapidly due to the advance of technology. It is crucial to learn how marketing practices change in order to fit the online space in a bid to grasp contemporary communication channels. The emergence of the Internet has shifted the paradigms of traditional marketing towards more interactive and effective methods, including viral marketing. Scholars have discovered through successful campaigns that there are factors such as campaign structure, product features, information in the message, and the peer-to-peer information channel as critical drivers of engagement and sharing. These findings not only enhance consumer conduct intelligence but also show the influence of how the communication channels of the digital world can bolster messages in an effective manner, thereby transforming market landscapes and expectations (Bourlakis et al., 2008). Lastly, summarizing the main points on the communication channels shows the fusion of health legislation and marketing tactics within an ever-evolving digital world. IHR (2005) field findings on issues in implementation underscore the importance of coordinated resource and strategy to ensure protection of public health, particularly at travel points in the global range. Concurrently, evolving viral marketing dynamics mark the revolutionary impact of Internet technologies on promotional strategies. Convergence of these topics supports a broad understanding of how communication channels serve as conduits of information flow that shape the attitudes and behaviors of the public in health and marketing disciplines. Engagement with such important perceptions can be supplementary to emerging debate on improving communication approaches across various disciplines (Bakari et al., 2013; Bourlakis et al., 2008).

REFERENCES

1. Ab Rahman, A., Basir, S. A., & Ibrahim, P. (2010). *Sustainable Economic Development: Concept, Principles and Management from Islamic Perspective.* https://core.ac.uk/download/19451582.pdf (accessed on 18th September 2025).
2. Abbas, R., Ahmad, I., Ali, S., Amuru, S., et al. (2020). 6G white paper on machine learning in wireless communication networks. http://arxiv.org/abs/2004.13875 (accessed on 18th September 2025).
3. Adelantado, F., Martinez, B., Melia, J., Tuset-Peiro, P., et al. (2017). Understanding the limits of LoRaWAN. http://arxiv.org/abs/1607.08011 (accessed on 18th September 2025).
4. Agrell, E., Alvarado, A., Bayvel, P., Lavery, D., et al. (2015). Replacing the soft FEC limit paradigm in the design of optical communication systems. https://core.ac.uk/download/79504498.pdf (accessed on 18th September 2025).
5. Alhakami, W., Mansour, A., & Safdar, G. A. (2014). Spectrum sharing security and attacks in CRNs: A review. https://core.ac.uk/download/29822044.pdf (accessed on 18th September 2025).
6. Alonso Zárate, J., Alonso Zárate, L. G., Kalalas, C., Laya, A., et al. (2016). Goodbye, ALOHA! https://core.ac.uk/download/81571287.pdf (accessed on 18th September 2025).
7. Alouini, M. S., & Yilmaz, F. (2011). Novel relations between the ergodic capacity and the average bit error rate. http://arxiv.org/abs/1201.1278 (accessed on 18th September 2025).
8. Andersson, M., Arzen, K. E., Chan, C. P., Colandairaj, S., Xia, F., et al. (2008). Cross-layer adaptive feedback scheduling of wireless control systems. https://core.ac.uk/download/18966374.pdf (accessed on 18th September 2025).
9. Andrew, L. L. H., Nazarathy, Y., Selen, J., Vu, H., et al. (2013). The age of information in gossip networks. https://core.ac.uk/download/577024468.pdf (accessed on 18th September 2025).
10. Andrews, J. G., Buzzi, S., Choi, W., Hanly, S. V., Lozano, A., Soong, A. C. K., & Zhang, J. C. (2014). What will 5G be? *IEEE Journal on Selected Areas in Communications, 32*, 1065–1082. https://doi.org/10.1109/jsac.2014.2328098 (accessed on 18th September 2025).
11. Apteker, C., Clark, F., Ford, G., Garrand, M., Ghinea, G., et al. (2006). A cognitive approach to user perception of multimedia quality: An empirical investigation. https://core.ac.uk/download/333158.pdf (accessed on 18th September 2025).

12. Arnau, J., & Kountouris, M. (2017). Delay performance of MISO wireless communications. http://arxiv.org/abs/1707.08089 (accessed on 18th September 2025).
13. Arrazola, J. M., Li, H., Liu, W. Z., Liu, Y., et al. (2017). Experimental unconditionally secure covert communication in dense wavelength-division multiplexing networks. http://arxiv.org/abs/1709.06755 (accessed on 18th September 2025).
14. Autant-Bernard, C., Manca, F., Suriñach, J., Massard, N., & Moreno, R. (2025). *The Diffusion/Adoption of Innovation in The Internal Market.* https://core.ac.uk/download/pdf/6970423.pdf (accessed on 18th September 2025).
15. Bakari, E., & Frumence, G. (2013). Challenges to the implementation of International Health Regulations (2005) on preventing infectious diseases: Experience from Julius Nyerere International Airport, Tanzania. https://core.ac.uk/download/pdf/19331445.pdf (accessed on 18th September 2025).
16. Barros, J., Pinto, P. C., & Win, M. Z. (2010). Techniques for enhanced physical-layer security. http://arxiv.org/abs/1008.3705 (accessed on 18th September 2025).
17. Boche, H., Jung, P., Strohmer, T., Wunder, G., et al. (2015). Sparse signal processing concepts for efficient 5G system design. http://arxiv.org/abs/1411.0435 (accessed on 18th September 2025).
18. Bondarenko, G. I., Korobenkov, P. M., Bondarenko, G. I., Bondarenko, P. M., et al. (2014). *Optimization of Corporate Network Communication Channel Capacity.* https://core.ac.uk/download/47234527.pdf (accessed on 18th September 2025).
19. Bourlakis, M., Li, F., Papagiannidis, S., Woerndl, M., et al. (2008). Internet-induced marketing techniques: Critical factors in viral marketing campaigns. https://core.ac.uk/download/29140551.pdf (accessed on 18th September 2025).
20. Bradley, T. T., & Suda, T. (1989). Survey of unified approaches to integrated-service networks. https://core.ac.uk/download/287623360.pdf (accessed on 18th September 2025).
21. Browne, T. D., Gilmore, J. S., & Heller, T. S. (2025). *Technology Transfer–A Selected Bibliography.* https://core.ac.uk/download/pdf/85235519.pdf (accessed on 18th September 2025).
22. Cassuto, Y., Hashemi, M., & Trachtenberg, A. (2015). Fountain codes with nonuniform selection distributions through feedback. http://arxiv.org/abs/1504.01802 (accessed on 18th September 2025).

23. Caviglione, L., Mazurczyk, W., Meier, M., Wendzel, S., et al. (2014). Hidden and uncontrolled–On the emergence of network steganographic threats. http://arxiv.org/abs/1407.2029 (accessed on 18th September 2025).
24. Chen, Z., Fang, J., Li, S., Mei, Y., et al. (2018). Physical layer service integration in 5G: Potentials and challenges. http://arxiv.org/abs/1707.01064 (accessed on 18th September 2025).
25. Coble-Neal, G., & Madden, G. G. (2025). *Internet Economics and Policy: An Australian Perspective.* https://core.ac.uk/download/pdf/7301255.pdf (accessed on 18th September 2025).
26. Dal Falco, F., & Vassos, S. (2017). Museum experience design: A modern storytelling methodology. https://core.ac.uk/download/153363274.pdf (accessed on 18th September 2025).
27. Didkowsky, R. M. (2013). Information transmission methods with through the use of stochastic signals. https://core.ac.uk/download/47228356.pdf (accessed on 18th September 2025).
28. Donohue, L. K. (2014). FISA reform. https://core.ac.uk/download/70375221.pdf (accessed on 18th September 2025).
29. Duong, Q. T., Nguyen, L. M. H., & Tran, T. M. T. (2025). Persuasive techniques in out-of-home advertising: A comparative study. https://core.ac.uk/download/666448064.pdf (accessed on 18th September 2025).
30. Effros, M., Goldsmith, A., & Liang, Y. (2007). Distortion metrics of composite channels with receiver side information. https://core.ac.uk/download/4879449.pdf (accessed on 18th September 2025).
31. Fomichev, M., Gardner-Stephen, P., Hollick, M., Steinmetzer, J., et al. (2017). Survey and systematization of secure device pairing. http://arxiv.org/abs/1709.02690 (accessed on 18th September 2025).
32. Hanzo, L., & Jiang, M. (2007). Multiuser MIMO-OFDM for next-generation wireless systems.
33. Imrie, J., Richens, J., & Weiss, H. (2003). HIV risk: Is it possible to dissuade people from having unsafe sex? https://core.ac.uk/download/1669845.pdf (accessed on 18th September 2025).
34. Karumanchi, S., Mancini, S., Winter, A., Yang, D., et al. (2016). Classical capacities of quantum channels with environment assistance. https://core.ac.uk/download/pdf/80187068.pdf (accessed on 18th September 2025).
35. Kim, J., Kim, S., & Lee, J. (2011). Adaptive duty cycling MAC protocols using closed-loop control for wireless sensor networks. https://core.ac.uk/download/323384018.pdf (accessed on 18th September 2025).

36. Knockaert, J., Rogier, H., Stevens, C., Van Torre, P., et al. (2014). Synchronous wearable wireless body sensor network composed of autonomous textile nodes. https://core.ac.uk/download/55846708.pdf (accessed on 18th September 2025).

37. Lachmayer, F., & Čyras, V. (2012). Multisensory legal machines and legal act production. https://core.ac.uk/download/14524802.pdf (accessed on 18th September 2025).

38. Lee, A., Malik, M., Cortesi, S., & Gasser, U. (2012). *Youth and Digital Media: From Credibility to Information Quality.* https://core.ac.uk/download/71362031.pdf (accessed on 18th September 2025).

39. Mao, Y., You, C., Zhang, J., Huang, K., & Letaief, K. B. (2017). A survey on mobile edge computing: The communication perspective. *IEEE Communications Surveys & Tutorials, 19*, 2322–2358. https://doi.org/10.1109/comst.2017.2745201 (accessed on 18th September 2025).

40. Missingham, R. (2016). Understanding information needs of Australian business organizations. https://core.ac.uk/download/156698448.pdf (accessed on 18th September 2025).

41. Mukherjee, A., Deb Gupta, P., & Ahuja, P. (2025). Indo-U.S. FTA: Prospects for audiovisual services. https://core.ac.uk/download/pdf/6604773.pdf (accessed on 18th September 2025).

42. Quan, Y. (2010). Cooperative symbol-based signaling for networks with multiple relays. https://core.ac.uk/download/70419449.pdf (accessed on 18th September 2025).

43. Serrat, O. (2012). *Communications for Development Outcomes.* https://core.ac.uk/download/16269461.pdf (accessed on 18th September 2025).

44. Slavkovsky, A. (2013). *Information in the Context of Philosophy and Cognitive Sciences.* https://core.ac.uk/download/159074078.pdf (accessed on 18th September 2025).

45. Wilks, L. (2016). Communicating an arts foundation's values: Sights, sounds and social media. https://core.ac.uk/download/82982537.pdf (accessed on 18th September 2025).

46. Yonis Buswig, Y. M. (2011). Development of a DC-AC power conditioner for wind generator by using neural network. https://core.ac.uk/download/12007862.pdf (accessed on 18th September 2025).

CHAPTER 7

ERROR CONTROL CODING

Contents

INTRODUCTION

The field of error control coding is essential in contemporary communication systems, as it offers the reliability and integrity of data transfer over various channels. As digital communications are on the rise, strong techniques that can detect and correct errors based on noise, interference, or other distortions are essential. Error control coding encompasses a range of techniques for enabling the sender and the receiver to detect and correct errors likely to compromise information received. Application of forward error correction (FEC) and automatic repeat request (ARQ) techniques enables communication systems to achieve the best possible balance between error tolerance and throughput and consequently enhance overall performance. Understanding these methods underlying principles sheds light on their critical applications across numerous industries, ranging from telecommunications to data storage, basically defining how information is received and stored today.

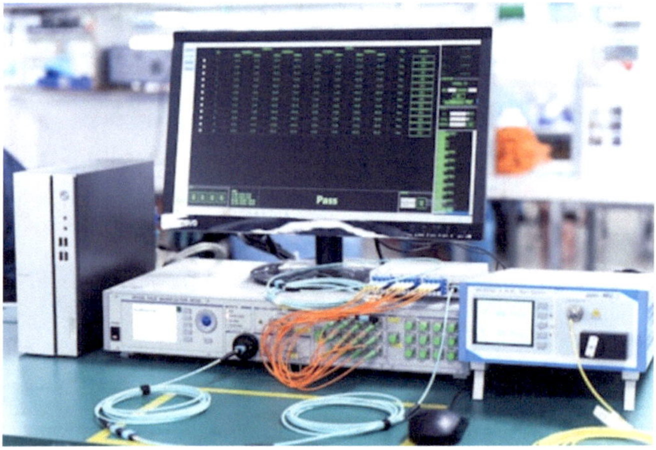

Figure 7.1. *Error control coding.*

Source: https://images.unsplash.com/photo-1682561477064-44fb2dfd82bf?crop=entropy&cs=tinysrgb&fit=max&fm=jpg&ixid=M3w3MDgyMDR8MHwxfHNlYXJjaHwxfHxlcnJvciUyMGNvbnRyb2wlMjBjb2RpbmclMkMlMjBzaWduYWwlMjBwcm9jZXNzaW5nJTJDJTIwZGF0YSUyMHRyYW5zbWlzc2lvbiUyQyUyMGRldGVjdGlvbiUyMGFuZCUyMGNvcnJlY3Rpb258ZW58MHwwfHx8MTc0OTkxOTM3OXww&ixlib=rb-4.1.0&q=80&w=500&?utm_source=numberanalytics&utm_medium=referral.

Mainly built into the core of error control coding is the principle of redundancy, employed for ensuring original data can be reconstructed even in the case of errors. Redundant bits, introduced into messages through coding schemes, serve as checks to allow receivers to identify mismatches when the data are sent. Various types of codes, like Hamming codes and Reed-Solomon codes, are employed based on the specific requirements of the communication channel. These codes provide not only a safety net to data transmission but also enhance the quality of service by bandwidth and scheduling changes, which is critical in dynamic environments like wireless networks. This versatility is warranted, particularly as it ensures that communication systems can adapt their error control schemes based on changing conditions, such as channel quality and traffic intensity, further substantiating the role of error control coding in achieving maximum system performance (Havinga et al., 2000). In addition to error detection and correction, error control coding frameworks also intersect with information processing mechanism and prediction explanation theories. For instance, how interoceptive information and prediction errors interact with each other can be likened to the way an error correction behaves in a communication system. Similarly, predictive models steer cognitive response to stimulus and emotional state, error control coding employs predictive algorithms in an effort to enhance the dependability of data transmission. Integration of these models, both cognitive neuroscience and communication theory, provides a comprehensive framework for describing how error control coding works. This multi-disciplinary perspective is essential to the further advancement of plans that not only focus on technical efficacy but also consider the wider implications of information processing in artificial and natural systems, thereby offering added insights to the nature of complexity of communication and data integrity (Anil et al., 2012) (Figure 7.1).

Definition of Error Control Coding

Error control coding is very vital in modern communication systems via mechanisms for detecting and correcting errors that can be experienced during data transmission or storage. Error control coding is simply the procedure of adding redundancy to the data so that the receiving system can detect and correct errors without the need for retransmission. This redundancy is designed based on specific coding techniques, classified into two broad categories: block codes and convolutional codes. While block codes handle fixed-length blocks of data and insert redundancy while encoding, convolutional codes process data streams in a stream form and utilize convolution to supply error correction along the transmission path. Actually, error control coding ensures the integrity of data, ensuring fidelity of communication over various channels and compensating for the impact of noise—and, therefore, is essential in the provision of reliable data exchange within digital networks. Error control coding theorems underlying them can be critiqued based on theorems such as the reliability function and the Shannon capacity of communication channels. The error exponent of reliability quantifies the biggest error exponent achievable under certain bandwidth and noise constraints, which is essentially plotting the limits of error correction capabilities in practical situations. As an example, one of the relations obtained in communications theory indicates the optimal error exponent to be $\max\{0, B(1-R(D)/C)\}$ with R(D) being the rate-distortion function for the source and C being the channels Shannon capacity. These theoretical frameworks contribute to the knowledge base of when and how to utilize particular error control coding methods, providing information on maximizing data transmission efficiency in certain situations and emphasizing proper selection of encoding methods in a given application in digital communications (Tan et al., 2019). Furthermore, error control coding has experienced new technologies used in applications beyond traditional communications, such as signal processing and machine learning. For instance, low-rank approximations, which form the core of many signal processing tasks, can be complemented by progress in error-correcting codes, where their structural aspects are utilized in the pursuit of enhanced computational efficiency. Recent work has explored how code matrices might be applied to achieve optimal approximations and solution of high-complexity problems, such as least squares regression, with reductions in randomness orders of magnitude and improved performance in upholding geometric factors of vector subspaces. These findings emphasize the variety of error control coding outside data integrity proper, showing its intrinsic role among the broader spectrum of computational methods and its ability to address modern challenges in technology and data analysis (Mazumdar et al., 2017).

Importance in Communication Systems

In communication networks, the role of error control coding cannot be overstated, particularly due to the implication of transmission errors. The errors can be brought about by a number of factors including noise, interference, and signal degradation, particularly in wireless networks. Detection and correction of the errors are important in data integrity and ensuring reliable communication. With the ongoing advancement of wireless technology, especially with the advent of mobile computing systems, the focus on energy-efficient error control mechanisms has become increasingly critical. The analysis of different techniques of error correction illustrates that balancing the energy consumption without compromising communication reliability is a fine art requiring sophisticated techniques of coding design and implementation. This emphasizes that merely resolving technical problems is not sufficient; one must incorporate the energy requirements enabled by the wireless interface into the overall plan of efficiency (Havinga et al., 1999). The creation of Ultra-Reliable Communication (URC) within modern systems such as 5G particularly emphasizes the importance of error control coding. URC aims to provide communication services at almost error-free levels of reliability, a critical requirement for safety-critical domains like industrial automation and vehicle coordination. Good understanding of concepts of error control coding becomes necessary as URC demands new methods of encoding transmitted information as well as metadata for control. By refreshing traditional encoding methods, communication protocols can achieve required levels of reliability by URC contexts. This change calls not only for innovation in coding techniques but also calls for a thorough analysis of reliability impairments on communication. Thorough investigation is required for appropriately solving the engineering challenges brought about by URC in new communication technologies (Popovski et al., 2014). The impact of effective error control coding extends beyond the realm of technology to affect other aspects of society, economy, and technological advancement. Reliable communications systems allow for trust and seamless running of various applications from telemedicine to vehicle networks. With the growing reliance on wireless communication, there is an escalating need for robust error correction schemes that can differentiate between errors acceptable and errors catastrophic. Understanding the importance of error control coding within a communication system allows engineers and researchers to design frameworks that are not only resilient but also dynamic to unfolding issues. The integration of strategies that are specialized based on specific reliability needs will eventually yield more effective communication infrastructures, which will further improve the overall reliability and competence of international communication systems regardless of the increasing demands and complexities (Figure 7.2).

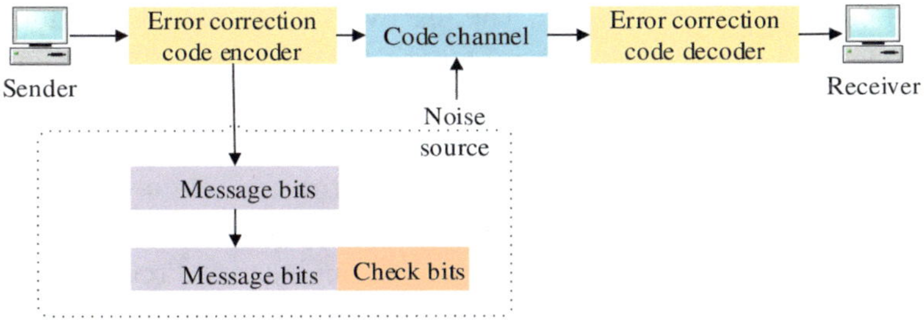

Figure 7.2. *Error control coding in communication systems.*

Source: https://encrypted-tbn0.gstatic.com/images?q=tbn:ANd9GcRRKdMHtO5o i5Qz5ZRqkXfJjBwIojF1yNLO5w&s.

7.1. HISTORICAL BACKGROUND

Error control coding history illustrates its rich development in parallel with the evolution of communication systems. The early 20th century saw the first emergence of error detection through simple mechanisms, whose primary purpose was to ensure data transmission via telephone lines to be secure. However, it was only in the mid-1940s that more sophisticated techniques, such as Hamming codes, were found. They not only identified errors but also provided ways of correcting them, which formed a paradigm shift in coding theory. Researchers began working on the mathematical underpinnings of coding, and the key principles that would become the building blocks of modern-day error control coding methods were established. It laid the groundwork for how it was possible to counter errors within data communications, which became increasingly important with the advent of digital data communication technologies. Capitalizing on early research, great error control coding achievements were developed during the 1950s and the 1960s, driven by the expanding need for robustness against errors in military and industrial applications. Scholars like Claude Shannon capitalized on information theory's theoretical base, proving it was possible to communicate over noisy channels with high reliability through sufficient redundancy. This period witnessed the development of convolutional codes and block codes, away from previous techniques by means of better data transmission. Researchers dabbled in trade-offs between performance and redundancy, resulting in advanced coding schemes to achieve maximum data integrity. Later, error control coding became an integral component of various systems, influencing the design of satellite communications to computer networks. The advent of modern error control methods in the latter half of the 20th century transformed the subject even more, particularly with the advent of digital communication systems. The discovery of advanced algorithms like Turbo codes and Low-Density Parity-Check (LDPC) codes was a significant milestone, as they attained very close to optimum performance under practical conditions. These were preceded by the increasing requirement for quick data transfer and the requirement for good error correction in applications such as cellular communications and data storage. In addition, the feedback systems used previously, such as the Software Engineering Process Simulation (SEPS), were important in enhancing project outcomes through the capability of providing good judgments and sound decisions in software development. Ongoing development of error control coding illustrates sustained interest in addressing the challenge of communication in an increasingly digital age, underlining its indispensable role in contemporary technology.

7.1.1. Early Developments in Coding Theory

The roots of coding theory stretch back to the mid-20th century when researchers sought to make information transmitted through noisy channels more reliable. The first efforts were directed toward the mathematical foundations of error detection and error correction, required for combating the effects of noise on transmitted data. Pioneers such as Claude Shannon paved the way with channel capacity as a theory, in a sense uncovering how the math of encoding can influence fidelity of the data during transmission. His work provided a theoretical base from which later advances in coding schemes and mechanisms for maintaining data integrity could expand. The union of combinatorics and algebraic structure became pervasive as scientists began experimenting with different encoding methods, driving the evolution of error control coding to meet the increasing demands of telecommunications and data storage industries. Building on the work of Shannon's seminal work, Ralph Hartley and others further expanded information theory concepts, further demystifying the role of redundancy in coding systems. Early coding theory breakthroughs involved basic parity checks and checksums, which were early error-detecting techniques. These basic methods signaled the detection of errors but did not have systematic ways for correcting errors.

Figure 7.3. *Code theory.*

Source: https://external-content.duckduckgo.com/iu/?u=https%3A%2F%2Fthumbs.dreamstime.com%2Fb%2Fcoding-theory-word-cloud-concept-grey-background-88647902.jpg&f=1&nofb=1&ipt=6e1985375b67f4fe02dddda36448363960bc5fed2cb653634ef49ff6d131fd40.

As technology progressed, the coding techniques grew more complex, resulting in the creation of block codes and convolutional codes. Block codes split data into fixed-sized blocks to allow for more intricate patterns when encoding, and convolutional codes used shift registers to generate a stream of data. This was the time that witnessed the crucial transition towards more powerful error correcting measures, leading to the creation of feasible applications in digital communication systems. The engineering significance of these early advances only increased with the advent of digital computers and the need for reliable data transmission. The invention of error correction codes transformed communications technology into a completely new field, allowing lost or corrupted data to be recovered in real-time applications. For instance, techniques such as Reed-Solomon coding and Hamming codes became essential to improving data integrity, particularly in environments where signal loss was prevalent. Moreover, advances in coding theory paved the way for advancements in data storage technologies, making error-free read and write processes possible in various media. Accordingly, early breakthroughs in coding theory not only enhanced the precision of data transmission but also facilitated a broad variety of technological innovations that would follow later in telecommunications, data processing, and more complex coding systems required in today's digital age (Anil et al., 2012; Ingels et al., 2025) (Figure 7.3).

7.1.2. Key Contributors and Their Contributions

In the area of error control coding, several significant figures who have advanced far are there who have shaped its theoretical foundations and applications in reality. Claude Shannon, also referred to as the father of information theory, opened the door to modern day coding practice with his research work on the Shannon-Hartley theorem. It quantifies the carrying capacity of a communication channel and serves as a benchmark to gauge error correction codes. Shannon also conceptualized source and channel coding, which is critical to coding data effectively with little interference by noise. His work revolutionized telecommunications and data transmission in its very nature, a watershed moment to influence a tremendous amount of subsequent innovation in error control coding, setting high expectations for the error-resilience of digital communications. Another prominent figure present here is Richard Hamming, whose contributions led to the development of Hamming codes. Hamming codes are of the type that can detect and correct single-bit errors while data is being transferred. Hamming's contribution was out of his desire to build more trustworthy communication systems, and he developed the concept of error detection and correction in one code framework. This two-way ability enabled systems to not just report errors but also correct them in real time, thereby enhancing the integrity of overall data transmission. Hamming's work is famed as a seminal innovation in error control technology development and continues to be used in numerous varied applications in digital communications. Additionally, Éric T. D. R. Turan's work should not be forgotten while talking about the history of error control coding's. Turan was an expert in designing low-density parity-check (LDPC) codes, which were a major achievement to achieve near-capacity performance on communication channels. The existence of these codes was a big step forward since they are efficient error correctors by utilizing sparse matrices, thus making error correction in high-speed data transmission simpler. Furthermore, the resurfacing of LDPC codes in the 21st

century, particularly through their becoming part of standards like Wi-Fi and digital TV, shows the relevance of such codes in contemporary communication systems. Turans work is one instance of the power of theoretical outcomes to give strong, useful applications that maintain data integrity within different networks.

7.1.3. Evolution of Error Control Techniques

The history of error control techniques dates back through a series of techniques that have been evolved in accordance with the growing requirement for error-free communication systems. Parity bits were the basic coding techniques used initially, which had very little error detection. However, as more and more complex systems were developed, more sophisticated techniques were investigated, and block and convolutional codes evolved. These methods allowed error correction of many errors and not only detection, significantly increasing the precision of data communications over noisy channels. As communication technology evolved, particularly digital transmission, and therefore the necessity for more robust techniques, there followed the emergence of modern methods such as Turbo codes and Low-Density Parity-Check (LDPC) codes that employ intricate mathematical structures to optimize performance while maintaining computations minimal. The unification of error control coding and information theory has motivated the invention of these techniques, particularly towards the solution of real-world problems such as multi-user systems and non-binary alphabets. One such recent example is proving polar codes to be optimal under various constraints, such as encoding and decoding complexities over numerous users. This new approach not only helps in achieving error-free data transmission but also maintains a polynomial relationship with the block size and implementation complexity, thus expanding coding's range of applications across different scenarios (Abbe et al., 2014). The generality of these codes to different communication environments represents a fundamental improvement in error control techniques, which is robust and flexible as the demand for efficient coding increases. Parallel to the progress in coding schemes, research into hardware solutions has also made great strides, particularly in settings where components are not trustworthy. Research in this area has suggested the ENCODED scheme, which functions successfully in coping with computational errors caused by faulty hardware by keeping error probabilities always low. This methodology is particularly relevant in probabilistic and permanent fault contexts, illustrating how constraints enforced by physics impact coding efficiency (Grover et al., 2017). Secondly, investigating dynamic voltage scaling and error control identifies a paradigm shift toward energy-efficient computation without degradation in performance. These advances serve further to underscore the interplay between coding theory and hardware, and to affirm a sustained emphasis upon enhancing reliability in the context of technological advancement, further entrenching the applicability of error control techniques to modern telecommunications (Figure 7.4).

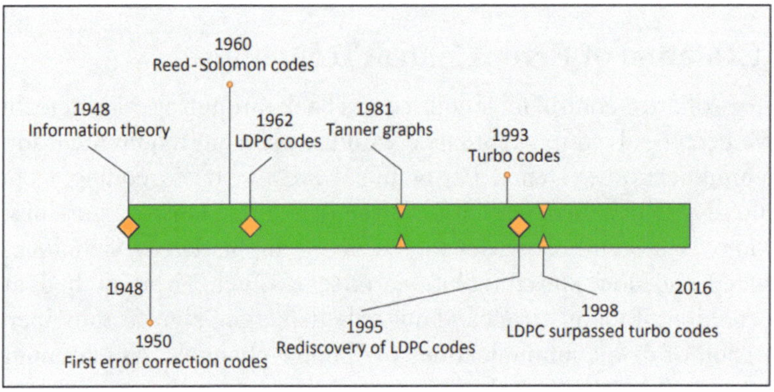

Figure 7.4. *Evolution of error control techniques.*

Source: https://www.researchgate.net/publication/307878339/figure/fig1/AS:403681959792662@1473256687759/Historical-milestones-in-error-correction-coding-theory.png.

7.2. TYPES OF ERROR CONTROL CODES

The topography of error control coding is diverse, consisting of several fundamental types of codes that are tailored to address specific issues in data transmission. Block codes are one of the primary groups; these codes function by dividing data into fixed-length blocks and adding redundant bits to each block to recover from errors. Hamming codes are some of the most well-known examples. These correct single-bit errors and detect two-bit errors by clever positioning of parity bits. A clear understanding of the underlying principles of block codes is important since their applications can result in dramatic enhancement of data transmission reliability in most communication media. Because the communication channels have the potential to bring about errors more than once, a proper block coding scheme must be put into practice to minimize such threats and ensure errorless recovery of data (Kirse et al., 2014). Convolutional codes are another critical class among error control codes that work differently from the block codes by processing data in a continuous stream rather than blocks. These codes utilize a sliding window technique to encode data, and hence are best fit for applications of real-time processing like cellular phones. Convolutional codes work on the principle of producing encoded bits using shift registers, and additional improvement can be done through techniques like soft-decision decoding. This allows adaptability to increase the capability of error correction, especially in signal degradation scenarios. Secondly, there have been numerous publicized examples of convolutional coding such as turbo codes and low-density parity-check codes that have proved to be excellent in theory as well as practice, demonstrating the versatility and use of convolutional coding methods in error correction (Goodman et al., 1989). Apart from block and convolutional codes, there has also been expert coding technique research to manage error correction in multilevel storage media such as Random Access Memories (RAM). Multi-bit RAMs, storing multiple bits per cell, introduce complexities due to their greater chances to be vulnerable to errors from various sources, such as alpha particle radiation and read ambiguities. In-chip error correction coding (ECC) becomes imperative in these scenarios, as it needs to effectively handle soft errors, hard errors, and read errors. Coding methods under such a situation may vary from bitwise methods, for example, parity checks to more advanced cellwise methods, for example, Reed-Solomon codes. Choosing a coding method depends on the environment of errors and RAM's functional performance requirements, with the implications of setting error control schemes to meet the complexity of data storage technology (Goodman et al., 1989) (Figure 7.5).

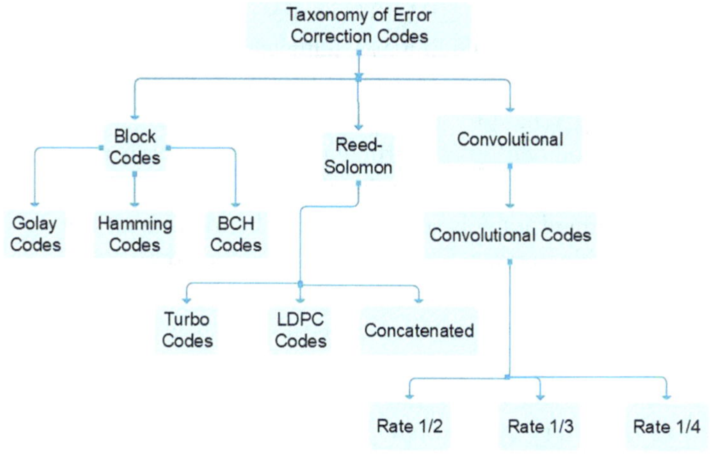

Figure 7.5. *Types of error control codes.*

Source: https://www.researchgate.net/publication/375503833/figure/fig4/AS:11431281203936131@1699533396055/Taxonomy-of-Error-Correction-Codes.jpg.

7.2.1. Block Codes

Applications of block codes are critical in error control coding precisely because of the systematic nature of their data encoding. Block codes operate by partitioning the input data into fixed-size blocks, and then adding redundancy through codeword formation. This added redundancy has a critical function: error detection and correction that can be introduced during data transmission. Block code performance can be assessed using their error detecting and correcting capabilities, which take firm cues from parameters such as block length and redundancy bit count. Inadequate redundancy could make block codes unable to correct errors, highlighting the need for careful design of coding systems to supply the necessary robustness for varying communication channels. A distinguishing characteristic of block codes is their compatibility with modulation diversity in wireless communication systems. Recent work has showcased how algebraic rotations can be used to construct block codes that achieve modulation diversity, especially in environments where conditions are poor such as fading channels. Employing low-complexity decoding strategies in conjunction with single algebraic rotations results in codewords whose projections are equidistant and, therefore, yield optimum performance under both unquantized and quantized receiver conditions. This design versatility not only enhances robustness against errors but also simplicity of decoding processes, which makes block codes viable for use in a wide range of applications across modern-day communications systems (Chockalingam et al., 2011). This reflects the ongoing advancement in methodology to error control coding as scientists attempt to improve block code techniques. Moreover, problems that fixed-length block codes face in terms of achieving increasing reliability through feedback mechanisms have caused a review of coding methodologies. Particularly, research indicates that feedback cannot accelerate the rate of error probability decay for fixed-length coding in the majority of channel models, propelling the boundaries of traditional block codes to more flexible options. By moving to variable-length coding and adding error-erasure decoding, higher reliability can be achieved with lower complexity. Specifically, the interplay between block code and feedback techniques highlights the importance of adaptive coding schemes in order to manage error rates effectively while keeping the overall system performance at acceptable levels (Nakiboğlu et al., 2011). Accordingly, research into these adaptive techniques is essential in order to maximize the actual-world usage of error control coding technologies in contemporary digital communications.

7.2.2. Convolutional Codes

Convolutional codes are a major tool in the field of error control coding, particularly due to their ability to efficiently correct errors during data transmission. Convolutional codes do this by encoding data streams based on a memory-based approach to incorporate information from past input bits in the construction of present encoded bits. This convolutional coding aspect is more beneficial than block codes because it allows for flexibility and efficiency optimization. Convolutional encoding generates output sequences that are a function of the present input and previous inputs, and it is therefore efficient in dealing with noise and interference when transmitting. Hence, this property upholds the integrity of communications for a variety of applications, from satellite communications to deep space missions, where reliability of data is crucial.

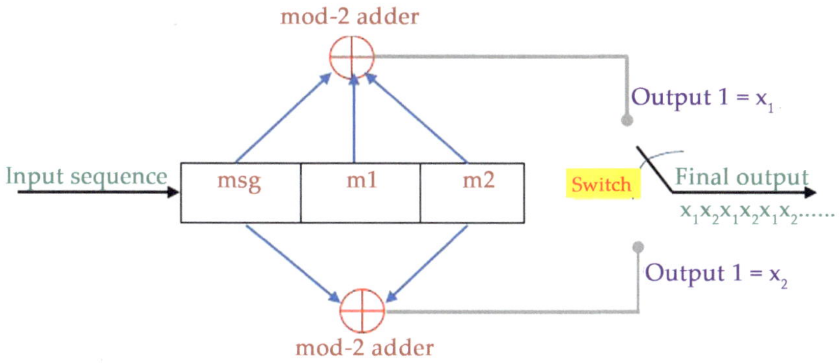

Figure 7.6. *Convolutional codes.*

Source: https://engineerstutor.com/wp-content/uploads/2020/09/convolutional-encoder-212.jpg.

Convolutional codes are generally decoded using algorithms with the ability to process the computational complexity of convolutional codes. A popular technique is the Viterbi algorithm that is optimum for maximum likelihood decoding of sequences of data coded by convolutionally. Operating on the principle of dynamic programming, the algorithm sequentially compares the input sequence with all possible states of the encoder, determining the most probable path that describes the input data observed while minimizing errors. This is particularly beneficial in highly impaired channels, allowing error correction efficiently while maintaining the computational load within reasonable limits. Furthermore, advances in soft decision decoding algorithms have been shown to have improved performance metrics, specifically in minimizing the bit error rate (BER), as these algorithms utilize past and future information to enable more accurate decoding (Hassibi et al., 2004). In recent developments, the marriage of convolutional coding with iterative decoding techniques has been shown to be beneficial for overall improvement of communication system performance. Experiments have indicated that convolutional codes, when paired with advanced decoding methods such as turbo coding, in certain conditions achieve performance close to the Shannon limit. The rationale for the efficiency of such a system is the fact that they can share soft information, which is utilized in iterative decoding iterations very efficiently. Furthermore, with advancements in networks from multiantenna to diversity mechanisms, convolutional codes combined with soft iterative decoding can also significantly minimize error propagation and channel fading issues, improving system robustness and throughput (Choi et al., 2007). Such adaptability of convolutional codes in dynamic networks of communications for the future speaks to their ongoing usefulness in error control coding schemes, projecting future-proof digital communication systems (Figure 7.6).

7.2.3. Turbo and LDPC Codes

For error control coding, Turbo codes and Low-Density Parity-Check (LDPC) codes are significant contributions for their ability to approach the channel capacity bounds theoretically. Turbo codes that were introduced in the mid-1990s consist of a pair of two or more convolutional codes and iterative decoding. This causes them to perform very well in bit error rates, particularly where signal-to-noise ratios are low. Turbo decoding, being iterative, enables soft information to be shared, which results in improved performance, particularly for applications requiring high reliability, for example, ultra-reliable low latency communication (URLLC) within 5G systems. This history depicts the code benefits of Turbo codes in meeting the challenge of high-tech communication needs and their utility as a main ingredient in the design of future error control techniques. By comparison, LDPC codes first designed in the 1960s but rediscovered in the 1990s are characterized by the sparsity of their parity-check matrices. This sparsity allows for efficient encoding and decoding operations that are friendly to iterative algorithms, similar to those of Turbo codes. LDPC codes have performed extremely well on a vast variety of communication channels, at times being as close as possible to the Shannon limit. The flexibility of design makes them capable of being designed for specific applications, and they are highly beneficial in a vast array of environments, ranging from satellite communications to wireless networks. Further, recent studies illustrate that coupled with state-of-the-art decoding algorithms, LDPC codes possess the ability to surpass traditional coding schemes in terms of complexity as well as error performance, further enhancing their status as an inspiration for future coding designs (Abbas et al., 2018). The integration of Turbo and LDPC codes into contemporary coding architectures testifies to their prominence in responding to the needs of future communication systems. As demands for latency and reliability continue to rise, particularly in tele-surgery and industrial automation applications, the performance trade-offs become increasingly important. Recent research reinforces the necessity of new coding technologies focusing on short block lengths wherein conventional strategies are vulnerable because conventional solutions are handicapped by the inherent limitations associated with retransmission and additional processing time. The advances in soft iterative decoding, particularly for Turbo and LDPC systems, show that capacity can be fully utilized with a low decoding complexity (Hassibi et al., 2004). The study of these codes in ultra-reliable and high-speed communication environments reflects their core significance in conceptualizing the future of error control coding.

7.3. ERROR DETECTION TECHNIQUES

In the domain of error control coding, some schemes of error detection play a pivotal role to ensure data integrity during transmission. One extremely famous method is the use of parity bits, where one bit is assigned to a sequence of data for making it even or odd. The method is very simple and works effectively to detect a single bit error but falters in more advanced scenarios where there are multiple errors and cannot distinguish among them. Sophistication in detecting errors rises with methods such as checksums and cyclic redundancy checks (CRCs), which apply mathematical algorithms to assess data integrity more securely. With the use of checksums, the sum of the data segment can be calculated and compared to a value transmitted, effectively sensing mismatches. In the same manner, CRC techniques rely on polynomial division, which yields high-quality error detection functions that are essential in modern digital communications where low rates of errors are of paramount importance to system performance and user satisfaction. The more advanced communication systems are, the greater is the imperative to employ advanced error detection functions, particularly in high data rate applications and complex signal environments. Techniques such as multiuser detection (MUD) emerged, particularly with multi-input multi-output (MIMO) systems. MUD effectively combats interference and minimizes error rates by decoding multiple user signals at the same time and therefore enhancing throughput as well as reliability. This is particularly critical in situations where channel conditions are rapidly changing, for instance in future-generation wireless systems. While traditional detection methods falter in rank-deficient circumstances—where there are more users than the system can handle—MUD is an advance in error detection technology. Based on recent findings, the use of genetic algorithms in MUD implementations has been observed to have potential for optimization of performance via achievement of near-optimal performance with reduced computational complexity, hence the inherent dynamic interplay between technology advancement and fault management mechanisms in modern communications (Hanzo et al., 2007). Despite the impressive merits of such sophisticated approaches, challenges do exist, more so with adaptive systems whose very nature compels them to constantly adapt to changing signal conditions. Vibrations of high-speed channel quality variation can significantly lower the efficiency of fixed-mode systems, necessitating the use of adaptive modulation schemes. For instance, BbB adaptive modems adjust parameters adaptively with respect to real-time channel conditions, thereby maximizing throughput and minimizing errors. Other diversity techniques such as space-time block coding can be assisted by adaptability, as it diminishes the effects of interference and fading. However, these techniques have their limitations, particularly in multiuser interference or correlated fading environments where it becomes increasingly difficult to maintain a low bit error ratio. The challenge is bringing these various techniques together to support robust performance in a large range of different transmission environments, highlighting the need for ongoing research and innovation in the field of error control (Choi et al., 2007) (Figure 7.7).

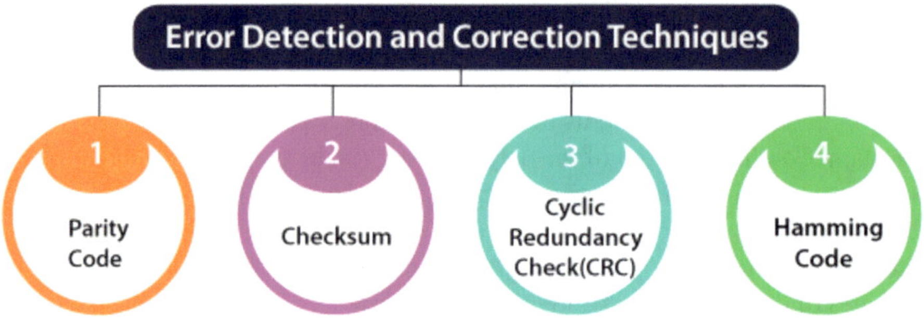

Figure 7.7. *Error detection techniques.*

Source: https://cn.edurev.in/ApplicationImages/Temp/1611942_11916d25-ab19-4efd-bac7-c18b55e00975_lg.png?w=360&dpr=2.6.

7.3.1. Parity Checks

Parity checks form part of the mechanism in error control coding, primarily for the detection of errors during transmitted information. Parity checks essentially mean the addition of an extra bit, known as the parity bit, to a group of bits. This parity bit is calculated based on the number of ones present in the original sequence; if the total is odd, the parity bit is set to one, whereas if it is even, it is set to zero. This simple yet effective method allows for the detection of single-bit errors, enabling systems to identify discrepancies on receipt. While valuable, parity checks have their own weaknesses, particularly in their inability to correct errors. Therefore, while capable of showing that there is indeed an error, other types of coding schemes would most likely be necessary to enact actual error correction in data communication systems. One of the biggest strengths of applying parity checks in error control coding is that they are extremely efficient and place virtually no overhead. Because the parity bit occupies only a single additional bit of space, the scheme is highly data-efficient in terms of transmission and hence very suitable to be utilized in applications involving strict bandwidth constraints. Finally, parity checks are simple to implement and involve minimal computational overhead which is actually particularly highly useful for low-power or embedded systems. However, with improvements in data communication, error and noise rates have necessitated stronger systems. For instance, in wireless communications, where signal loss is an inevitability, parity checks are perhaps not sufficient, as the need for complementary schemes like those presented in recent advances such as S-PRAC, which offer additional overall error detection and correction capabilities for noisy channels (Cabrera et al., 2019). In contemporary uses, parity checking is often an integral part of more complex error control coding structures. As a primitive scheme, parity checks are readily integrated into multi-level coding structures and improve their functioning as part of a larger system. In these new models, multiple coding techniques are used together to ensure increased reliability and immunity from data corruption. For example, pairing parity checks with advanced coding techniques such as polar codes has shown promise to enable error-free transmission rates near the channel capacity. This pairing is an important milestone in the progress of coding theory that enables systems not only to detect errors but correct them, overcoming challenges presented by exclusive use of traditional parity checks (Hassani et al., 2018). In general, while parity checks may be an easy way to identify errors, they do nonetheless form a part of the overall error control coding scheme in ensuring data integrity in the growing complexity of communications environments.

7.3.2. Checksums

In error control coding, checksums are employed to ensure data integrity when data is being transmitted over potentially erroneous channels. A checksum functions by generating a fixed-length bit string from a region of data, which is also transmitted with the data. Upon arrival, the receiver computes the checksum for received data and verifies whether it matches the transmitted checksum. If both values are equal, the data is complete; otherwise, such a mismatch suggests potential errors that were introduced in the process of transmitting data. The effectiveness and ease of checksums make them widely applied in numerous communication protocols, though they are far from being ideal. Its effectiveness is largely a function of the nature of the data to be transmitted and the condition of the communication channel, which emphasizes the necessity for continual refinements in the methods of detecting errors (Eggert et al., 2008). Additionally, traditional implementations of checksum often apply a single checksum across the whole of the data frame and this leads to inefficiencies in use, particularly in error-prone networks. Each time errors are found, there is a usual practice to resend the entire frame, though perhaps only a few bits could have been wrong. This action causes huge bandwidth wastage and increased transmission time, particularly in high-error-rate channels. To prevent such inefficiency, researchers propose a novel approach involving multiple checksums distributed over little data blocks within a frame. Only when errors are discovered within these smaller fragments do they have to be resent individually, optimizing recovery and total efficiency. This serves to illustrate the potential worth of sophisticated error control techniques and calls for the creation of new techniques to maintain pace with increasing needs for data transmission (Willig et al., 2004). Additionally, matching checksum techniques with dynamic channel conditions can optimize effectiveness of error control. Variable chunk sizes provide for a more responsive approach to error correction, adjusting retransmission activity based on real-time decisions about the link quality. Such flexibility is not only beneficial for errors currently being managed by the checksum method but also for errors in the future through dynamic chunk size adjustment where necessary. Experimental results demonstrate that goodput can be significantly improved using this method, particularly under suboptimal conditions where standard checksums are ineffective. Integration of adaptive algorithms into checksum design is a significant breakthrough in error control coding, with enhanced reliability and effectiveness assured for contemporary communication systems. By utilizing these mechanisms, the integrity of data could be more assured in more intricate digital environments, a huge breakthrough in error control policy (Willig et al., 2004).

7.3.3. Cyclic Redundancy Checks (CRC)

Cyclic Redundancy Checks (CRC) is a significant tool to ensure data integrity in digital communication networks. CRC works on the principle of identifying unwanted changes to raw data by computing a fixed-size checksum of the binary data to be transmitted. The CRC algorithm operation is done by converting data into a polynomial, dividing it by a predetermined polynomial divisor, and sending out a remainder as the CRC value. It gives a simple validation facility on the receiving side; if the CRC calculated at the receiving end is identical with the received CRC, then data is taken to be uncorrupted. Although good against normal errors, such as burst errors, CRCs are not flawless and their effectiveness depends on a correct choice of the polynomial used in generation and detection procedures. As handy as CRCs are, the dynamic nature of communication technology demands a reconsideration of CRCs' role. According to research, in modern satellite networks, such as those using DVB-S and DVB-S2 standards, utilization of advanced Forward Error Correction (FEC) methods may be employed to enhance error-detection capabilities beyond what is possible with CRCs.

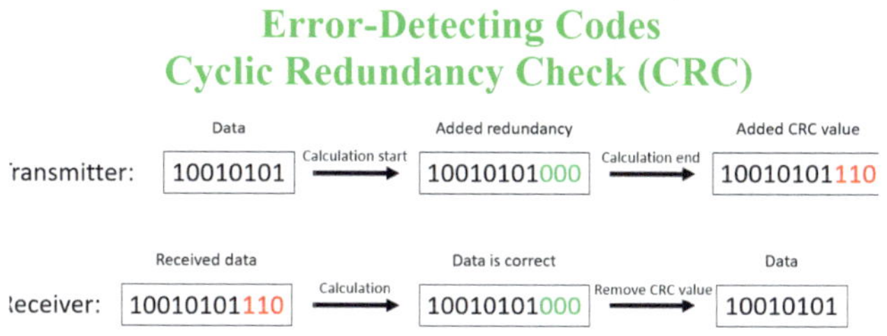

Figure 7.8. *Cyclic redundancy check.*

Source: https://www.knowelectronic.com/wp-content/uploads/2021/12/cyclic-redundancy-check-1000x448.png.

For instance, outer block codes Reed-Solomon and BCH can provide reliable error-detection information without putting excessive processing demands. Such effectiveness means that, after FEC settings are optimized for detection and correction of errors, the rate of uncorrected occurrences after FEC decoding reduces significantly. Thus, there is significant rationale for minimizing reliance on CRCs, or even eliminating CRCs in certain encapsulation methods, thereby reducing error control methods to a streamlined process in satellite communications (Buret et al., 2006). Furthermore, the relationship among error detection, data integrity, and security is underscored by the broader context of cryptographic application. Maintaining data integrity within contemporary digital transmissions is of paramount significance, especially in situations threatened by malicious interference or denial of service attacks. Techniques employed in cryptography, for example, private and public key techniques, extrapolate the fundamental principles of CRCs to deal with not just random errors, but deliberate alterations. Their application within encryption protocols is to provide authenticity and data integrity within information sent over insecure media. Therefore, while CRCs are a valuable component of the error control coding toolkit, their application must be balanced against other technologies and methods advancing enhanced world data security and integrity to an increasingly digital environment (Patra et al., 2007).

7.4. ERROR CORRECTION TECHNIQUES

Error-correcting methods play a significant role in maintaining data integrity during transmission in most communication networks. Key to these methods is the error detection and correction principle that occurs as part of the transmission process. Various coding strategies such as Hamming codes, Reed-Solomon codes, and low-density parity-check (LDPC) codes effectively enhance robustness against errors. All coding methods apply different methods to insert redundancy into the data being received, so that errors can be detected and corrected at the receiver end. Hamming codes, for example, make use of parity bits in such a way that single-bit errors can be corrected with little effort. In addition, Reed-Solomon codes are particularly helpful in cases like CDs and DVDs, where burst errors occur frequently. The versatility of these techniques across different environments is a gauge of their value in the delivery of consistent communication over multiple technological environments. The use of error correction techniques is most applicable to mobile computing environments, where the challenges of energy efficiency and network reliability are coincidental. In wireless networks, for instance, variations in signal strength and interference may result in higher error rates.

As provided in energy-efficient network interface studies, adaptation of the error correction protocol is vital to have the best performance with an assurance of quality of service (QoS) within such dynamic conditions (Bos et al., 2000). Not only does such adaptation conserve battery power for portable devices, but it also ensures that data exchange continues to be reliable, irrespective of wireless communication uncertainties. Through the combination of error correction techniques and energy-efficient designs, developers facilitate improved user experiences and seamless connectivity. Innovating improvements in error correction techniques also tackle new access techniques that can improve bandwidth efficiency while maintaining error rates. An emerging technology is Galois-Division Multiple Access (GDMA), which leverages finite field transforms to offer effective bandwidth demands (de Oliveira et al., 2015). This approach facilitates more effective allocation of available channels, which is necessary in virtually all contexts for communication, including satellite and mobile networks. Through incorporation of GDMA into present digital frameworks, there arises a new paradigm for data transfer, where multiple users can easily utilize shared resources with reduced errors and improved overall performance. With growing bandwidth requirements, research into more advanced methods for error correction is an essential field of study, with the potential to increase the reliability and efficiency of next-generation communication systems.

7.4.1. Hamming Codes

Hamming Codes, which were created by Richard Hamming during the 1950s, are a fundamental innovation in the field of error control coding. Hamming Codes are used to detect and correct for single-bit errors that occur when moving data. By using redundancy through the inclusion of parity bits, Hamming Codes ensure that a receiver can ascertain whether data had been altered. The peculiarity of Hamming Codes is that they have a mathematical basis, which utilizes linear algebra and binary operations in order to establish the minimum distance between the codewords. The minimum distance is important because it establishes the error-correcting capability of the code, which allows it not only to detect the errors but to locate them efficiently. As a direct consequence, Hamming Codes are implemented in numerous applications, including computer memory and digital communication systems, where data integrity is essential. Encoding and decoding in Hamming Codes are systematic and efficient. To encode data, the pre-determined number of parity bits is produced and combined with the original message bits according to specified positioning rules. This rigorous process allows for the identification of errors on reception of data; an algorithm checks the parity to identify any discrepancy. On identifying a discrepancy, the system is capable of correcting the error by inverting the erroneous bit based on the output of the checks for parity. Their efficiency has been underscored by various studies, which have demonstrated their use in a number of varying contexts, from satellite communications to index coding problems in networks. As shown, the flexibility and efficiency of Hamming Codes also underscore their use in contemporary error control schemes (Byrne et al., 2016). Moreover, the flexibility of Hamming Codes is also in the way they are able to accommodate complex instances of noise and data loss. Among the recent advancements are the use of Hamming-like methods within wiretap codes and index coding with side information. The use of Hamming Codes in these environments has also been explored by researchers, highlighting not only their historical value but also their adaptability to modern needs. As today's literature has described, error correction in noisy channels can be greatly improved with the application of Hamming Code principles, allowing secure communication in even the worst of conditions (Gomes et al., 2017). Systems are made capable of strengthening their immunity to all types of interference and deliver consistent data integrity over a wide range of applications with the application of these codes. This evolution highlights the persistent applicability and promise of Hamming Codes in the ever-changing realm of error control coding (Figure 7.9).

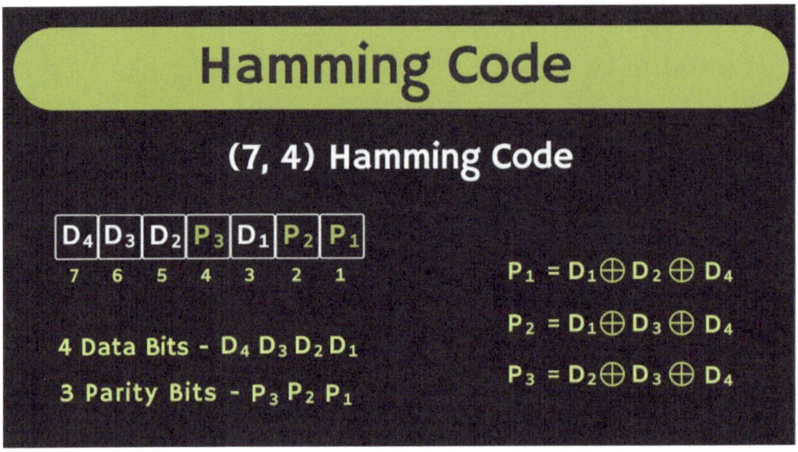

Figure 7.9. *Hamming codes.*

Source: https://i.ytimg.com/vi/t4kiy4Dsx5Y/maxresdefault.jpg.

7.4.2. Reed-Solomon Codes

Reed-Solomon codes are a powerful error control coding mechanism, commonly used in digital communication systems to efficiently prevent data corruption. Dating back to the 1960s, the codes utilize the polynomial algebra field over finite fields to correct and detect over one symbol error in the data transmitted. The effectiveness of Reed-Solomon codes is in their ability to create maximum-distance separable (MDS) codes, or simply put, maximize the distance between words. This is crucial for applications from compact disc storage to satellite communications, where data integrity is of prime concern. Exactly, mathematical design of these codes allows the recovery of the message as initially transmitted even in the case of severe errors, making them indispensable in applications where data loss may have disastrous effects. In implementation, Reed-Solomon codes show a sophisticated structure, integrating encoding and decoding techniques to provide the best performance while maintaining processing to a minimum. The process of encoding itself involves the computation of polynomials at roots of given primitive elements of a finite field, thus giving a solid basis for error correction. Among such developments is the development of fast decoding algorithms that accelerate the computation of Reed-Solomon codes, particularly in satellite communication where instant recovery of information is crucial. Current research emphasizes the use of various algorithms enhancing these codes' decoding performance, with mention of dual-basis approaches that guarantee reliability while optimizing resource management in transmission systems (Lin et al., 2025). Such flexibility not only illustrates coding theory advances but also emphasizes the promise to vastly improve existing technologies. Further, the versatility of Reed-Solomon codes extends beyond conventional error correction to new applications in information dispersal and peer-to-peer file systems. Due to developments in contemporary computing techniques, novel constructions of MDS Reed-Solomon erasure codes have been developed. For instance, utilizing the Fermat Number Transform (FNT) has been shown to be able to facilitate practical encoding and decoding algorithms with less computational complexity. These advancements enable data streams to keep speeds of 150 Mbps even with the codeword length increasing to thousands of symbols (Lacan et al., 2009). Thus, Reed-Solomon codes are not used solely for error control; they are at the hub of new data distribution techniques, encompassing the art of error control coding in a constantly evolving era of technology (Figure 7.10).

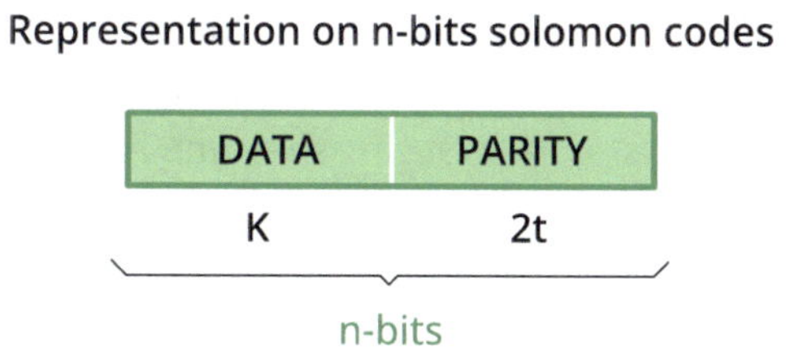

Figure 7.10. *Reed–Solomon code.*

Source: https://media.geeksforgeeks.org/wp-content/uploads/20220210103827/Representationonnbitssolomoncodes.png.

7.4.3. Low-Density Parity-Check Codes

Low-Density Parity-Check (LDPC) codes represent a revolutionary development in error control coding that enhances communication through noisy channels considerably. LDPC codes are characterized by their sparse bipartite graph representation, where the encoding and decoding procedures leverage a systematic approach to dealing with redundancy. By employing a large ratio of parity-check equations to codeword length, LDPC codes can provide high error-correcting capacity at low computational complexity. The balance between the code length and redundancy allows LDPC codes to approach the Shannon limit and, therefore, find applications in high-performance applications involving data transmission, such as digital TV and data storage devices. Their inherent adaptability to alternative channel models underscores their value in modern-day communication systems and makes them an essential part of coding theory today. Recent advances have spurred a resurgence of interest in the design and application of LDPC codes specifically in certain communication environments, such as transmission of short-data packets over Additive White Gaussian Noise (AWGN) channels. These channels present unique challenges, particularly when state information of the channel (CSI) is not readily accessible to receivers and transmitters. For these cases, authors have examined the trade-offs in mismatched decoding as a method to optimize performance under short block lengths. Introduction of new analysis methods has brought to the forefront the strength of LDPC codes under such situations, demonstrating that not only are they better under traditional metrics but also provide robust advantage when used in conjunction with preambles that permit channel estimation. This combined capability becomes important when dealing with short packets, where error correction has to be maximized against channel conditions, as identified by performance tests between competing schemes and LDPC codes (Chiani et al., 2017). In addition, the flexibility of LDPC codes exceeds mere error control capabilities to also focus on security applications, particularly with wireless data transmission media. The union of error correction and encryption is an advanced coding technique that is designed to enhance reliability as well as confidentiality in data communications. Non-Binary LDPC codes, particularly, offer improved performance parameters over their binary counterparts, particularly during concurrent error correction and encryption processes. This new approach holds that encryption use will have no negative effect on bit error rates—a concern often cited when unbundling traditional methods into discrete application. The synergy of such technologies highlighting the efficiency and usability of LDPC codes places them at the apex of coding technology as it answers growing demands for security, consistent data transfer in many applications (Bonadero et al., 2023).

7.5. APPLICATIONS OF ERROR CONTROL CODING

Applications of error control coding have key roles to enhance the reliability of data communications in a wide range of communication systems. One of the most important areas of application is in wireless communication networks, where signal decay due to interference and atmospheric conditions can lead to significant data loss. Error control coding techniques, such as convolutional and turbo codes, allow for error correction without retransmission and thereby improve overall system efficiency. Such is particularly significant in multimedia applications, where maintaining quality at low latency is critical. To these demands, network coding (NC) is an emerging paradigm that integrates error control schemes directly into the nodes of a network, making networks increasingly resilient to transmission errors and changing network conditions (Frossard et al., 2012). As a result, the robust nature of error control coding not only corrects quality defects but enhances user experience in real-time data transmission settings. A further prominent use of error control coding exists in peer-to-peer multimedia streaming, in which large data files are transferred over decentralized networks. In such applications, error control schemes can go a long way towards minimizing visual artifacts that could result from data loss during the course of streaming, so users can receive the best possible quality. Network coding improves this process by allowing the forwarding of packet combinations rather than single message forwarding, raising the throughput and lessening the likelihood of further mistakes. This method is particularly important in high client heterogeneity and changing network scenarios, so the adaptive aspect of error control coding is even more relevant (Frossard et al., 2012). Moreover, with multimedia services increasingly demanding reliability and urgency, the implementation of advanced error control coding techniques has to be implemented in order to satisfy the stringent criteria of the current digital communication. In the case of wireless networks, error control coding must meet the varying bandwidth and quality demands introduced by different types of traffic. The employment of these coding techniques on different layers of the network protocol stack maintains efficient data transmission even in poor conditions. The capability to achieve this is not just for error correction but also for Quality of Service (QoS) control by adjusting parameters of communication based on real-time feedback. This enables applications to stream their data to conform to current network conditions with the added advantage of conserving energy and performance to the maximum extent. In this regard, the use of adaptive error control coding structures not only enhances the reliability of wireless network data transmission but also contributes to having more efficient and greener modes of communications that closely follow the dynamic demands of contemporary digital systems.

7.5.1. Telecommunications

In telecommunications, error control coding can't be overemphasized. Data transmission integrity is heavily dependent on the ability to detect and correct errors that occur because of some range of causes such as noise, interference, and signal weakening. Advanced coding mechanisms such as Low-Density Parity-Check (LDPC) codes have come in to increase the effectiveness of error correction. These codes are efficient in working on sparse parity-check matrices and thus enable efficient encoding and decoding. With evolving improvements in telecommunication, particularly with the shift towards 5G networks, the need for the implementation of robust error control mechanisms is imperative to ensure seamless service and high data integrity. Thus, aligning coding techniques with evolving communication protocols is the need of the hour to counter the increasing demand for fast and reliable data transfer. Incorporation of adaptive systems in telecommunication has shown significant promise in mitigating the impact of unstable channel conditions.

Figure 7.11. *Error control coding in telecommunications.*

Source: https://external-content.duckduckgo.com/iu/?u=https%3A%2F%2Flearncodingusa.com%2Fwp-content%2Fuploads%2F2023%2F10%2FThe-Beginners-Guide-to-Debugging-in-Python-1.jpg&f=1&nofb=1&ipt=c1f861e29ee276ccb357e10e04ab34cb0e2b560bdfaa6a3df373198d8077407b.

The advent of Bursts-by-burst adaptive schemes is a strategic step in countering sudden variation in channel quality, which can seriously degrade overall system performance. For instance, adaptive orthogonal frequency division multiplexing (OFDM) modems can adaptively modify their operating parameters, thereby achieving improved throughput and reliability. Besides, diversity techniques such as Rake receivers and space-time block coding (STBC) serve to further boost system resistance to fading and interference. However, the interplay between channel conditions and error correction mechanisms is subtle; while diversity techniques can be used to advance error performance under perfect conditions, their efficiency is lessened when operating within specific correlated fading channels. Thus, continuous optimization of such coding techniques has to be done in order to ensure maximum performance in real-world telecommunication applications, particularly in environments with unfavorable conditions. Finally, the integration of pragmatic techniques for simplicity reduction in error control coding solutions is crucial in modern telecommunications. With smaller systems and fewer resources, innovative coding architecture design methodologies are necessary in order to sustain functionality while minimizing the consumption of resources. The capability to replace many quasi-cyclic LDPC codes with a single code offers significant memory efficiency and implementation simplicity on devices like System on Chips (SoCs). The simplification is not only alleviating memory loads but can even render coding structures more viable for Software Defined Radio (SDR) platforms. By simulating the performance of altered LDPC structures compared to traditional designs, scientists can ascertain the feasibility of such innovations. As such, with growing developments in the field, the emphasis on practical applications emphasizes the need to balance theory-driven advancements with real application in telecommunications (Figure 7.11).

7.5.2. Data Storage

The intricate interaction between data storage and error control coding is crucial in modern-day computing systems, specifically the way information is retrieved and stored across different systems. Since data volumes continue to grow exponentially, data storage mechanisms employed must be flexible so that data will be intact and accessible. Error control coding is then an essential technique in this setting, providing error detection and correction that can emerge in data storage or transmission. In software-as-a-service applications, the condition is even more complicated, as locally stored data are updated sporadically. In such cases, maintaining the latest version while possibly even holding older versions necessitates efficient coding techniques that achieve a balance between compression and modifiability. Consequently, the intersection issue of error control coding and storage remains a promising research area, with the significant trade-offs that come with having flexible and efficient data handling (Goyal et al., 2016). Secondarily, data storage techniques remain more reliant on the effectiveness of error control codes, particularly for networked and distributed data systems. When data is replicated or transmitted over error-prone mediums, ensuring that the original data can be recreated in a proper state is the central engineering issue. The binary erasure channel (BEC) and binary defect channel (BDC) are two methods in which error control codes can be employed to reverse certain types of data loss. Knowledge of the duality between these channels enhances our insight into channel capacities and the highest limit of successful data recovery. Techniques such as Reed-Muller codes and polar codes have shown promise at attaining optimal performance under such paradigms, establishing a core principle for the application of error correction in data storage systems (Kim et al., 2016). This dichotomy is representative of the complex relationship between theoretical limits and practical implementation in error control coding. Beyond current error correction methods, locally writable codes (LWC) offer a new paradigm for increasing data storage in resistive memories. The codes have the ability to significantly increase system endurance and power efficiency, working to meet growing demands for resilient memory systems in a world of growing data needs. LWC is the counterpart of locally repairable codes (LRC) for distributed storage, emphasizing the significance of flexibility with regard to data management strategy. Through the principles of error control coding, these innovations enable more reliable and efficient data storage technologies, enabling organizations with the ability to tackle issues regarding data integrity effectively. As researchers move deeper into these connections, the need for adding error control coding to data storage devices will become more and more obvious, providing new avenues for innovation and data management effectiveness (Kim et al., 2016).

7.5.3. Wireless Communication

Wireless communication is an evolutionary leap in the exchange and reception of information via numerous media. Wireless communication relies on data transmission via electromagnetic waves, giving room for cable-free connectivity. Strong error control coding mechanisms are pivotal to the success of wireless communication. These mechanisms are required to counteract the brutal impacts of noise, interference, and signal attenuation that typically torment wireless channels. Using efficient error control coding, the integrity and reliability of data being transmitted can be improved so that vital information is delivered to where it should go without any damage. With increased dependence on wireless networks creeping into sectors ranging from telecommunication to health care, the significance of error control coding grows by leaps and bounds since it directly impacts end-user experiences and the overall effectiveness of the communication systems. The evolution of wireless communication technologies, particularly with the advent of 5G systems, emphasizes the need for advanced error control techniques. Unlike previous generations, 5G is accompanied by features such as Ultra-Reliable Communication (URC), which aims to provide communication services with significant reliability most of the time. Such reliability is highly critical in applications where continuous connectivity is paramount, such as autonomous vehicle coordination and industrial automation. To achieve this, there is a requirement to look back at encoding control information and data so that every aspect of the communication channel is fault-tolerant in terms of potential failures. This review highlights the importance of error control coding not just as an extra protection scheme but as a fundamental component of wireless communication in the future, driving innovation and improving service provision in hostile environments (Popovski et al., 2014). One of the dilemmas of wireless communication is the reconciliation of theory and practice, particularly in network coding. As traditional networks primarily transmit messages without in-network processing, the emerging trend of network coding introduces the potential for combining packets in intermediate nodes for better throughput and efficiency. Not only does this reduce delay, but also enhances transmission error robustness inherent in wireless channels. Multimedia-specific application studies have gained momentum, highlighting the way network coding can address various network conditions and client diversity in an optimal manner. Through the use of methods incorporating error control coding with concepts of network coding, the industry can look forward to more reliable wireless communication systems capable of serving the sophisticated requirements of next-generation multimedia applications (Frossard et al., 2012) (Figure 7.12).

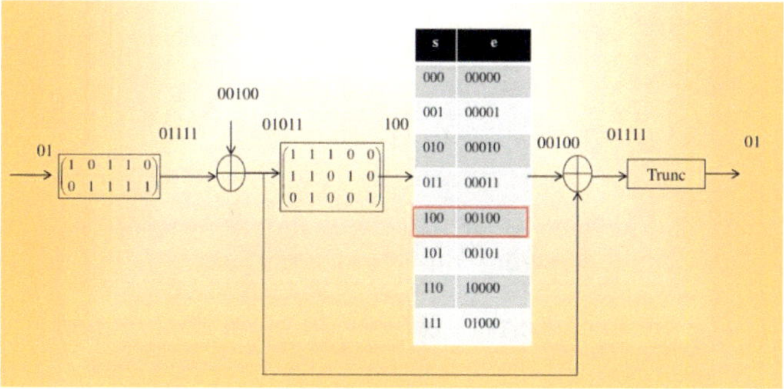

Figure 7.12. *Error control coding for wireless communication technologies.*

Source: https://external-content.duckduckgo.com/iu/?u=https%3A%2F%2Fimage1.slideserve.com%2F2404481%2Fthe-coding-scheme-l.jpg&f=1&nofb=1&ipt=23a30fe529f13b2baa13256a2170c7b68fc16f1a09af4d061312d794b8653b64.

7.6. CHALLENGES IN ERROR CONTROL CODING

In the rapidly changing communications technology landscape, error control coding is the linchpin to data integrity in varied transmission channels. However, one of the fundamental issues here is the added complexity introduced by newer designs such as Cloud-RAN. While mobile networks change from point-to-point dedicated fronthaul connections to packet-based multi-hop networks, reliability and latency reduction become a sine qua non (Mahmoodi et al., 2018). This evolution requires new error control coding techniques that can cope with the uncertain nature of packet-based systems. Moreover, users not only need high throughput but also ultra-reliable low-latency communication, particularly in use cases such as the Tactile Internet. Consequently, error control coding must strike a delicate balance between these conflicting requirements in order to provide efficient performance.

Figure 7.13. *Challenges in error control coding.*

Source: https://www.21kschool.com/in/wp-content/uploads/sites/4/2023/05/A-kids-guide-how-to-find-common-coding-language-errors.png.

The application of advanced error control coding in contemporary networking is also under threat due to the amount of multimedia content that is invading various platforms. With the dissemination of Networked Media continuing to rise, the size of the data to be encoded and sent continues to expand at an exponential rate (Aggoun et al., 2008). This growth necessitates robust coding mechanisms to manage various types of media efficiently without compromising quality and preventing transmission errors. Utilization of multiple coding methods, such as erasure and convolutional coding, offers a solution but adds more complexity. Switching in compliance with evolving network conditions and user requirements can thus lead to issues of encoding efficacy and data recovery mechanisms, which underscore the need for flexible and dynamic coding systems in error control. The other important challenge to error control coding is ongoing advancements in mobile and wireless technology. With the users requiring more sophisticated services like virtual collaboration and personalized media interaction, coding schemes must also evolve accordingly

(Aggoun et al., 2008). Real-time computation and low delay transmission are demands that often conflict with the overhead of employing more sophisticated error correction schemes. Combining various coding schemes for enhancing reliability without affecting performance is a daunting task. In addition, while the network architecture environment evolves, error control coding must also endure interference and channel variability in mobile environments. Overcoming such challenges will be pivotal in ensuring the sustainability of error control coding in future networks (Figure 7.13).

7.6.1. Trade-Offs Between Complexity and Performance

Complexity-performance interaction is a crucial consideration in error control coding methods. As technology advances to accommodate increasing data rates and complicated channel conditions, the coders have to deal with creating coding schemes that provide the best performance at minimal cost without contributing disproportionately to computational overhead. For instance, traditional modulating techniques such as Orthogonal Frequency Division Multiplexing (OFDM) offer higher data rates, but they have a high peak-to-average power ratio (PAPR) that reduces power amplifier efficiency. It achieves this at the expense of a trade-off because it raises throughput when employed but demands more sophisticated hardware and higher power consumption, translating into greater operational costs and complexities. Lastly, effective error control coding must tread a fine line between system reliability in adverse conditions and resource utilization optimization. One innovative solution to reconcile the complexity-performance dilemma is the novel flexible multi-group single-carrier (FMG-SC) modulation scheme, an innovative mixture of OFDM and Single-Carrier Frequency Domain Equalization (SC-FDE) that puts them into one package for greater flexibility. By orthogonal grouping of frequency subcarriers in relation to channel gains, the FMG-SC method facilitates parallel transmission of multiple data streams in a single framework. The system maximizes the maximum sum-rate of all groups by utilizing low-complexity grouping methods that closely realize optimal performance of higher-complexity approaches. These innovations demonstrate how error control coding can be engineered to improve the rate-PAPR trade-off without too much complexity, hence being a practical avenue for modern communication systems to maximize data throughput while system efficiency and reliability are ensured (Lie et al., 2017). Other applications of error control coding trade-offs include practical implementations in the form of Code Division Multiple Access (CDMA) in optical communications. In this instance, error correction complexity is balanced with enhanced performance. For instance, optical CDMA techniques enable transmission of data for many users simultaneously, promoting efficient utilization of system bandwidth.

However, strict performance constraints, such as low bit error rates, translate into complicated coding procedures for suppression of unwanted effects like intercode crosstalk and electronic noise. Using sophisticated error correction codes like Reed-Solomon coding, it is possible to reduce error rates drastically, even though it means higher computational overhead. This example shows how even as performance targets become increasingly complex, judicious coding decisions can enable high efficiency while dealing with the operational complexity that comes with state-of-the-art communication systems (Gagliardi et al., 2025) (Figure 7.14).

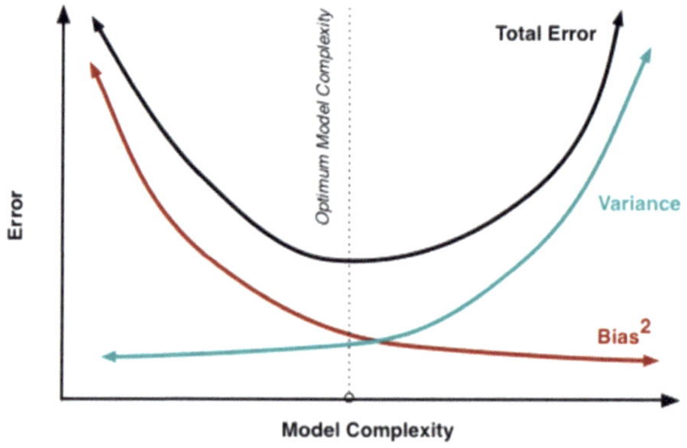

Figure 7.14. *Trade-offs between complexity and performance in error control coding.*

*Source: https://miro.medium.com/1*qqFreoBD9OrSE4NdLiCUZw.png.*

7.6.2. Limitations of Current Techniques

With increasing reliance on error control coding techniques in data transmission systems, it is important to identify some of the limitations that restrict their efficiency. One major limitation is the trade-off between error correction capability and working performance. Many conventional coding techniques utilize enormous bandwidth, which also leads to increased delay and poor overall performance, especially in constrained environments such as wireless networks. While newer codes like Low-Density Parity-Check (LDPC) and Turbo codes are error resilient, their high complexity necessarily leads to bigger computations, which in turn adversely affects viable implementation within practice. This need for highly sophisticated hardware or software to facilitate the computations involved further diminishes these methods' viability in most real-world applications, pointing towards an imperative direction to continue development and research for balanced error management techniques. A second pressing limitation concerns the evolving nature of modern communication environments. Today's error control coding techniques have a tendency to utilize constant channel conditions as an assumption, rather than adapting in real-time to levels of interference and signal degradation. For wireless communication, where channel quality is likely to vary dramatically, such assumptions are certain to lead to suboptimal resource utilization as well as less than optimal performance. As noted in newer studies, the adaptability of bandwidth scheduling and error control has to be efficient in order to achieve energy efficiency and a proper quality of service within such environments (Bos et al., 2000). Present approaches tend to be less responsive, however, and this can lower the reliability of data transmission as a whole. This restriction is addressed by the development of more sophisticated algorithms that dynamically alter error correction parameters based on real-time feedback from network quality. Another important restriction of error control coding methods is their scalability as network sizes grow. As the Internet of Things (IoT) and increasingly networked devices emerge, the modern error correcting methods must be able to handle a vast and diverse set of traffic types with high efficiency. However, many traditional schemes are unscalable, leading to detection and correction delays that degrade data integrity. Current work has begun to explore the adaptability of network protocols to address these problems, such that error control is able to provide adequate support for diverse types of traffic (Havinga et al., 2000). Integration of schemes across the entire protocol stack, however, is still an even bigger challenge, requiring further innovations able to effectively blend reliability and scalability into error control coding.

7.6.3. Future Directions in Research

With the information technology environment still evolving, future directions in research on error control coding continue to drive significant advances. An especially intriguing area of exploration is integrating machine learning algorithms with error correction strategies. These algorithms can enhance the performance of traditional coding standards by dynamically adjusting coding schemes according to shifting channel conditions. For instance, the application of sparse coding and reinforcement learning concepts can enhance perception/action loops in coding models. The combination is likely to enhance coding efficiency, particularly in impoverished environments. As researchers develop systems that dynamically enhance their error-control capability, outcomes may determine an integrated interconnection between the dynamic data environment and coding techniques providing maximum efficiency and reliability, pushing the frontier of error control coding even further (Shi et al., 2014). Further research on non-orthogonal multiple access (NOMA) is another future pillar area of research in the field of error control coding. With the emergence of NOMA technologies as integral components of next-generation wireless networks, their integration with Multi-Input Multi-Output (MIMO) systems presents great promise. Solving the technical challenges inherent in MIMO-NOMA, for example, power allocation and beamforming, presents rich areas for methodological progress. Researchers need to improve the spectral efficiency of such systems and, therefore, optimize the network performance. Additionally, the issue of stability in successive interference cancellation (SIC) still matters. An insight into the impact of using attainable rates as performance metrics could lead to more robust coding strategies that minimize interference while maintaining data integrity (Huang et al., 2018). This intersection of coding techniques with future-generation access technologies represents a significant subject for future research studies. Also of note in error control coding research is the development of advanced coding schemes for emerging communication paradigms, such as millimeter-wave and massive MIMO systems. These technologies promise to revolutionize data transmission by allowing increased bandwidth and lower latency. Reliable error correction codes for such high-frequency applications are challenging to design, though. Future research must include how codes may be designed that are capable of sustaining performance in the presence of enormous amounts of data that these technologies can make feasible. Further investigation into hybrid solutions that draw upon aspects of both conventional and innovative coding approaches can build robust frameworks able to address the complicated nuances presented by developing communication systems. Such a multi-disciplinary approach has the potential to drive the next generation of error control coding innovations, as researchers struggle with ensuring reliable communication under increasingly difficult operating conditions (Shi et al., 2014; Huang et al., 2018).

SUMMARY

Error control coding has evolved as a key factor in making digital communication systems more reliable. Increasing technology demands more efficient data transmission, and error control coding through various coding techniques such as convolutional and block codes helps communication networks minimize errors due to noise and interference, maintaining the integrity of transmitted data. This emphasizes the requirement for further research in the field, particularly in developing more sophisticated coding schemes with the capacity to operate under constraints of low channel capacities, as proposed by Alexander et al. (2007). The use of such sophisticated coding schemes not only increases performance but also facilitates the maximum utilization of data traffic in current networks. In addition, problems arising in current networking conditions, especially concerning data class heterogeneity and congestion, require stringent solutions in error control frameworks. The realization of Asynchronous Transfer Mode (ATM) reflects an attempt towards maximizing network performance under such constraints. Flexibility and efficiency inherent in ATM networks guarantee that they are a viable beginning solution for the complex issues of traffic handling and error minimization, as stressed in Bae et al. (1991). Directions for future research must focus on examining these system vulnerabilities and counteracting future bottlenecks with innovative coding methodologies, contributing to a more efficient and resilient communication infrastructure. Lastly, the effects of error control coding extend far beyond simple detection and correction of errors; they influence the fundamental structure and complexity of communications networks themselves. With emerging new standards and existing protocols being optimized, the role of error control coding will soon become increasingly indispensable. Experts and researchers need to strive to comprehend and master versatile coding techniques that can adapt to meet the evolving demands of digital communication. Continued focus to this area not only assures greater provision of service but also ensures that new technologies can provide the consistent flow of information in an increasingly connected world, solidifying the central role of error control coding in modern communication systems.

The landscape of error control coding is characterized by the high stake that it has in ensuring the consistency of information communication through various channels. It employs techniques like forward error correction (FEC) and automatic repeat request (ARQ) to correct and identify errors which may be present in information transmission. FEC allows the data to be encoded such that the receiver is able to reconstruct the original message even in the presence of errors, thereby enhancing efficiency if retransmitting the data would be costly or inconvenient. ARQ, however, operates on the principle of sender and receiver acknowledging error detection followed by retransmission of the lost or corrupted data. This is a double-edged attention that calls for the selection of appropriate coding techniques depending on provided applications, which has been stressed in previous studies on the effectiveness and efficiency of various coding schemes for applications in various environments. Apart from the fundamental methods of error control coding, the increasing complexity of communication systems requires investigations into their vulnerabilities. Security attacks, particularly in wireless mesh networks (WMNs), are major challenges because WMNs are exposed to multiple malicious attacks at the network, link, and physical layers. Recognizing these vulnerabilities, researchers have begun to suggest comprehensive defense techniques designed to ensure data integrity and user privacy. For instance, effective key management protocols are required for secure connection establishment in WMNs. The demand for reputation and trust-based systems, which are ever gaining popularity as they are utilized in complementing security systems, also highlights the importance of adopting solid security practices. These guarantee that the transmission is safe, thereby tightening the general objectives of error control coding and its application in today's communication systems (Gupta et al., 2013). Looking to the future, ongoing growth in information systems will have far-reaching impacts on error control coding policy. Technology evolution will surely introduce increased computational capacity, automation, and network involvement, which in turn will contribute to improved performance from error control devices. Most notably, human-interface technology and data transport technique advances will enable smoother and more efficient error-correction processes. In addition, as illustrated in more contemporary architectural specifications for systems such as the Deep Space Network (DSN), incorporation of open-systems standards is crucial to achieving interoperability and enhancing data security (Atkinson et al., 1992). These advances will usher in the next generation of error-control coding techniques that are not only more resilient to errors but also more secure against attack, ensuring the integrity and reliability of data transmission in an increasingly interconnected world.

REFERENCES

1. Abbas, R., Han, G., Johnson, S. J., Li, Y., et al. (2018). Short block-length codes for ultra-reliable low-latency communications. http://arxiv.org/abs/1802.09166 (accessed on 18th September 2025).
2. Abbe, E., & Liu, J. (2014). Polynomial complexity of polar codes for non-binary alphabets, key agreement and Slepian-Wolf coding. http://arxiv.org/abs/1405.0776 (accessed on 18th September 2025).
3. Abdel-Hamid, T., Lin, C. Y., & Sherif, J. S. (1992). Software-engineering process simulation (SEPS) model. https://core.ac.uk/download/pdf/42812624.pdf (accessed on 18th September 2025).
4. Aggoun, A., Amon, P., Arbel, I., Chernilov, M., et al. (2008). Multimedia delivery in the future internet. https://core.ac.uk/download/335423.pdf (accessed on 18th September 2025).
5. Atkinson, D. J., Beswick, C. A., Cooper, L. P., Crowe, B. J., et al. (1992). Deep Space Network information system architecture study. https://core.ac.uk/download/pdf/10438278.pdf (accessed on 18th September 2025).
6. Bae, J. J., & Suda, T. (1991). Survey of traffic control schemes and error control schemes for ATM networks. https://core.ac.uk/download/287623208.pdf (accessed on 18th September 2025).
7. Barker, P., Hughes, L., Jacobs, G., Kernohan, D., et al. (1996). Reviews. https://core.ac.uk/download/6873.pdf (accessed on 18th September 2025).
8. Bonadero, J. C., Castiñeira Moreira, J., Coppolillo, L., Liberatori, S., et al. (2023). An encryption and error-control coding scheme based on non-binary LDPC codes. https://core.ac.uk/download/574357969.pdf (accessed on 18th September 2025).
9. Bos, M., Havinga, P. J. M., & Smit, G. J. M. (2000). Energy-efficient adaptive wireless network design. https://core.ac.uk/download/pdf/11454370.pdf (accessed on 18th September 2025).
10. Buret, I., Cantillo, J., & Lacan, J. (2006). A CRC usefulness assessment for adaptation layers in satellite systems. https://core.ac.uk/download/12041369.pdf (accessed on 18th September 2025).
11. Byrne, E., & Calderini, M. (2016). Error correction for index coding with coded side information. http://arxiv.org/abs/1506.00785 (accessed on 18th September 2025).
12. Cabrera, J. A., Fitzek, F. H. P., Irianto, K. D., Nguyen, H. Q., et al. (2019). S-PRAC: Fast partial packet recovery with network coding in very noisy wireless channels. http://arxiv.org/abs/1902.10645 (accessed on 18th September 2025).

13. Chiani, M., Durisi, G., Liew, S. C., Liva, G., et al. (2017). Short codes with mismatched channel state information: A case study. http://arxiv.org/abs/1705.05528 (accessed on 18th September 2025).
14. Chockalingam, A., Hong, Y., Mohammed, S. K., Viterbo, E., et al. (2011). Modulation diversity in fading channels with quantized receiver. http://arxiv.org/abs/1104.1155 (accessed on 18th September 2025).
15. Choi, B. J., & Hanzo, L. (2007). Near-instantaneously adaptive HSDPA-style OFDM versus MC-CDMA transceivers for WIFI, WiMAX, and next-generation cellular systems.
16. de Oliveira, H. M., & Miranda, J. P. C. L. (2015). On Galois-division multiple access systems: Figures of merit and performance evaluation. http://arxiv.org/abs/1502.03698 (accessed on 18th September 2025).
17. Dobos, J., Farkas, P., Palenik, T., Rakus, M., et al. (2015). Analysis of minimal LDPC decoder system on a chip implementation. https://core.ac.uk/download/30312872.pdf (accessed on 18th September 2025).
18. Eggert, L., & Fairhurst, G. (2008). Unicast UDP usage guidelines for application designers. https://core.ac.uk/download/19915619.pdf (accessed on 18th September 2025).
19. Falaschetti, D. (2025). Can voting reduce welfare? Evidence from the US telecommunications sector. https://core.ac.uk/download/pdf/9312525.pdf (accessed on 18th September 2025).
20. Fradkov, A. L., Andrievsky, B., Blekhman, I. I., Andrievsky, B., De Persis, C., Demidovich, B. P., Dudnik, E. N., et al. (2007). Controlled synchronization of one class of nonlinear systems under information constraints. http://arxiv.org/abs/0712.0636 (accessed on 18th September 2025).
21. Frossard, P., Magli, E., Markopoulou, A., Wang, Y., et al. (2012). Network coding meets multimedia: A review. https://core.ac.uk/download/pdf/11426964.pdf (accessed on 18th September 2025).
22. Gagliardi, R. M., Mendez, A. J., & Park, E. (2025). Optical multiple access techniques for on-board routing. https://core.ac.uk/download/pdf/42812397.pdf (accessed on 18th September 2025).
23. Gomes, M. A. C., Harrison, W. K., Pfister, J., Vilela, J. P., et al. (2017). Quantifying equivocation for finite blocklength wiretap codes. http://arxiv.org/abs/1701.05484 (accessed on 18th September 2025).
24. Goodman, R. M., & Sayano, M. (1989). On-chip ECC for multi-level random access memories. https://core.ac.uk/download/84317825.pdf (accessed on 18th September 2025).

25. Goyal, V. K., Kusuma, J., & Varshney, L. R. (2016). Malleable coding for updatable cloud caching. https://open.bu.edu/bitstream/2144/39228/1/cutpaste_JSAC_v7.pdf (accessed on 18th September 2025).
26. Grover, P., Kar, S., & Yang, Y. (2017). Computing linear transformations with unreliable components. http://arxiv.org/abs/1506.07234 (accessed on 18th September 2025).
27. Gupta, A., Mishra, A., Naveed, A., Perrig, A., Shamir, A., Franklin, A. A., Wood, A. D., et al. (2013). Security and privacy issues in wireless mesh networks: A survey. http://arxiv.org/abs/1302.0939 (accessed on 18th September 2025).
28. Hanzo, L., & Jiang, M. (2007). Multiuser MIMO-OFDM for next-generation wireless systems.
29. Hassani, S. H., Mondelli, M., & Urbanke, R. (2018). How to achieve the capacity of asymmetric channels. http://arxiv.org/abs/1406.7373 (accessed on 18th September 2025).
30. Hassibi, B., Kailath, T., & Vikalo, H. (2004). Iterative decoding for MIMO channels via modified sphere decoding. https://core.ac.uk/download/4871837.pdf (accessed on 18th September 2025).
31. Havinga, P. J. M. (1999). Energy efficiency of error correction on wireless systems. https://core.ac.uk/download/pdf/11454366.pdf (accessed on 18th September 2025).
32. Havinga, P. J. M. (2000). Energy-efficient wireless communication. https://core.ac.uk/download/pdf/11460969.pdf (accessed on 18th September 2025).
33. Haworth, L. A., & Newman, R. L. (2025). Test techniques for evaluating flight displays. https://core.ac.uk/download/pdf/42808248.pdf (accessed on 18th September 2025).
34. Huang, Y., Jing, Y., Wang, J., Yang, L., et al. (2018). Signal processing for MIMO-NOMA: Present and future challenges. http://arxiv.org/abs/1802.00754 (accessed on 18th September 2025).
35. Ingels, F. M., & Shahsavari, M. M. (2025). *A Study of Data Coding Technology Developments in the 1980–1985 Time Frame, Volume 2.* https://core.ac.uk/download/pdf/42871670.pdf (accessed on 18th September 2025).
36. Kim, Y., & Kumar, B. V. K. V. (2016). Duality between erasures and defects. http://arxiv.org/abs/1602.03536 (accessed on 18th September 2025).

37. Kirse, H. (2014). *Studies in Error Correction Coding.* https://core.ac.uk/download/58825524.pdf (accessed on 18th September 2025).
38. Lacan, J., & Soro, A. (2009). FNT-based Reed-Solomon erasure codes. https://core.ac.uk/download/12041389.pdf (accessed on 18th September 2025).
39. Lie, J. P., Yang, Y., Zhang, R., Zhang, S., et al. (2017). Flexible multi-group single-carrier modulation: Optimal subcarrier grouping and rate maximization. http://arxiv.org/abs/1710.10001 (accessed on 18th September 2025).
40. Lin, S. (2025). *Coding for Reliable Satellite Communications.* https://core.ac.uk/download/pdf/42846530.pdf (accessed on 18th September 2025).
41. Mahmoodi, T., Mountaser, G., & Simeone, O. (2018). Reliable and low-latency fronthaul for tactile internet applications. https://core.ac.uk/download/195272646.pdf (accessed on 18th September 2025).
42. Mazumdar, A., Saad, Y., & Ubaru, S. (2017). Low rank approximation and decomposition of large matrices using error correcting codes. http://arxiv.org/abs/1512.09156 (accessed on 18th September 2025).
43. Nakiboğlu, B. (2011). Exponential bounds on error probability with feedback. https://core.ac.uk/download/4429020.pdf (accessed on 18th September 2025).
44. Patra, N., & Sankar, S. S. (2007). Data reduction by Huffman coding and encryption by insertion of shuffled cyclic redundancy code. https://core.ac.uk/download/53186854.pdf (accessed on 18th September 2025).
45. Popovski, P. (2014). Ultra-reliable communication in 5G wireless systems. https://core.ac.uk/download/60611226.pdf (accessed on 18th September 2025).
46. Seth, A. K., Critchley, H. D., & Suzuki, K. (2012). An interoceptive predictive coding model of conscious presence. https://core.ac.uk/download/9427700.pdf (accessed on 18th September 2025).
47. Shi, B. E., Triesch, J., Zhang, C., Zhao, Y., et al. (2014). Intrinsically motivated learning of visual motion perception and smooth pursuit. http://arxiv.org/abs/1402.3344 (accessed on 18th September 2025).
48. Tan, V. Y. F., & Truong, L. V. (2019). The reliability function of lossy source-channel coding of variable-length codes with feedback. http://arxiv.org/abs/1710.07419 (accessed on 18th September 2025).
49. Willig, A. (2004). Intermediate checksums for improving goodput over error-prone links. https://core.ac.uk/download/35465152.pdf (accessed on 18th September 2025).

CHAPTER 8

INFORMATION THEORY AND CHANNEL CAPACITY

Contents

INTRODUCTION

In the realm of information theory, the concept of channel capacity is important in grasping the limit of communication systems. Channel capacity, or the greatest achievable rate of error-free communication over a given channel, is a notion inherent to quantifying the efficiency of data transmission schemes. The Shannon-Hartley theorem describes such a relationship by establishing one for channel capacity and channel bandwidth as well as for the signal-to-noise ratio. This theorem has profound implications in communication system design and development, guiding engineers and researchers to select appropriate modulation schemes and encoding techniques to optimize data transmission. In addition to this, advances in coding theory, especially in the construction of low-density parity-check (LDPC) codes, also allow us to approach channel capacity further, with the possibility of error correction being important for error-free communications. Research into various coding techniques, for instance, spatially-coupled LDPC codes, has revealed substantial progress in our understanding of channel capacity. In particular, recent studies by Kudetar et al. show that spatial coupling can increase the threshold for MAP decoding by belief propagation over the BEC to be the same as that of the underlying codes' MAP threshold. This result is a departure from the traditional LDPC codes, which have a tendency to require higher column and row weights for parity-check matrices in a bid to achieve comparable performance.

The relevance of these findings extends beyond theory, in the sense that they provide empirical evidence demonstrating that codes like the Hsu-Anastasopoulos (HA) and MacKay-Neal (MN) codes can indeed be effectively used in spatially-coupled structures. The channel capacity research is thus inextricably linked with the real world, underlining the necessity of pioneering coding techniques. Moreover, information theory research does not only encompass data transmission but also ensuring secure channels for communication. To this end, research into secret key agreement protocols, particularly over correlated Gaussian sources, illustrates the delicate interaction between communication rates and security limitations. The findings of the study present a closed-form expression describing the optimal compromise between key generation rates and public communication limits. This separation of discrete and continuous sources is done to illustrate the complexity of information theory when viewed within applied contexts, whereby the requirement for effective coding techniques becomes an overriding issue. Through continued examination of information theory's environments and channel capacity, it seems increasingly that these related concepts not only advance the theoretical realm but also enable us to optimize more effectively the design of robust and secure communication systems under the perpetually shifting technological landscape.

Definition of Information Theory

Information theory, as a branch of applied mathematics and electrical engineering, is a theory for the storage, processing, and transmission of information. It was formulated in the mid-20th century by Claude Shannon, and the theory formally defines the concept of information using such quantities as entropy, which is a measure of the uncertainty of a random variable. It describes the fundamental limitations of data transmission and compression, establishing principles that determine how information can be conveyed effectively across various channels. Central to this field lies the idea of channel capacity, a building block that determines the maximum point in terms of which rate of information can be sent over a communication channel without any error. This work of foundation not only revolutionized telecommunication but opened the door for countless applications in computer science, cryptography, and network engineering. By exploring channel capacity, information theory reveals profound implications for classical and quantum communication systems. Shannon's seminal theorem supplies the bounds of what is achievable with a given amount of noise and bandwidth for classical channels.

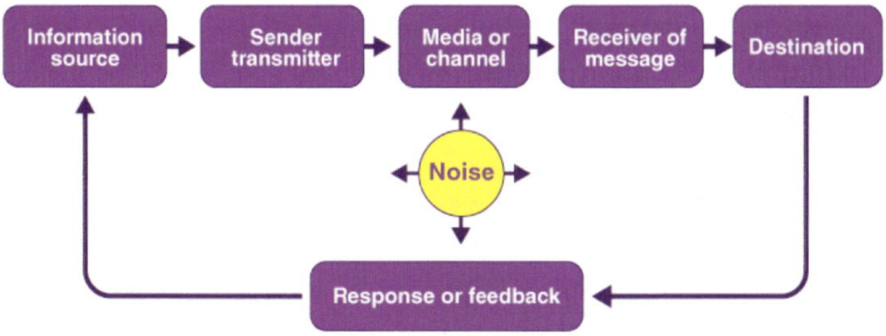

Figure 8.1. *Information theory.*

Source: https://cdn1.byjus.com/wp-content/uploads/2022/07/information-theory.png.

The concept of channel capacity is a boundary in which, in attempting to transmit information across this boundary, there is an increase in error levels, so error-correcting codes are required to ensure fidelity. Quantum information theory adds complexity over classical thinking, such as the locking effect highlighted by recent research. This phenomenon allows the encoding of information beyond classical limits, particularly when employing quantum channels. According to Guha et al. (2013), locking capacity is the maximum locked information which may be retained against eavesdroppers using quantum resources and an arbitrarily small secret key. Its compatibility with real-world applications makes its role even more important in modern telecommunications and computing. Recent advancements, particularly in optical communications, reveal the necessity to maximize channel capacity under various conditions. The evidence indicates that under static point-to-point channels, maximizing mutual information with limited transmit power yields better performance, as revealed in Agrell et al.

(2014). This phenomenon applies not only to straightforward information transfer but has broader implications for interference channels, wherein joint optimization of input distributions can yield significant increases in achievable sum-rate capacities. These findings serve to illustrate the dynamic nature of channel capacity and sensitivity to operational constraints, which is required for uncovering insight into the design and optimization of communication systems within a rapidly evolving technological setting (Figure 8.1).

Importance of Channel Capacity

Channel capacity is a key information theory concept that determines the maximum amount of information that may be reliably transferred through a communication channel. Channel capacity is extremely significant in the optimization of a wide range of various communications systems, from cellular networks to quantum key distribution. As engineers design these systems, they must remember channel attributes such as noise, bandwidth, and modulation techniques. Channel capacity implications are not limited to mere rates of data transmission; they also impact energy efficiency and the sustainability of communication networks. Ayanoglu et al. (2014) say that the traditional wireless networks were developed without energy efficiency as a major consideration, leading to massive carbon footprints. The significance of increasing channel capacity becomes more pivotal as data transmission demand rises, with the need for enhanced design efficiency without compromising performance while reducing environmental exposure. Moreover, the significance of channel capacity assumes greater importance with current communication protocols, such as quantum key distribution (QKD). Here, channel capacity determines how fast secure keys can be produced between two parties without betraying information to possible attackers. The efficiency of secret-key generation depends directly on the efficient reconciliation protocols used in QKD, as noted in Alléaume et al. (2009). As scientists improve more sophisticated approaches, the potential for enhanced secret key rates demonstrates that channel capacity is not only a theoretical principle but an actual reality used as a standard of measure for enhancing communication security and efficiency. The relationship between channel capacity and these new approaches demonstrates its paramount importance in the creation of secure communications technology. Lastly, the pursuit for more channel capacity lies at the core of addressing growing issues in today's communication networks. As mobile data traffic keeps expanding, the inefficiencies of existing network designs must be overcome to handle rising demands for throughput and reliability. By concentrating on channel capacity enhancement, network designers can develop solutions that allow for greater densities of users and better service quality, which is vital in today's interconnected universe. Through the investigation of multi-tier networks, as discussed in Ayanoglu et al. (2014), one can see how effectively managed energy efficiency can bring about immense improvements in channel capacity. Lastly, channel capacity prioritization and its comprehension, not only maximizes the efficiency of communications systems but also facilitates bigger purposes toward developing sustainable technology culture in a rapidly evolving digital landscape (Figure 8.2).

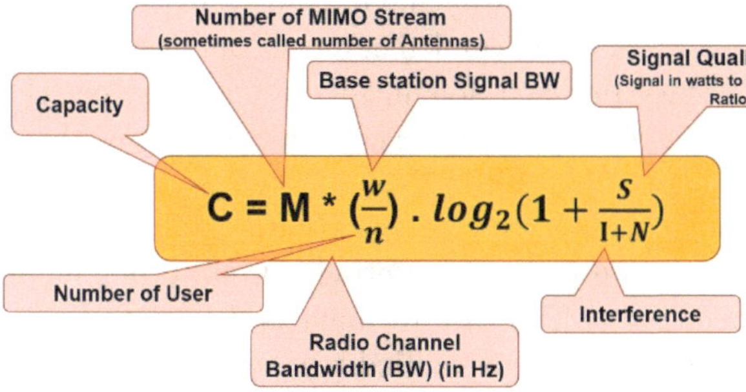

Figure 8.2. *Channel capacity.*

Source: https://media.licdn.com/dms/image/v2/C4D12AQHzau6xCLoZeA/article-inline_image-shrink_1000_1488/article-inline_image-shrink_1000_1488/0/1635483159477?e=1760572800&v=beta&t=REgcdC5EpFG-44JigmLIvE5pv_Mk-DeWv2kPQaVM8PGQ.

Overview of Key Concepts

In exploring information theory's nature and channel capacity, one has to contend with the central concepts constituting the foundation of our understanding of data transmission and communication systems. Foremost among these is channel capacity as the maximum rate at which information may be transmitted over a communication channel without error. This concept is elementary in the sense that it imposes an upper bound on data transmission efficiency, quantifying the interaction between bandwidth, noise, and signal power. In order to increase this transmission capability, various techniques have been devised, such as advanced modulation schemes and error detection codes, which are employed to optimize utilization of given bandwidth as well as counteract the influences of noise inherent in communications systems. In addition, the evolving domain of wireless communication has also led to the exploration of physical-layer security as a means of data protection while in transit. As illustrated in existing research, secure communication and high data rates are major objectives for future wireless systems, particularly the fifth generation (5G) networks (Chen et al., 2018). Physical-layer security becomes increasingly attractive in this regard, as it can provide robust security simply as a function of the wireless medium properties. Leveraging these fundamental properties, researchers propose integrative ideas such as physical-layer service integration (PHY-SI), which integrates multiple coexisting transmission services into a single framework to maximize spectral efficiency without degrading information security and reliability. In addition, the strategic implications of these ideas go beyond theoretical applications into concrete uses, particularly in marketing and organizational performance. The literature of strategic management points out that the principles of information theory also draw on organizational capabilities by expanding product and service delivery in changing environments. Specifically, organizations can draw upon strategic marketing models to expand the insights provided into economic drivers that inform consumer behavior, which in turn impacts the relative fine-tuning of communication strategies towards specified performance levels (Akinyele et al., 2011). As information theory converges with business applications ever more directly, the implications for both fields serve to emphasize the value of building upon cornerstone concepts in order to stimulate innovation and optimize communication efficiencies in an increasingly globalized world.

8.1. HISTORICAL BACKGROUND

The origins of Information Theory have profound roots in the historical development of communication and signal processing technologies. The very first paradigms for communication analysis are traced to mathematicians and scientists trying to quantify the transmission of knowledge and information. Cross-fertilization between the development of telecommunication systems and the evolution of entropy and redundancy concepts provided the foundation upon which modern information theory exists. Notably, at the Arab-Islamic Golden Age, astronomy and mathematics played significant roles with significant contributions, where scientists employed systematic approaches in calculations and measurements to enhance their understanding of mechanisms of transmission. The period underscored the popularization of knowledge, a factor that was key in group innovation in science. Such historical contexts enlighten how cross-cultural exchange and intellectual efforts pave the way for contemporary theory, describing the evolution toward methodology defining channel capacity today (Lelli et al., 2015). In the pre-20th century decades, the intertwining of telecommunications and information theory became more and more unavoidable. Claude Shannons seminal work in the 1940s, wherein he discovered the mathematical foundation of information theory, occurred against this context of rising technological refinement. His pioneering work conveyed the theory of maximum data transmission and the concept of channel capacity, quantitatively described as the utmost rate of information transfer within a communication channel. Shannon's research not only crossed the theoretical limits of data transmission but also addressed practical limits present within existing communication systems. This period in history was probably a paradigm shift, bringing theoretical concepts into applied designs for electronic and digital communications systems. As a result of the fast-moving developments of the period, the significance of Shannon's work cannot be overstressed, as it pushed forward later innovation in coding and modulation schemes that underpin modern telecommunications (Hanzo et al., 2007). Moreover, the evolution of channel capacity continues to be a reflection of past trends in technology and society. Multiple-Input Multiple-Output (MIMO) integration is a great case of how historical problems have conditioned present practice. Researchers have learned how employing sophisticated multicarrier structures can optimize capacity and performance in dense communication environments. It is a result of precursor experiments on signal processing, which already grappled with the complexities of multiple sources of signals vying for bandwidth. The achievements to date reflect the ongoing interplay between ancient practices and modern technological remedies, citing the cyclical pattern of knowledge evolution in this discipline. With escalating demands for effective communication in a globalizing world, history adds value to our appreciation of today's methodologies and innovations in channel capacity (Hanzo et al., 2007) (Figure 8.3).

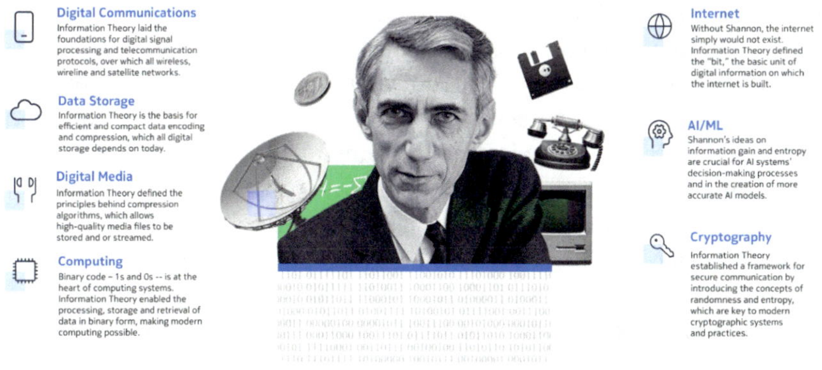

Figure 8.3. *Historical background of information theory.*

Source: https://www.nokia.com/sites/default/files/2024-11/infographic-information_theory_shannon.original.jpg.

8.1.1. Development of Information Theory

The origin of information theory can generally be traced back to the early work of Claude Shannon in the mid-20th century when he developed the basic principles of data transmission and communication. Shannon's seminal paper, A Mathematical Theory of Communication, was the very first to introduce the idea of a communication channel and the mathematics behind its capacity. This capability, as the information transmission rate that could be sent reliably to its limit, was central in acquiring the limits of data transmission protocol. The applicability of Shannon's work goes beyond theory; it gave shape to digital communication systems which pervade contemporary times. His findings on the ways in which noise distorts signal quality and what techniques can be utilized to encode information frugally prove not only his profound understanding of the mathematical nature of information but also their utilitarian application, thereby making him the father of information theory. Shannon's work was later extended by other scholars, and the range of information theory expanded to embrace more and more areas and implications. Scientists began to study complex systems and their behavior when they are subjected to information processing constraints, pointing out how various forms of entropy can be applied in informing decision-making in these systems. Integrating psychological and cognitive models into information theory is an intriguing avenue in human information processing. As described in current studies, biological cognitive systems are plagued by limitations of the sort that accrue to technological systems, where increased requirements lead to a critical point of performance. Such incidents represent the interplay between uncertainty of information sources and the surrounding environment, which shows us that processes linking human cognition and machine communication are very closely related. These forays indicate that an appreciation of information theory requires a multi-disciplinary approach to its broad-reaching applications across disciplines. In addition, the development of information theory has raised significant questions about the impact of structural dynamics on information flow, particularly in systems functioning in a state of stress. Information-processing systems, whether biological or mechanical, suffer issues synonymous with Rate Distortion Theorem constraints under which entropic forces can precipitate ruinous collapse, as in the case of biological systems in disease or breakdown. The analogy that follows that biological processes have lessons for us in terms of technological implements is increasingly true, especially when we observe that mechanism-based models conventionally employed are prone to overlook the nuance of more dynamic biological processes. As suggested, the complexities of biological function and their malfunctions may not just demand creative solutions but also mandate a holistic appreciation like that utilized in psychiatric treatment. This view supports the case for individualized system design in the field of information theory, proposing that custom approaches will better allow for effectiveness and robustness against system stress (Wallace et al., 2008; Wallace, 2010).

8.1.2. Key Figures in the Field

The evolution of information theory and channel capacity has been led in considerable measure by key players in these fields whose research has determined how they are currently understood. Claude Shannon is arguably the most prominent among them, referred to often as the father of information theory. His landmark work provided the foundation for measuring information, inventing the terms of entropy and channel capacity. Shannon's revolutionary 1948 paper, A Mathematical Theory of Communication, introduced the concept of measures of the limits of information transmission via noisy channels, a concept that remains fundamental to modern telecommunication. The fundamental ideas established by Shannon not only revolutionized communications engineering but have since been applied to numerous domains, from data compression to cryptography. Hence, familiarity with his work is essential to the understanding of current advancement in information theory. Subsequently, the topic has received more contributions from some other notable individuals, such as David Slepian and John K. Wolf, who elaborated on Shannon's original ideas regarding channel capacity. They came up with advanced concepts regarding communication systems that operate with indefinite or incomplete information, especially in their work on the blind index coding (BIC) problem. This problem illustrates a case in which a sender must transmit varying messages to various users without complete access to side information known to each user. Slepian and Wolfs work, which emphasizes the pivotal position of channel capacity in the context of uncertain environments, has implications for current use in wireless networks and cache systems. By creating hybrid coding schemes that make use of side information, they have evolved methods that optimize transmission rates with the least interference, greatly impacting research trends today. Further, the spread of multilateral aid structures mirrors the increasing complexity inherent to communication networks, which relates back to information theory fundamentals. Looking at the difficulties of multi-bi aid, often earmarked for specific purposes, highlights the necessity of robust data management and robust communication tactics within many networks. As outlined in recent evaluations, improving the effectiveness of aid involves overcoming fragmentation and alignment with recipient-country priorities, drawing parallels with difficulties experienced in channel capacity when dealing with streams of data. Through the application of information theory principles to optimize channels of communication, stakeholders can optimize legitimacy and efficiency in aid. Thus, the integration of information theory and real issues like multi-bi aid identifies the importance of such impactful personalities in leading not only theoretical frameworks, but also real-world applications, which are still evolving in today's interconnected world.

8.1.3. Milestones in Channel Capacity Research

Channel capacity milestones form a fundamental aspect of the entire information theory framework and have a deep effect on present-day communication systems. Having the ability to understand the maximum amount of information that can be communicated via a certain channel without any errors is the very core of determining the most ideal data transmission method. Claude Shannons contribution in 1948 was revolutionary in presenting the concept of channel capacity as a foundation stone for telecommunications. His formulation of the Shannon-Hartley theorem provided a mathematical expression that calculates channel capacity based on bandwidth and signal-to-noise ratio, enabling scientists and engineers to develop systems that maximize data rates with minimized errors. This was a milestone moment not only in laying the groundwork for future technological advancements but also in highlighting the necessity of a systematic exploration of how different parameters influence communication efficiency. Other advances in channel capacity research have been focused on various communication contexts, adapting Shannon's theorems to meet modern applications. These achievements involved studies in multi-user contexts, where the development of concepts such as multiple-input multiple-output (MIMO) technology greatly improved capacity without offering additional bandwidth. This achievement demonstrated the potential of increasing channel capacity through leveraging spatial dimensions, confirming that the capacity of a channel is not fixed but rather a dynamic variable that can be improved by innovation. The use of advanced coding techniques directly aligned with reliability and efficiency in communication systems, further confirming the assertion that constant research is warranted. As scientists continue to advance the frontiers of new approaches and new mediums, the pursuit of further capacity remains a prime concern area, signaling the relevance of continued research in this field. Beyond theory development, practical applications in channel capacity work have permeated industries, redefining the manner in which information is exchanged. Theoretical frameworks of new programs in public health interventions, such as behavior change campaigns, have relied on information transmission efficiency knowledge to most efficiently use outreach efforts, demonstrating the broader social relevance of channel capacity research findings. Such cross-disciplinary applications demonstrate how channel capacity research may transcend disciplinary boundaries, applying to subjects as diverse as environmental science and robotics. By researching the convergence of information theory and real-world communication challenges, researchers seek to push the physical limits of data transfer higher, opening the doors to future progress and globalisation. As shown by the efforts in fields like microbial fuel cell technology and robotics, the use of channel capacity principles can lead to remarkable innovation in energy-efficient designs and optimum communication systems, showing how much this research has impacted a lot of aspects of life today (Figure 8.4).

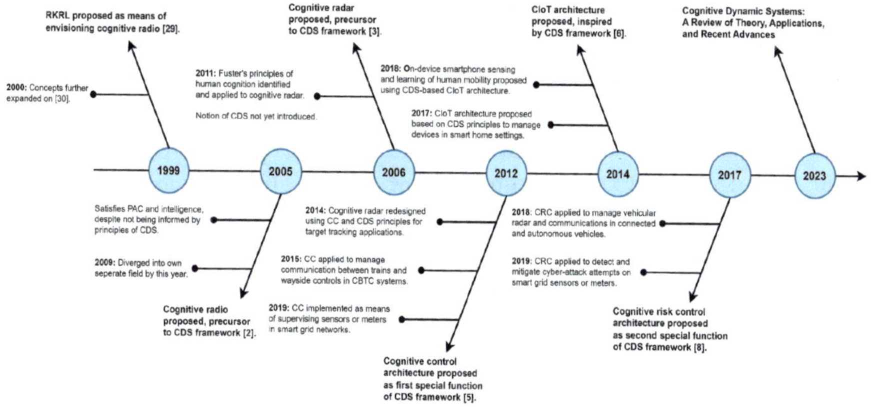

Figure 8.4. *Milestones in channel capacity research.*

Source: https://www.researchgate.net/publication/370835126/figure/fig1/AS:11431281159388676@1684345857026/Timeline-of-the-major-milestones-and-advancements-of-the-field-of-CDS-as-surveyed-in.jpg.

8.2. FUNDAMENTAL CONCEPTS

In the realm of information theory, fundamental concepts serve as the building blocks for understanding complex communication systems. Central to these concepts is the idea of entropy, which quantitatively measures the uncertainty or information content inherent in a random variable. By applying this principle, one can derive the maximum amount of information that can be transmitted over a communication channel without error. Entropy not only informs the design of efficient coding schemes but also provides insights into the limits of message transmission under various conditions. Emphasizing the relevance of these fundamental concepts allows researchers to explore more intricate dimensions of information transfer, laying a foundation for innovations in data compression and error correction, essential in today's digitally-driven society.

This theoretical groundwork is crucial as it guides practitioners in navigating the challenges of real-world communication scenarios. The infinite diversity of information systems highlights the importance of channel capacity, defined as the maximum rate at which information can be reliably transmitted over a given channel. Understanding channel capacity involves considering various factors such as bandwidth, signal-to-noise ratio, and modulation techniques. The Shannon-Hartley theorem provides a fundamental limit, outlining the relationship between these factors and channel capacity, emphasizing the need for effective strategies in optimizing communication performance. However, channel capacity is not merely a theoretical abstraction; it directly informs the practical design of systems, from wireless networks to satellite communications. By exploring these core principles, one can grasp the various constraints that affect information transmission, providing a deeper appreciation of how engineers and scientists strive to maximize data throughput while minimizing errors in increasingly complex environments. Advancements in quantum information theory have also enriched the discourse surrounding fundamental concepts in communication. Notably, quantum Gaussian states play a pivotal role in expanding traditional information paradigms. These states are not only manipulable through existing technologies but also facilitate various quantum tasks such as teleportation and cryptography, as highlighted in recent studies (Hayashi et al., 2008). By extending the study of quantum states to encompass continuous variables, researchers are uncovering new opportunities for optimizing channel capacity. Additionally, the interplay between radar and communication systems illustrates the potential for cooperative signaling, transforming how we perceive interference in these domains. The investigation into joint systems reveals that effective communication and radar operations can, under appropriate conditions, enhance overall performance, emphasizing the relevance of fundamental concepts in developing integrated signaling strategies (Bliss et al., 2014). This holistic perspective underscores the significance of innovative approaches in navigating the complexities of modern information systems.

8.2.1. Entropy and Information Content

Entropy in information theory is a basic principle that quantifies the uncertainty of a random variable. Entropy is actually the probabilistic character of information, and it was by this definition that Shannon gave a mathematical framework that makes information content measurable. Specifically, entropy is commonly defined as the mean quantity of information produced by a stochastic data source. This parameter not only informs information systems of the efficiency of data coding but also of the limitation caused by the source's inherent uncertainty. With various coding schemes being deployed, familiarity with the entropic nature of information becomes crucial in order to maximize channel capacity, ensure information transmission fidelity under varying conditions, and optimize overall system performance by maximizing bandwidth and resource utilization (Cheng et al., 1995). Information content, deeply related to the concept of entropy, is the amount of uncertainty reduction at message reception. This amount is extremely important in encoding systems because it specifies the least number of bits to be used in the transmission of information without loss.

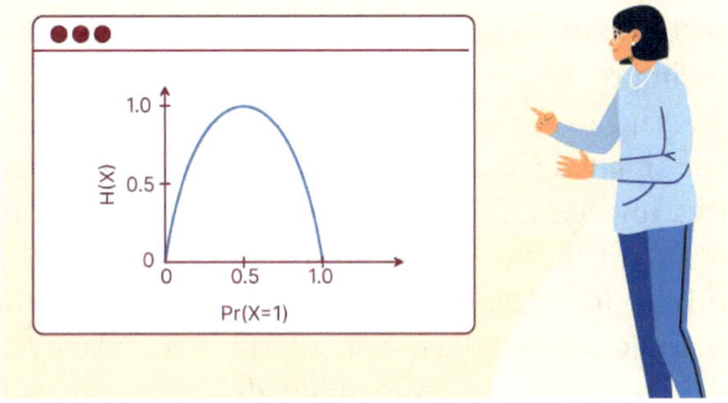

Figure 8.5. *Info entropy in information theory.*

Source: https://images.shiksha.com/mediadata/ugcDocuments/images/wordpressImages/2023_02_MicrosoftTeams-image-156.jpg.

The entropy-information content relationship highlights the significance of efficient coding schemes using this relationship in the elimination of redundancy and optimization of transmission efficiency. By developing advanced coding techniques, including those explained in both classical and quantum systems, researchers are better able to deal with the complexities of error control and signal integrity. For instance, recent studies on position-based coding highlight its promise in error bounding and advancing our understanding of quantum channel communication, hence adding more insight to the analysis of information content and its practical applications in various scenarios (Qi et al., 2017). In information theory usage, entropy and information content are both critical tools for determining channel capacity—the rate at which information can be communicated reliably. Their comprehension provides insight into the limits of communication systems and directs the strategy for reaching beyond them. For example, techniques in the conventional approach to determining channel capacity concentrate on the need for precise encoding that reduces error rates and maximizes throughput. The inclusion of concepts such as rate distortion theory and multiuser systems is a sign of the ongoing attempt to generalize such fundamental concepts into more complex situations. As technologies advance, the ability to study entropy and information content is a valuable tool to support diverse communication needs, ultimately deciding telecommunications, storage, and secure communication advances (Cheng et al., 1995) (Figure 8.5).

8.2.2. Redundancy and Noise

In the field of information theory, redundancy and noise are the basis of the theory of channel capacity. Redundancy refers to the inclusion of extra bits in a transmission that do not bring new information but have a function of error detection and the improvement of the reliability of communication. Channels become more resistant to the harmful effect of noise—unwanted signals that distort or mask the signal by including redundancy in the transmitted messages. Noise can be of various types, that is, electronic noise or environmental noise, and holds significance in the impact on information integrity. Where redundancy is effectively utilized, it assists receivers in differentiating the intended signal from the noise so that decoding occurs properly. Redundancy serves not only to increase the data reliability transmitted but also serves to contribute towards achieving maximum overall channel capacity, which is a significant factor in communication systems. The connection between noise and redundancy is pertinent to actual applications, particularly coded transmission plans. A case in point is the fact that recent advances in low-density parity-check (LDPC) coding have shown how multiple levels of error protection can be used to distinguish confidential from public messages in noisy environments. As highlighted in the work assigned for the Gaussian broadcast channel with confidential messages, two approaches can neutralize redundancy: unequal error protection (UEP) and non-UEP LDPC codes. By adjustment of the redundancy to provide higher protection to more important bits of information while maintaining overall transmission efficiency, these methods operate positively to neutralize noise impacts. This capability is crucial in instances where unauthorized receivers may intercept part of the information, thereby establishing that strategic redundancy use has a direct capacity to increase the reliability and capacity of communication channels effectively counteracting the interaction with noise. In addition, the study of feedback in channel capacity again emphasizes the employment of redundancy for the regulation of noise with minimal delay. The findings of more recent work demonstrate that feedback will be able to significantly extend the method of channel capacity, allowing systems to adaptively correct errors due to noise. For example, the use of rate-compatible sphere-packing (RCSP) analysis in hybrid automatic repeat requests (ARQ) systems illustrates that redundancy can be dynamically adjusted, with significant reductions in delay while maintaining high information transfer rates. Numerical optimizations show that, with efficient redundancy control, channels could achieve 92% of capacity with extremely short block lengths, illustrating that efficient focus on redundancy and noise control could produce drastic efficiency gains as well as reliability. This blending of redundancy methods with noise reduction not only illuminates theoretical models but highlights real-world applications in real communication systems, ultimately furthering our knowledge of information theory and channel capacity.

8.2.3. The Role of Probability in Information Theory

Probability is information theory's underlying construct that affects how information is encoded, quantified, and transmitted. Essentially, information theory is trying to quantify the amount of uncertainty present in messages, and probability provides the means for such a quantification. Channel capacity, another central information theory building block, is heavily grounded in probability models that explain how information travels through different channels as well. They also factor in noise and interference, which can greatly affect communication quality and efficiency. With probabilistic methods, researchers are able to plot the limits of data transmission and how one would go about optimizing these limits in order to convey as much information as possible (Hassibi et al., 2001). This stochastic model not only embodies the intricate dynamics of multiple communication channels but also opens up possibilities for wireless communication innovations and capacity improvement. In channel capacity assessment, probability also illuminates the significance of mutual information, a fundamental measure in information theory. Mutual information quantifies the measure of information carried by one random variable about another, forming a critical connection in decoding and error analysis. Probability and mutual information together dictate the amount of information that can be sent with certainty through a channel, particularly under conditions such as noise and fading (Jafar et al., 2012). With multiple antennas, for instance, reliance on probability becomes essential while computing the density of received signals, thereby making it simpler to approximate mutual information under varying configurations. This intricate interrelationship emphasizes the manner in which probability not only provides for the theoretical demarcation of communication but also prods practical approaches to enhancing data transfer in actual implementations. On top of this, probability's application extends into error correction and network topology in communication systems. With probabilistic modeling, engineers can forecast data transmission behavior and design techniques for overcoming errors caused by noise and interference. This is especially relevant in heavy communication networks, where there are several users who attempt to transfer data simultaneously. When this occurs, understanding the interference probability and strength of various links facilitates better resource allocation. For example, research has established that awareness of equal strength interference links as bottlenecks can allow optimal capacity strategies in phase-fading interference networks (Jafar et al., 2012). By applying probabilistic analysis, information theorists can construct robust systems capable of dealing with a broad variety of communication issues while trying to achieve full channel capacity (Figure 8.6).

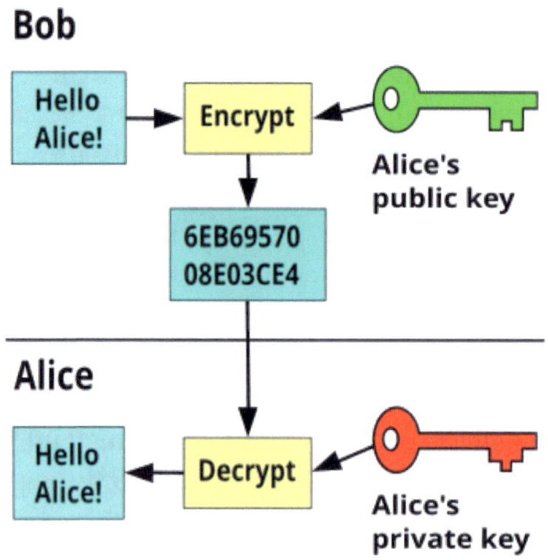

Figure 8.6. *Probability theory and information theory.*

Source: https://storage.googleapis.com/static.prod.fiveable.me/search-images%2F%22Random_variables_and_distributions_in_cryptography%3A_probability_theory_uncertainty_key_generation_and_decryption_outcomes.%22-375px-Public_key_encryption.svg.png.

8.3. MATHEMATICAL FOUNDATIONS

The discipline of Information Theory highly engages with mathematical structures that allow scholars to model and analyze communication systems efficiently. Fundamentally, Information Theory provides fundamental measures such as entropy, which quantifies the uncertainty in a message, hence allowing scholars to compute the optimal encoding needed in transmission across noisy channels. The interaction between the mathematics and applications in real life is the focus, especially in systems that must operate under varying constraints and operating conditions. This type of strict mathematical modeling is at the core of addressing today's issues in telecommunication and computer science, especially when accuracy and dependability are required. As highlighted by ongoing research, mathematical constructs like the ones derived from Blum–Shub–Smale (BSS) computability theory shed light on the theoretical and practical boundaries of information exchange (Boche et al., 2025). Furthermore, ongoing innovation in state-of-the-art communication systems relies heavily on the mathematical underpinning that derives channel capacity. The movement toward "Faster-than-Nyquist" (FTN) transmission modes is a prime example of this, where conventional capacity limits are met and being pushed by innovative methodologies designed to maximize data rates in the presence of inevitable intersymbol interference. Description of these systems requires advanced mathematical models that can reveal insight into achievable specific capacities and error rates under given conditions. The use of techniques such as time resolution theory and pulse amplitude modulation (PAM) methods offers new avenues for theoretically calculating error performance and capacity, serving as useful tools for engineers. By practicing finding modified mathematical models, one can assess realistic constraints while enhancing the overall sense of channel capacity (Lerner et al., 2024). A deep understanding of the mathematical foundations of Information Theory not only enhances theoretical studies but also inspires innovation in practical applications. The development of rigorous mathematical models constitutes a basis upon which valid predictions and analyses can be inferred, leading ultimately to directing the design and construction of efficient communication systems. Future research in Information Theory will likely be reliant on such sound mathematical foundations, as researchers and professionals alike strive to address the growing demand for more rapid and efficient means of communication. As technology advances, continued refinement of these mathematical models remains imperative in uncovering new potential and in ensuring that systems can handle the challenges of more complex settings. As analog computing advances and new signal processing techniques emerge, the promise of revolutionary innovations in communication technology hangs in the balance, spearheaded by principles established through mathematical profundity and creativity.

8.3.1. Shannon's Theorems

Shannon's Theorems are the pillars of Information Theory, describing the data transmission limits of information through various communications channels. The channel capacity theorem, the first theorem, describes a maximum transmission rate, referred to as channel capacity, at which information may be sent with an error probability arbitrarily close to zero when there is sufficient signal strength and time. This concept transforms the manner in which theorists and engineers cope with communication systems by emphasizing the importance of good quality codes that are of low error rates and yet optimize the throughput. Through establishing these limits, Shannon's theorems create a crucial basis for the evaluation and optimization of communication protocols in numerous applications, thus achieving advancements in wireless communication and data networking. In order to examine the significance of Shannon's Theorems, one has to consider their use in real-life scenarios, particularly within the domain of layered coding and multicast/broadcast communications. The second theorem extends Shannon's by considering the performance of codes for various channel conditions, showing that it is possible to achieve reliable communication even if error probability does not approach zero.

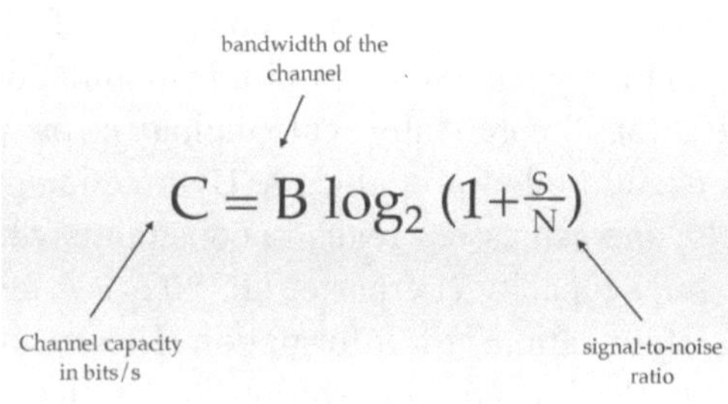

Figure 8.7. *Shannon's theorems.*

Source: https://files.speakerdeck.com/presentations/9c716593a1974b3bac610c3b95c07515/slide_69.jpg.

This is particularly relevant for systems with multiple users, in which users experience different channel conditions. Methods like successive refinement (SR) and superposition coding make an adaptive approach to data transmission possible, where power allocation for layered signals can be optimized to address the multi-perspective challenges caused by time-varying channel conditions and maximize total throughput. These methods illustrate how Shannon's axioms may guide the development of high-level algorithms to address the demands of modern communication networks (Shi et al., 2009). In addition, Shannon's Theorems have a key role in supporting theoretical research, particularly with regard to asymptotic estimates under non-vanishing error probabilities. This is an improvement on acknowledging communication limits beyond absolute asymptotic studies towards addressing cases where error rates are limited to non-zero bounds. Its use is observed in contemporary network information theory, where researchers study optimal coding rates and how to achieve such with realistic assumptions about channel diversity and signal distortion. For instance, studies concentrate on various communication strategies, including the use of multiple description coding and superposition coding to constrain distortion in the face of multi-user settings. Meanwhile, these studies continue, scientists are creating robust frameworks that build on Shannon's early theorems, which will tackle harder communications challenges and inform future innovations (Tan et al., 2015) (Figure 8.7).

8.3.2. The Noisy Channel Coding Theorem

The Noisy Channel Coding Theorem is a cornerstone of information theory that solves, in a general manner, the problem presented by noisy channels. According to the theorem, regardless of what type of noise degrades a channel, it is possible to transmit information at some rate, referred to as the capacity of the channel, with extremely low probability of error. The development of this theorem truly shifted the focus from merely studying the physical aspects of communication channels to understanding the mathematical basis for information transmission. It truly provides a mathematically sound method for encoding data such that data can be effectively transmitted over an erroneous channel, thereby allowing communication systems to achieve optimal performance even at sub-optimal conditions. At the core of the Noisy Channel Coding Theorem is the concept of error correction. With the help of various coding techniques, data can be structured so that it can mask potential errors due to noise. The theorem provides means by which redundancy may be employed effectively to identify and correct errors at the receiving end without requiring the retransmission of the original data.

This theorem is the foundation for most contemporary coding techniques utilized in digital communications, such as convolutional codes and turbo codes. Further research also finds newer protocols such as the channel-state coding protocol extending the boundaries of this theorem even further to allow for more powerful forms of coding techniques optimized for new quantum communication paradigms (Hsieh et al., 2010). New work in network coding has also helped to extend the importance of the Noisy Channel Coding Theorem by demonstrating that it is not exclusive to traditional environments. Specifically, the integration of new schemes for application within noisy networked environments demonstrates how the operations of the theorem can be mapped to provide greater communication effectiveness within more challenging environments. These innovations widen the capability of multi-source communication where the coding strategies can offset the particular difficulties of different noisy channels, increasing network capacity and reliability overall. For instance, recent relay channel schemes minimizing interference and successfully splicing signals unveil more understanding into capacity limits in disparate network structures, demonstrating theorems sustained relevance in contemporary information theory (Chung et al., 2010).

8.3.3. Capacity Formulas and Their Applications

The finding of capacity equations in information theory is at the core of determining the quality with which the information can be transmitted through various types of channels. Capacity, or the maximum achievable rate of transmission of information with reliability, is strongly dependent upon the nature of the channel, including noise and signal distortion. In investigating channel models, there is a need to derive precise capacity formulas that take into account various constraints and conditions. The derivation of random-coding exponents and their relation to channel types, for instance, compound discrete memoryless channels (CDMC), provides important information about error rates in transmission. This material forms the basis of real-world use like secure communication protocols and data encoding techniques with maximum capacity, which play a fundamental role in contributing to the evolution of telecommunications and data storage systems. In real life, formulas of capacity are extensively used across various industries, particularly in designing reliable communication systems. For example, the use of capacity formulas assists in defining watermarking and data hiding protocols for information integrity maintenance against malicious attacks. Moreover, such formulas assist in guiding the encoder and decoder to improve the process of transmitting data in cases of partially known channel states, thereby enhancing robustness against unknown conditions. One of the most significant contributions in this direction is the stacked binning scheme that comes together with maximum penalized mutual information decoders to achieve greater random-coding exponents. This work enlightens mechanisms for building systems that maximize throughput as well as minimize error probabilities and is crucial for application in real-time communications as well as multimedia streaming. In addition, channel capacity results have wider implications within other theoretical and practical models relevant to high-dimensional statistical inference. Within that environment, eigenvalue distributions and associated finite models, for example, non-central and correlated zero mean Wishart distributions, become most significant. These advanced mathematical techniques explain the behavior of large data sets and assist scientists in dealing with problems where asymptotic analyses are irrelevant. Creating a deep understanding of such relations with the aid of algorithmic methods enables significant progress in channel decoding and channel design. The union of free probability theories with information transfer creates a deeper understanding of capacity-aware approaches, thereby widening the horizon for development in adaptive communication systems, as noted in recent researches on both fundamental and application problems in information theory (Moulin et al., 2006; Debbah et al., 2010) (Figure 8.8).

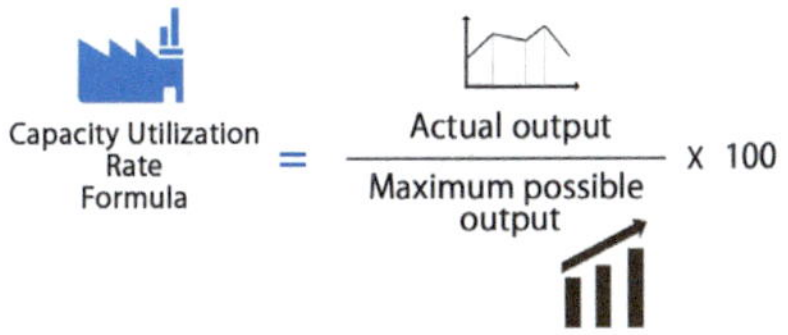

Figure 8.8. *Capacity formulas.*

Source: https://cdn.educba.com/academy/wp-content/uploads/2018/12/Capacity-Utilization-Rate-Formula.jpg.

8.4. TYPES OF CHANNELS

The classification of channels in information theory is significant in defining the transmission of data over various mediums. Channels are principally classified into a variety of types, including noisy and noiseless channels, discrete and continuous channels, and more specialized channels like finite-state channels. Noisy channels are characterized by the introduction of errors during transmission, where external influences can alter the signal. In comparison, noiseless channels operate in ideal conditions, preserving the integrity of information transmitted. Understanding these distinctions is important in designing communication systems to effectively deal with the complexities and limitations offered by different types of channels. Advances in information theory have also continued to shed more light on channel behaviors, particularly through channel capacity formulation that takes into account both noise and the specific properties of the medium of transmission. In the framework of finite-state channels, there is yet another subdivision that emphasizes the demands put upon feedback mechanisms. Such channels are particularly noteworthy for the capacity to allow time-invariant feedback, which can greatly enhance reliability in communication. Research has indicated that feedback does not necessarily enhance the capacity of channels, particularly compound channels like the Gilbert-Elliot channel, for which there is no measurable benefit from the availability of feedback compared to non-availability. The development of universal decoders for finite-state channels is a best example of developments in the field that ensure efficient decoding regardless of the specific properties of the channel employed. The developments are a showcase of the necessity of understanding the dynamics of the underlying channels for the best information transmission and ultimately improved performance in practical applications. Further study of quantum channels has provided a new perspective of classical channel frameworks. Quantum channels, driven by quantum mechanics theory, have concepts such as entanglement-assisted classical capacity running parallel to Shannon's theory classically. This added factor enables the examination of channel behaviors under quantum constraints and determination of the relationship between entanglement and information capacity. Specifically, convergence rates of entanglement-assisted codes have been investigated, and it has been demonstrated that the rates converge towards capacity for increasing channel uses. This helps to highlight the intricate nature of quantum channels and their distinct properties, showing how they stretch classical communication theory. As research in this field advances, the effect on classical and quantum communication systems will keep driving inventive developments in information transmission and processing techniques, closing the gap further between theory and real application in the evolving technology of today.

8.4.1. Discrete Versus Continuous Channels

The distinction between continuous and discrete channels is the foundation for knowledge of information theory and channel capacity. Information is transmitted via a finite number of symbols in discrete channels, which are highly suitable for systems like digital communication systems. The signal can be exactly specified and quantified, and optimal schemes of coding, for example, error-correcting codes, may greatly enhance data transmission reliability. Conversely, continuous channels handle an infinite collection of signal values, typically in conjunction with analog communication systems, such as video or audio signals. With continuous channels, the issue is managing noise, which greatly degrades signal quality. Thus, knowing the capacities of these channels at different levels of noise is significant for degradation minimization and optimal performance in real-world applications. In channel capacity, discrete channels have easier computational models, with closed-form capacity expressions. To illustrate, the binary symmetric channel, a simple model of discrete transmission, has an easy derivation of capacity and allows exact prediction of optimal rates of reliable communication. Continuous channels are made difficult to analyze by the infinite dimensional signal space. For example, the capacity of an additive white Gaussian noise (AWGN) channel heavily relies on the signal-to-noise ratio and possesses smooth characteristics which take high-level mathematical tools such as functional optimization and Monte-Carlo simulations to identify capacity bounds in an effective way. This refers to the intricacy of continuous channels, whose capacities are determined by numerous variables like noise behavior and channel conditions, making it hard to create efficient communication systems. Discrete versus continuous channel use is also better understood by research into their secrecy capacity, particularly in wiretap channels. In all these cases, the analysis will reflect that the secrecy capacity of individual channels can be suitably bounded and analyzed in terms of traditional lower bounds, as explained in work that extends theory to arbitrary sets of uncertainty. On the other hand, the settings of continuous channels, including those in the presence of Gaussian noise, demonstrate the necessity of optimal signaling techniques, which can offer a compound capacity equivalent to the worst-case capacity. This elucidates the necessity of tailored signaling for continuous channels to achieve reliable performance limits with the guarantee of secrecy in communication. Understanding these differences has a direct impact on how theorists and engineers address problems like channel design, error correction, and the development of secure communication protocols in various environments, hence the importance of understanding the differences between discrete and continuous channels in information theory and channel capacity studies (Figure 8.9).

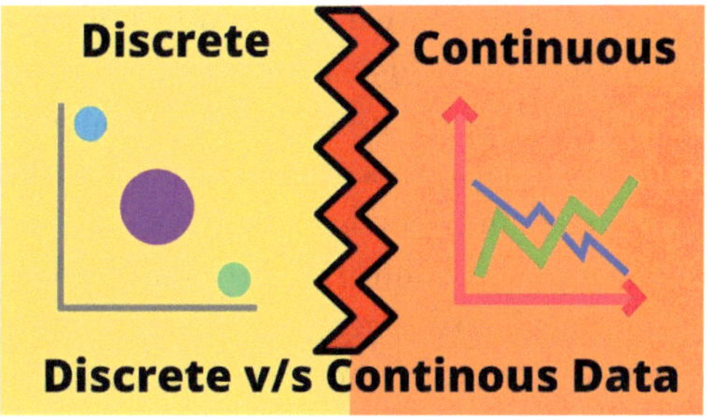

Figure 8.9. *Discrete versus continuous channels.*

Source: https://i.ytimg.com/vi/Cg0W6mod9Hw/maxresdefault.jpg.

8.4.2. Linear and Nonlinear Channels

The theory of linear and nonlinear channels is fundamental in establishing information theory's inherent limits of communication systems. Linear channels are characterized by the mere superposition of signals, allowing one to count on linear behavior when carrying information. The predictability is such that it guarantees that the capacity of such channels can theoretically be extended without bound with the energy of the input signal, a desirable aspect for a communication system designer. Nonetheless, with the development of information systems, the physical nature of nonlinear channels has to be realized. Such channels tend to exhibit more intricate behaviors that may stray far from linearity, incorporating a range of distortions and interferences that impact throughput and data integrity of information conveyed through these channels. As such, understanding the differences between linear and nonlinear channels is necessary to maximize channel capacity and enhance communication performance under practical operating conditions. While linear channels make computation of channel capacity straightforward, nonlinear channels present the challenge unique to their nature that can limit capacity even as power input is increased. In the case of optical fibers, for instance, the nonlinear propagation of light presents challenges of sustaining a growing capacity due to intricacies introduced. As in Mitra et al. (2000), nonlinear interactions among fibers create measurable physical bounds, which are in direct opposition to linear additive noise channel predictions. Thus, channel capacity in these nonlinear channels does not similarly increase without bound with power and instead reaches a peak value. This limitation demands innovative theoretical approaches, necessitating the reframing of thought in developing successful communication systems operating within the constraints of nonlinear channels to ensure that engineers and researchers are able to adequately address such issues. Compounding further the analysis is the manner in which various types of channels interact, particularly when integrating components of memory and interference. For example, memoryless vector channel performance shows that channel capacity is an increasing function of power, as in Agrell et al. (2014). This is due to the fact that power control is part of achieving maximum mutual information, particularly for interference channels which often are confronted with interfering signals that reduce overall performance. Nonlinearity in such contexts, however, specifically in wavelength-division multiplexing systems, also finds its use in exhibiting the significance of understanding such dynamics in detail. The relationship between power, interference, and channel intrinsic behavior compels a high-level understanding of linear and nonlinear models to aim for maximizing channel capacity in modern-day communication systems.

8.4.3. Real-World Examples of Communication Channels

The research of real-world communication networks reveals the richness of information flow through various circumstances. Telecommunications wireless networks, for instance, demonstrate the challenge introduced by interference and capacity limitations. Multiple simultaneous signals are managed by these networks, necessitating sophisticated techniques toward increased efficiency. One of these techniques is interference alignment where signals are intentionally aligned in order to minimize destructive interference. This method enables multiple nodes in the network to utilize a common frequency spectrum, leading to greater throughput data. Therefore, understanding the channel capacity in such scenarios does not only shed light on practical applications but also helps with theoretical work in information theory, as can be seen in the work on ergodic interference alignment and its implications on large wireless networks (Aldridge et al., 2011).

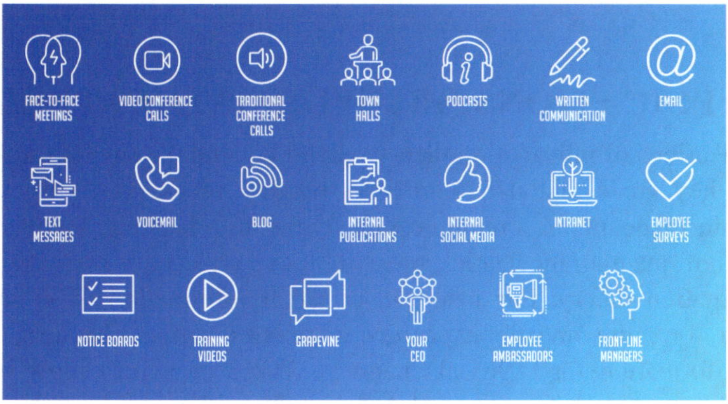

Figure 8.10. *Real-world examples of communication channels.*

Source: https://www.yourthoughtpartner.com/hubfs/Blog_images/Communication-channels-in-business.png.

Another rather fascinating case is found in optical communications, that of bosonic Gaussian channels. The latter exhibit an unusual quantum capacity behavior; quite fascinatingly, standalone channels may have zero capacity, yet pooling them can yield positive capacity through a mechanism known as super-activation. This strange characteristic only serves to highlight the complexities at play in quantum communications and demands further study into the resources required to attain the maximum channel performance. The need to compress, a quantum resource, makes it even harder to establish this relationship, and it conditions the structure and use of channels. These results not only have a profound impact on quantum technology applications but also offer increased knowledge of channel capacities in different conditions, and their implications are of far-reaching value to information theory (Giedke et al., 2013). Apart from wireless and optical channels, research into communication in sensor networks also provides further details regarding channel capacity and data transmission methods. These networks rely on multiple small, decentralized sensors transmitting messages to centralized systems, often in bandwidth-limited environments. To overcome this challenge, scientists have come up with techniques such as compressed sensing, which takes advantage of the sparsity of signals to reconstruct data from fewer samples than conventional methodology. Through such novel approaches, such systems are able to optimize information transmission while ensuring optimal utilization of available resources. Therefore, observation of these real channels of communication provides substantial information about the application of information theory, outlining how channel capacity can be optimized in different contexts (Figure 8.10).

8.5. APPLICATIONS OF INFORMATION THEORY

The applications of information theory extend beyond the simple transmission of data; they are essential to numerous modern techniques, particularly in the domain of quantum communication. Quantum channels, founded on the principles of quantum mechanics, present unique challenges and opportunities for information theory. For instance, coherent information is an important parameter for estimating the capacities of noisy quantum channels, just like mutual information for classical channels. These capacities must be found since different applications demand different channel capacities, which can be found tightly bounded by computing the properties of coherent information. The implications of such conclusions reinforce the use of information theory models to a larger degree so as to soak up the complexity that exists in quantum information systems, which would naturally yield greater fidelity and trustfulness in communication in quantum networks (Barnum et al., 1997). Apart from its use towards quantum communication, information theory has enormous applications toward creating secure communication systems, especially wireless networks. Information-theoretic security offers a solid foundation that exploits the randomness of communication channels to establish secure connections between transmitters and receivers. One of the viable methods is the development of the intrinsically secure communications graph (iS-graph), which enables the description and analysis of secure connectivity in intricate systems. From this foundation, authors have developed methods such as sectorized transmission and eavesdropper neutralization, significantly improving network security and connectivity. Such techniques not only enhance the robustness of secure channel codes but also provide opportunities for future wireless communication systems to securely guard against potential threats (Barros et al., 2010). Moreover, the complementarity of information theory and channel capacity has created new avenues of applications in data compression and error correction. Data compression is the encoding of data in a compact and efficient form, minimizing redundancy without losing the information that can be employed for faithful reconstruction. This is directly connected to channel capacity because the largest rate of error-free data transmission is bounded by the characteristics of the channel itself. In addition, error correction codes exploit the boundaries established by information theory to facilitate effective communication over noisy channels for reconstructing original messages even in the face of distortion. The fact that such applications demonstrate how fundamental notions of information theory can be extrapolated for solving problems in real-world environments and contribute to technology and communications optimization is evidence of their relevance.

8.5.1. Telecommunications and Data Transmission

Pioneers in telecommunication and data transmission technologies have been at the forefront of boosting the efficacy of information transmission through various media. Technologies such as cloud radio access networks (C-RANs) provide a revolutionary approach to fifth-generation (5G) wireless communications systems, radically transforming the manner in which data capacity is being provided. C-RANs reduce capital expenditures as well as operational expenses while maximizing spectral efficiency (SE) as well as energy efficiency (EE) in telecommunications. The fronthaul, as the transmission channel between remote radio heads and baseband units, must cater to high capacity but often encounters constraints that prevent it from optimal functioning. Compression, quantization, and synchronized processing are some of the strategies required in reducing such obstructions, thus highlighting the crucial role of efficient management of channel capacity in modern telecommunications networks (Lau et al., 2015). Aside from wireless communication innovations, power line communication (PLC) has been of particular interest as a competitive data transfer option. The technology makes communication possible over existing power wiring, thus eliminating the need for installing new wire-based infrastructures. Narrowband and broadband PLC systems provide universal solutions for a variety of applications, ranging from smart grid technologies to home automations. With researchers going deeper into PLC's physical layer performance and channel characterization, it is crucial to realize the effects on medium access and higher-layer specifications. Incorporating these systems into the network infrastructure makes having an efficient regulatory and standardization mechanism crucial in order to deliver flawless performance and reliability in data transmission. More study of PLC will become necessary in order to counter the existing problems and ensure its widespread application (Cano et al., 2016). Moreover, the interplay between various telecommunication inventions and information theory principles further educates us on channel capacity for data communication. Information theory, as pioneered by researchers like Claude Shannon, is a mathematical framework for calculating the amount of information that can be transmitted reliably through a given communication channel. The impact of channel constraints, such as interference and noise, necessitates the formation of sophisticated encoding and modulation techniques to permit maximum throughput. Furthermore, the interoperations of heterogeneous network environments, such as a combination of different technologies like PLC and C-RAN, call for an in-depth understanding of channel capacity to facilitate maximum performance. By integrating these concepts, researchers and practitioners can plan better for the application of telecommunications technologies in a way that data transmission is robust under varied conditions and according to the information theory guidelines (Figure 8.11).

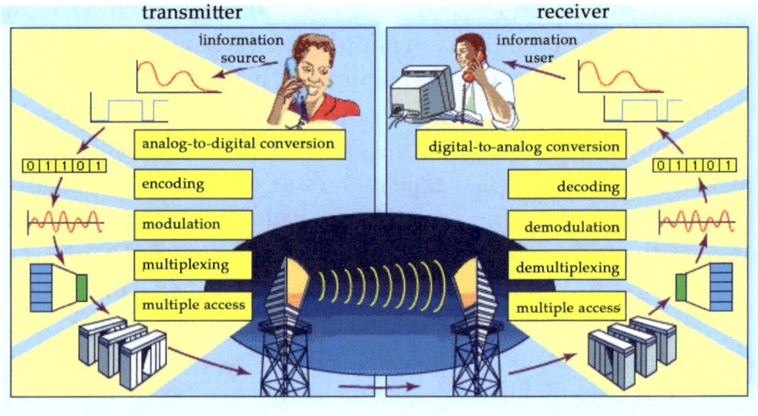

Figure 8.11. *Telecommunications and data transmission.*

Source: https://cdn.britannica.com/15/4615-004-75EA480A/Block-diagram-tele-communications-system.jpg.

8.5.2. Data Compression Techniques

Analysis of data compression techniques is of vital importance in information theory and channel capacity because they have a direct effect on data transmission efficiency and capacity. Data compression techniques are of two wide categories: lossless and lossy compression. Lossless techniques such as Huffman encoding and Run-Length Encoding (RLE) allow the original data to be reconstructed exactly from the compressed data. This is particularly important in applications where data integrity is essential, for example, executables or text files. On the other hand, lossy compression schemes such as JPEG for images or MP3 for audio sacrifice their compression ratio by shedding some non-vital information that they do not keep, causing the quality to deteriorate. This trade-off becomes increasingly important in high-bandwidth usage situations, where transmission speed is the highest priority. The advent of cloud radio access networks (C-RANs) as a contemporary solution to increase spectral and energy efficiency in wireless communications has rekindled the interest in data compression techniques. In C-RANs, the fronthaul links, which have the responsibility of connecting baseband units and remote radio heads, are typically capacity-constrained. Compression and quantization are the major techniques that assist in mitigating the degradation due to such constraints on system performance. By efficiently compressing data transmitted over such constrained links, C-RANs facilitate significant improvements in spectral efficiency (SE) as well as energy efficiency (EE) while maintaining the QoS of users. The integration of advanced data compression algorithms within such networks not only alleviates fronthaul bottlenecks but also assists in the optimization of resource usage, and thus pushes the frontiers of communication technology (Lau et al., 2015). Against the backdrop of evolving situations in digital technologies, the perennial issue of data piracy has necessitated the use of effective data compression methods within information theory. The transition from analogue to digital media has rendered traditional copy obstacles such as quality degradation moot and thereby amplified the necessity for robust digital rights management methods. One of the potential solutions is digital watermarking, where data-compression techniques are employed to embed information into digital material without making it visibly different. The development of wavelet-transform-based systems has been successful in balancing capacity and strength to prevent copying and maintain data integrity. Through the strategic application of powerful error correction and human vision models, these watermarking techniques prove the potential of new data compression techniques to address both effective data transmission issues and intellectual property right protection (Serdean et al., 2002).

8.5.3. Error Detection and Correction Techniques

Error detection and correction methods in information theory are important in offering reliable data storage and communication. The methods are designed to identify and correct errors that may be caused by noise in the communication channel or in computation. Simple error detection methods such as checksums and cyclic redundancy checks provide a simple means of ensuring message integrity in transmitted messages. As an illustration, a checksum is a computed value that helps to identify any data block errors by summarizing the binary form of the data. However, while such systems may identify errors, they need not necessarily correct them, and it is therefore necessary to include error correction algorithms. Such a two-pronged strategy adds to the strength of communication systems, especially when data integrity is of concern. Borrowing from rudimentary detection techniques, advanced error correction techniques employ coding techniques that not only detect but also correct errors. Reed-Solomon codes and Low-Density Parity Check (LDPC) codes are great examples of this robust methodology, providing frameworks for systems that enable safe data recovery.

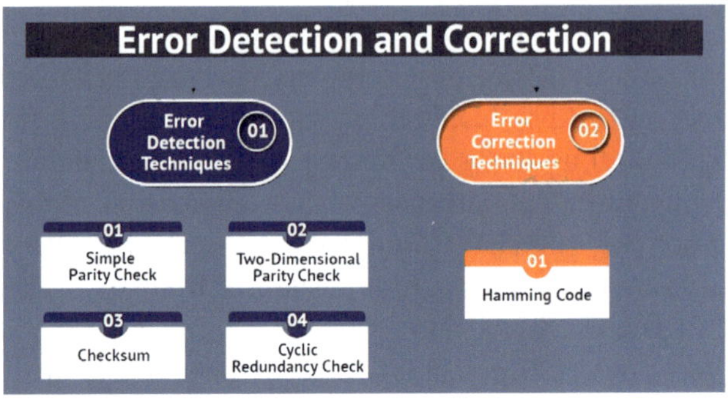

Figure 8.12. *Error detection and correction methods.*

Source: https://encrypted-tbn0.gstatic.com/images?q=tbn:ANd9GcSZzUo_bNiJ1f AFw5GEnSfBYP6lKdWbJceFiw&s.

Reed-Solomon codes are particularly notable for their capacity to correct burst errors, which are found regularly in digital communication networks. On the other hand, LDPC codes utilize a sparse parity-check matrix to achieve excellent performance near-channel capacity, as it is theoretically envisioned in information theory. The continually improving such methods demonstrate a sustained interest in incorporating reliability into data transmission within increasingly complex networks. Moreover, as digital technologies have evolved, the function of maintaining data accuracy has been increasingly more important and has necessitated innovative and efficient schemes of error correction (Serdean et al., 2002). Furthermore, the synergies between quantum information theory and error correction methods have promising research avenues. Quantum systems, as a result of their inherent characteristics, require particular protocols of error detection and correction, which differ from classical ones. Quantum error correction codes such as the Shors code and the surface code make use of entanglement in order to protect quantum information from decoherence as well as operational errors. With continuous-variable quantum information processing gaining momentum, work is done in parallel towards developing compatible error correction procedures that leverage Gaussian states as well as operations (Arthurs et al., 2011). This paradigm shift not only extends the operational lifespan of quantum computers but also provides entry to advanced applications in quantum cryptography and communication. Thus, advances in error detection and correction strategies remain a corner stone within the extended range of information theory and channel capacity, reflecting the need for robust frameworks suited to evolving technological landscapes (Figure 8.12).

8.6. CURRENT RESEARCH TRENDS

In the new era of Information Theory and Channel Capacity, current research trends indicate an increased focus on physical layer security, particularly in view of the evolving wireless technologies such as 5G. Information-theoretic approaches leveraging the intrinsic randomness of communication channels are increasingly being researched by scholars to enhance confidentiality of data. With the expansion of 5G networks, problems of grand scale have been achieved, which necessitate creative solutions to securing information at the physical level. This involves a thorough examination of various 5G technologies like physical layer security coding, millimeter wave communication, and non-orthogonal multiple access. These developments provide the promise of very significant advances in secure data transmission, which remain of top priority since the necessity for robust security measures increases with greater connectivity and data exchange between networks (Caire et al., 2018). Meanwhile, the energy efficiency of wireless communication is garnering much attention from researchers on economic as well as environmental grounds. There is a compelling need for development of methods that minimize energy use in all mobile networks, both the base stations and mobile terminals. Existing studies have evaluated the attainable rate region of additive white Gaussian noise (AWGN) broadcast channels using comparisons of various coding approaches, which include time-division, frequency-division, and superposition coding. Results show that superposition coding is best in energy efficiency under ideal circumstances; however, things alter after incorporating real-world parameters such as power of signaling.

Energy-efficient strategies are particularly critical as modern wireless systems experience rising energy demands. As such, researchers are aiming at optimizing performance given real-world metrics, balancing the use of energy against the need for rate fairness among users (Akbari et al., 2011). Besides that, an important trend in recent research is the convergence of energy efficiency and physical layer security in wireless communications. As data integrity threats become more and more important and power usage becomes a key issue, researchers are looking into integrated solutions that address these two key issues simultaneously. This convergence applies particularly to next-generation communication system development, where achieving secure transmission with energy management becomes a key engineering challenge. Researchers are innovating by combining techniques from both domains, thereby enhancing security without imposing excessive energy costs. As systems seek to achieve higher capacity and secure communication modes concurrently, understanding their interdependence becomes essential. This ongoing crossing of ground is a testament to the constant evolution of Information Theory and Channel Capacity, a stepping stone to future innovation in secure and trustworthy wireless communication (Caire et al., 2018; Akbari et al., 2011).

8.6.1. Quantum Information Theory Progresses

The quantum information theory research has greatly enhanced our understanding of channel capacity as well as quantum communication error correction. Underpinning these advances is the development of effective Quantum Error Correction Codes (QECCs), which are key to qubit stabilization and preservation of information integrity against the deleterious action of quantum decoherence. Through the application of a concatenated code design and iterative decoding techniques, researchers have been able to push QECC performance to extraordinary heights. A case in point is the Quantum Irregular Convolutional Code (QIRCC), which demonstrates extraordinary flexibility. This optimization allows the QIRCC to dynamically adjust to varying inner codes, achieving performance near theoretical hashing bound, thereby contributing to the knowledge of the field concerning the way quantum systems can effectively address information transmission and storage with the primacy of strong error correction codes in future quantum technologies (Alanis et al., 2015). In parallel with the development of error correction, research on channel capacity itself has also seen significant advance, particularly that related to nonlinear communication channels. Information theory models early on primarily addressed linear additive noise channels, but issues induced by channel nonlinearities such as optical fibers require a re-examination of these models. It seems that research indicates analyzing the way to reaching knowledge about nonlinear effects on information capacity can provide significant insight. For example, capacity can grow to infinity in the case of additive noise, whereas nonlinear channels possess an utmost capacity beyond which signal power is not raised by optimizing information transfer. All these findings also apply to quantum channels but have implications in other fields, like neurobiology, suggesting broader implications of nonlinear dynamics of information processing in different systems (Mitra et al., 2000). Furthermore, as quantum information theory evolves, it can not only enhance communications systems but also reshape computational architectures. Quantum algorithms that take advantage of the superposition and entanglement properties allow for greater processing capabilities than traditional systems. Consequently, as research in quantum channel coding and capacity advances further, it provides an underlying infrastructure on which future technologies can be developed.

Investigating the interplays between quantum mechanics and information theory illuminates paths to achieving record-breaking speeds and efficiencies. Synergies between advances in quantum error correction and channel capacity principles present crucial insights into realizing the practical applications of quantum technologies, leading to revolutionary breakthroughs in secure communications and complex computational operations (Figure 8.13).

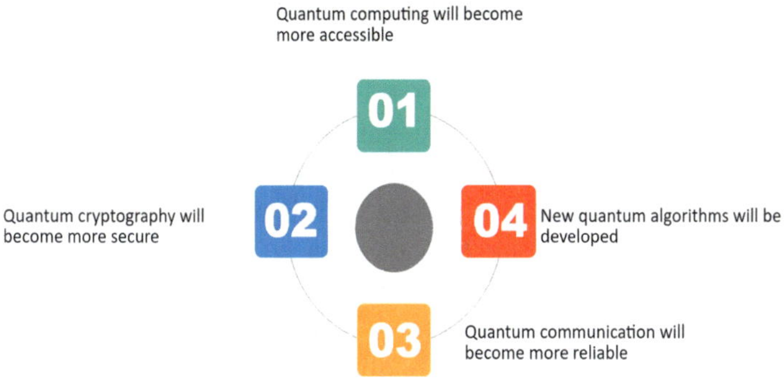

Figure 8.13. *Advances in quantum information theory.*

Source: https://fastercapital.co/i/Quantum-Information-Theory--Encoding-and-Decoding-Q--Future-of-Quantum-Information-Theory.webp.

8.6.2. Information Theory in Machine Learning

The integration of information theory and machine learning (ML) has yielded huge advances in channel capacity and the optimization of communication systems. Information theory, as initially developed by Claude Shannon, provides a theoretical foundation for measuring information and understanding how and when it is communicated over channels in an optimum way. In machine learning, the theoretical foundation is used to solve problems in algorithm training over complex data sets. By utilizing concepts such as entropy and mutual information, scientists are able to find out the optimal way to encode and transmit data so that the quality of information and its relevance are not compromised even in noisy and uncertain environments. This symbiosis identifies the importance of valuing channel capacity and its applicability in designing robust ML applications, particularly in facilitating reliable communication across different network structures, as discussed in advancements outlined by Chen et al. (2019). Minimization of quantization operations by means of communication channels is one of the beneficial applications of information theory in machine learning. For instance, there has been recent work showing algorithms to find best quantizers that maximize the mutual information between channel inputs and outputs and minimize the loss of data. Such utility becomes particularly highlighted in binary-input discrete memoryless channels, where information transfer is a high priority. The significance of such improvements comes from the fact that they can improve machine learning model performance by representing and passing on data effectively. The findings indicate that, although faced with complex networking scenarios, information theory concepts can be applied to maximize learning algorithms' effectiveness and accuracy, thus their applicability in the real world, as indicated in Kurkoski et al. (2014). Further, the application of machine learning to wireless network analysis identifies an era of transformation towards intelligent decision-making systems. Parallel to advancements in adaptive learning and parameter estimation, machine learning has been instrumental in coping with increased complexity in current communication systems, from IoT devices to cognitive radio networks. Specifically, the application of supervised, unsupervised, and reinforcement learning methodologies provides advanced insights into system performance and improves users' experience through facilitating high-rate, low-latency services. The analysis of these approaches with the limelight thrown in current studies highlights how in-depth understanding of information theory can propel the process of developing new communication application and services both militarily and in civilian life. This inquiry ensures that the unification of ML and information theory not only enhances the knowledge of channel capacity but also sets the ground for further studies on wireless technologies, as with Chen et al. (2019).

8.6.3. Challenges in Modern Communication Systems

The rapid development of communication networks, particularly with the advent of fifth-generation (5G) technology, has brought into focus a number of fundamental challenges in modern networks. One of these is the blending of high-transmission speeds and high security. As new generations of communication technology are developed, demands for faster and more reliable transmission of data intensify, and with it no less intense pressures for secure connections. In an environment like this, physical-layer security, where the characteristics of the wireless medium are used to ensure data integrity, is a possible solution. This system of security not only enhances the security of the transmission but also supports multiple simultaneous communication services being incorporated into individual channels, leading to greater spectral efficiency. Here, the concept of physical-layer service integration (PHY-SI) introduces an extremely interesting solution, embedding greater than one service such as multicast and secret transmissions for enhanced performance and to achieve the primary objectives of modern communication systems (Chen et al., 2018). Another major issue for modern communication systems is the prevalent issue of interference. In dense networks with greater than one transmitter being active at once, interference can significantly degrade signal quality. Interference alignment (IA) has been one of the strategic ways of addressing this issue by aligning transmitters in a way that their interference actually aligns at receivers for the application of simpler and more effective interference cancellation techniques. The new strategy has also been extensively researched and shown to achieve optimum degrees of freedom, which is an approximation to sum capacity. However, its performance relies on a number of practical considerations, including the reliability of channel state information at the transmitter and under realistic propagation environments. Managing these remains an issue in the practical application of IA to large systems and ensuring that existing systems have the ability to maintain the demand for high-capacity communication despite rising interference levels (Ayach et al., 2012). In addition, the amalgamation of newer communication technologies is a double-edged sword that complicates existing structures and challenges standardization. As different communication paradigms, such as millimeter-wave and ultra-dense networks, have appeared on the scene, there is a pressing need for unified standardization processes to ensure interoperability across different systems. This is further added to by the challenge of accommodating legacy systems while amalgamating newer solutions effectively. As multiple services compete for available bandwidth, the issue is not merely one of attaining maximum channel capacity but assuring service reliability with heterogeneous applications. To address these challenges needs innovative research and development in enhancing the resilience and efficiency of modern communication systems, thereby providing the avenue for future breakthroughs and maintaining the pace of technology advancements in a more globalized world.

SUMMARY

The exploration of information theory and channel capacity reveals critical insights into the parameters influencing communication systems. Through the examination of absolute judgments, as discussed in the work of Garner and Hake, it becomes evident that traditional interpretations associated with information transmission may overlook significant cognitive factors. The illusion of channel capacity, as argued, is largely a function of sampling bias and subjective interpretations of memory capabilities. This brings about the need for greater integration of cognitive ideas into models of information transmission, again stressing that quantified peaks in information are no guarantee of sound understanding of channel capacity in the classical sense (Nizami et al., 2011). In addition, the employment of intricate channel models, such as ones targeting multi-input multi-output (MIMO) systems, brings about an emphasis on distinctions inherent in conventional approaches. The Kronecker model, while well accepted, has been shown to be too optimistic in channel capacity estimation, particularly in practical applications. Recent advances indicate that a more precise examination of the sparse statistical degrees of freedom of channel behavior will lead to improved performance estimates. This paradigm change in modeling also aligns with empirical fact and requires a re-evaluation of existing practice in the framework of signal transmission and signal processing capability (Kotecha et al., 2008). Finally, the findings of this study extend beyond theoretical accounts to practical re-evaluations in technology and communications. As information theory continues to develop, it is important to build models that are specifically designed to account for cognitive limitations and channel variability. The implications of learning from critically assessing information transmission and the capacity of memory can inform future research and advancements in the discipline. By appreciating the complexities in channel capacity and other cognitively relevant constructs, we will be able to foster innovations better aligned with reality, presenting potential for more effective communication systems in an increasingly interconnected world (Nizami et al., 2011; Kotecha et al., 2008). Information theory explains the fundamental principles underlying communication systems, with specific emphasis laid on quantification and transmission of information. Channel capacity is an important one, which defines the maximum rate at which information can be transmitted effectively over a channel of communication. Capacity varies with factors such as noise, channel type, and coding techniques employed. Recent advances, particularly in 5G wireless communications, have underscored the need for continuous study of channel capacity to enable sufficient capacity to meet increasing needs for data transmission and security. Blending information theory with its practical application in telecommunications marks a revolutionary milestone towards a more secure communication technology that ensures data integrity under evolving threats and operational conditions in transmission environments. Physical layer security has gained significant research momentum in recent times. Drawing inspiration from information-theoretic principles, physical layer security aims to provide data confidentiality based on the usage of inherent randomness in communication channels. This becomes essential when the development of 5G wireless networks introduces new complexities to communication channels. Research in physical layer security has been classified as coding schemes and innovations in multiple-input multiple-output systems. Such developments not only strengthen data security but also call for a review of existing channel capacity constraints. The combination of these technologies is going to ensure that, as the demand for higher data rates rises, associated risks in information transmission are defied by more security screening and novel coding schemes to defy existing vulnerabilities and constraints in practice (Caire et al., 2018). In the study of relay channels, much has been achieved insofar as the upper bounds on capacity are concerned, with particular reference to discrete memoryless relay channels having three nodes: a destination, a source, and a relay. New research suggests a new inequality for tighter capacity bounds where traditional methods are insufficient, most notably at high encoding rates that are beyond the individual channel capacities of connected nodes. This involves a generalization of the blowing-up lemma that relates conditional entropy to decoding errors and hence enhances relay functioning comprehension. Examination of statistically degraded channel instances reveals that improved bounds involve specific computational pathways for further investigation. These findings are not only evidence of theoretic advancement but also have practical relevance, particularly in optimizing data rate transmission and minimizing error probabilities in current and prospective communications systems (Xue et al., 2012).

REFERENCES

1. Agrell, E. (2014). Conditions for a monotonic channel capacity. http://arxiv.org/abs/1209.2820 (accessed on 18th September 2025).
2. Akbari, A., Imran, M. A., & Tafazolli, R. (2011). Maximizing average energy efficiency for two-user AWGN broadcast channel. https://core.ac.uk/download/77601413.pdf (accessed on 18th September 2025).
3. Akinyele, S. T. (2011). *Strategic Marketing Management in The Nigerian Oil And Gas Industry: A Theoretical Framework.* https://core.ac.uk/download/18295049.pdf (accessed on 18th September 2025).
4. Alanis, D., Babar, Z., Botsinis, P., Hanzo, L., et al. (2015). The road from classical to quantum codes: A hashing bound approaching design procedure. http://arxiv.org/abs/1503.02372 (accessed on 18th September 2025).
5. Aldridge, M. (2011). Interference mitigation in large random wireless networks. http://arxiv.org/abs/1109.1255 (accessed on 18th September 2025).
6. Alléaume, R., Boutros, J., Elkouss, D., Leverrier, A., et al. (2009). Efficient reconciliation protocol for discrete-variable quantum key distribution. http://arxiv.org/abs/0901.2140 (accessed on 18th September 2025).
7. Arthurs, E., Bachor, H. A., Bennett, C. H., Bergou, J., et al. (2011). *Gaussian Quantum Information.* https://core.ac.uk/download/15143138.pdf (accessed on 18th September 2025).
8. Avestimehr, A. S., Kao, D. T. H., & Maddah-Ali, M. A. (2015). Blind index coding. http://arxiv.org/abs/1504.06018 (accessed on 18th September 2025).
9. Ayach, O. E., Heath., R. W., & Peters, S. W. (2012). The practical challenges of interference alignment. http://arxiv.org/abs/1206.4755 (accessed on 18th September 2025).
10. Ayanoglu, E., & Davaslioglu, K. (2014). Quantifying potential energy efficiency gain in green cellular wireless networks. http://arxiv.org/abs/1501.06099 (accessed on 18th September 2025).
11. Baldi, M., Chiaraluce, F., Maturo, N., Ricciutelli, S., et al. (2014). LDPC coded transmissions over the Gaussian broadcast channel with confidential messages. http://arxiv.org/abs/1404.2864 (accessed on 18th September 2025).
12. Barnum, H., Nielsen, M. A., & Schumacher, B. (1997). Information transmission through a noisy quantum channel. https://core.ac.uk/download/4870568.pdf (accessed on 18th September 2025).

13. Barros, J., Pinto, P. C., & Win, M. Z. (2010). Techniques for enhanced physical-layer security. http://arxiv.org/abs/1008.3705 (accessed on 18th September 2025).
14. Bliss, D. W. (2014). Cooperative radar and communications signaling: The estimation and information theory odd couple. http://arxiv.org/abs/1403.1476 (accessed on 18th September 2025).
15. Boche, H., Böck, Y. N., Deppe, C., & Fitzek, F. (2025). Remote state estimation and Blum–Shub–Smale machines—A computability analysis with applications to virtual twinning. *IEEE Transactions on Automatic Control, 70*, 3165–3180. https://www.semanticscholar.org/paper/90d50fa6a5b6793b8ceaf7586f24bc61d4a3a782 (accessed on 18th September 2025).
16. Caire, G., Gao, X., Khisti, A., Wong, K. K., et al. (2018). A survey of physical layer security techniques for 5G wireless networks and challenges ahead. https://core.ac.uk/download/158170707.pdf (accessed on 18th September 2025).
17. Cano, C., Dabak, A., Lampe, L., Malone, D., et al. (2016). State-of-the-art in power line communications: From the applications to the medium. https://core.ac.uk/download/297029547.pdf (accessed on 18th September 2025).
18. Chen, K. C., Hanzo, L., Jiang, C., Ren, S., et al. (2019). Thirty years of machine learning: The road to Pareto-optimal wireless networks. http://arxiv.org/abs/1902.01946 (accessed on 18th September 2025).
19. Chen, T. Y., Wesel, R. D., & Williamson, A. R. (2012). A rate-compatible sphere-packing analysis of feedback coding with limited retransmissions. http://arxiv.org/abs/1202.1458 (accessed on 18th September 2025).
20. Chen, Z., Fang, J., Li, S., et al. (2018). Physical layer service integration in 5G: Potentials and challenges. http://arxiv.org/abs/1707.01064 (accessed on 18th September 2025).
21. Cheng, J. F., Dolinar, S., Effros, M., et al. (1995). Data expansion with Huffman codes. https://core.ac.uk/download/4875561.pdf (accessed on 18th September 2025).
22. Chung, S. Y., Gamal, A. El, Kim, Y. H., et al. (2010). Noisy network coding. http://arxiv.org/abs/1002.3188 (accessed on 18th September 2025).
23. Datta, N., Tomamichel, M., & Wilde, M. M. (2016). On the second-order asymptotic for entanglement-assisted communication. https://core.ac.uk/download/520189547.pdf (accessed on 18th September 2025).

24. Debbah, M., Masucci, A., Ryan, Ø., & Yang, et al. (2010). Finite dimensional statistical inference. http://arxiv.org/abs/0911.5515 (accessed on 18th September 2025).

25. Eriksson, T., Khanzadi, M. R., Krishnan, R., et al. (2015). On the capacity of the Wiener phase-noise channel: Bounds and capacity achieving distributions. http://arxiv.org/abs/1503.08644 (accessed on 18th September 2025).

26. Giedke, G., Lercher, D., & Wolf, M. M. (2013). Standard super-activation for Gaussian channels requires squeezing. https://core.ac.uk/download/210617384.pdf (accessed on 18th September 2025).

27. Guha, S., Hayden, P., Krovi, H., et al. (2013). Quantum enigma machines and the locking capacity of a quantum channel. https://core.ac.uk/download/520189581.pdf (accessed on 18th September 2025).

28. Hanzo, L., & Jiang, M. (2007). *Multiuser MIMO-OFDM for Next-Generation Wireless Systems.*

29. Hassibi, B., & Marzetta, T. L. (2001). Multiple-antennas and isotopically random unitary inputs: The received signal density in closed form. https://core.ac.uk/download/4875021.pdf (accessed on 18th September 2025).

30. Hayashi, M., Hiroshima, T., Tomita, A., & Wang, et al. (2008). Quantum information with Gaussian states. http://arxiv.org/abs/0801.4604 (accessed on 18th September 2025).

31. Hsieh, M. H., & Wilde, M. M. (2010). Entanglement generation with a quantum channel and a shared state. https://core.ac.uk/download/520189641.pdf (accessed on 18th September 2025).

32. Jafar, S. A. (2012). The ergodic capacity of phase-fading interference networks. http://arxiv.org/abs/0902.0838 (accessed on 18th September 2025).

33. Kasai, K., & Sakaniwa, K. (2011). Spatially-coupled MacKay-Neal codes and Hsu-Anastasopoulos codes. http://arxiv.org/abs/1102.4612 (accessed on 18th September 2025).

34. Kotecha, J. H., Raghavan, V., & Sayeed, A. M. (2008). Why does a Kronecker model result in misleading capacity estimates? http://arxiv.org/abs/0808.0036 (accessed on 18th September 2025).

35. Kurkoski, B. M., & Yagi, H. (2014). Quantization of binary-input discrete memoryless channels. http://arxiv.org/abs/1107.5637 (accessed on 18th September 2025).

36. Lau, V., Peng, M., Poor, H. V., et al. (2015). Fronthaul-constrained cloud radio access networks: Insights and challenges. http://arxiv.org/abs/1503.01187 (accessed on 18th September 2025).

37. Lelli, G. (2015). Arab-Islamic reception of Hellenistic science. https://core.ac.uk/download/55799783.pdf (accessed on 18th September 2025).
38. Lerner, A. N., Khairullin, V. I., Il'in, G. A., & Garifullina (2024). Lower boundary capacity estimations of frequency-selective communication channels with PAM-n-signals achievable using resolution time theory. *Electromagnetic Waves and Electronic Systems*. https://www.semanticscholar.org/paper/9c7dd9342fd149498f36d5d030400f77923cc38f (accessed on 18th September 2025).
39. Loyka, S., & Schaefer, R. F. (2015). The secrecy capacity of compound Gaussian MIMO wiretap channels. http://arxiv.org/abs/1509.02182 (accessed on 18th September 2025).
40. Mitra, P. P., & Stark, J. B. (2000). Nonlinear limits to the information capacity of optical fiber communications. https://core.ac.uk/download/25299959.pdf (accessed on 18th September 2025).
41. Moulin, P., & Wang, Y. (2006). Capacity and random-coding exponents for channel coding with side information. http://arxiv.org/abs/cs/0410003 (accessed on 18th September 2025).
42. N/A. (2011). *Handwashing with Soap–Two Paths to National-Scale Programs: Lessons from the Field: Vietnam and Indonesia*. https://core.ac.uk/download/75783149.pdf (accessed on 18th September 2025).
43. Nizami, L. (2011). Memory model of information transmitted in absolute judgment. https://core.ac.uk/download/186332049.pdf (accessed on 18th September 2025).
44. Oohama, Y., & Watanabe, S. (2010). Secret key agreement from correlated Gaussian sources by rate-limited public communication. http://arxiv.org/abs/1001.3705 (accessed on 18th September 2025).
45. Permuter, H., & Shrader, B. (2007). Feedback capacity of the compound channel. http://arxiv.org/abs/0711.0705 (accessed on 18th September 2025).
46. Qi, H., Wang, Q., & Wilde, M. M. (2017). Applications of position-based coding to classical communication over quantum channels. https://core.ac.uk/download/520191746.pdf (accessed on 18th September 2025).
47. Reinsberg, B. (2017). Five steps to smarter multi-bi aid: A new way forward for earmarked finance. https://core.ac.uk/download/161770908.pdf (accessed on 18th September 2025).
48. Scott, G. (2025). Low power micro robotics utilizing biologically inspired energy generation. https://core.ac.uk/download/pdf/211016422.pdf (accessed on 18th September 2025).

49. Serdean, C. V. (2002). Spread spectrum-based video watermarking algorithms for copyright protection. https://core.ac.uk/download/29816683.pdf (accessed on 18th September 2025).

50. Shi, K. (2009). Optimal power allocation for a successive refinable source with multiple descriptions over a fading relay channel using broadcast/multicast strategies. https://core.ac.uk/download/144143726.pdf (accessed on 18th September 2025).

51. Tan, V. Y. F. (2015). Asymptotic estimates in information theory with non-vanishing error probabilities. http://arxiv.org/abs/1504.02608 (accessed on 18th September 2025).

52. Wallace, R. (2008). *Lurching Toward Chernobyl: Dysfunctions of Real-Time Computation.* https://core.ac.uk/download/pdf/87128.pdf (accessed on 18th September 2025).

53. Wallace, R. (2010). Protein folding disorders: Toward a basic biological paradigm. https://core.ac.uk/download/pdf/289091.pdf (accessed on 18th September 2025).

54. Xue, F. (2012). An upper bound on relaying over capacity based on channel simulation. http://arxiv.org/abs/1210.6740 (accessed on 18th September 2025).

CHAPTER 9

DETECTION AND ESTIMATION THEORY OF DIGITAL COMMUNICATION

Contents

INTRODUCTION

In communication technology, detection and estimation theory is the essential framework for increasing signal fidelity and reducing uncertainty in data transmission. Detection and estimation theory encompasses methods employed to derive useful information from signals received amid noise and interference. With the growing demand for quality communication networks, particularly in the face of emerging technologies, demand for effective detection and estimation is taking root. Through an examination of these techniques, researchers can address the inherent complexities of digital communication systems, from straightforward point-to-point communications to intricate networks characterized by numerous users and high-order modulation schemes. More recent advances in wireless communication systems, in the form of the introduction of technologies like 5G and the imminent arrival of 6G, put emphasis on multi-disciplinary solutions to detection and estimation. Researchers are now examining how such theories can be integrated with new paradigms such as Integrated Sensing and Communications (ISAC). This multi-disciplinary effort attempts to tap the dense infrastructure of future networks for allowing a richer embedding of sensing capabilities alongside traditional communication channels. As characterized via recent studies, the synergy between sensing and communication can potentially revolutionize network performance, with robust detection and estimation systems potentially playing a key role in leveraging these innovative technologies (Liu et al., 2022). In addition, IRS represents a groundbreaking platform to optimally design communication environments. Via adaptive reflection-optimizing signal forwarding, IRS can significantly enhance the overall performance of wireless networks. For effective deployment, the channel estimation and reflection optimization problems are fundamental research domains. IRS integration in existing communication systems demands robust detection algorithms and estimation methods that are powerful enough to adapt to dynamic environments and dynamic network structures. Accordingly, detection and estimation theory are not only fundamentals but also an essential tool to resolve real-world issues in future wireless communications (Wu et al., 2021). These principles will be discussed in this essay with the view of their contribution to creating digital communication systems.

Definition of Detection and Estimation Theory

Detection and estimation theory is a subfield of digital communication dealing with the methods of detecting and estimating signals from noisy measurements. Detection theory actually attempts to detect the presence or absence of a signal in the presence of interference and noise, and estimation theory attempts to quantify the parameters of the signal. This dual role is critical in various uses like telecommunications, radar, and remote sensing, whose effective data transmission has a high reliance on robust detection and estimation operations. By applying statistical methods, these theories provide methods to guarantee that communications systems are optimized, thus ensuring reliability and effectiveness in information delivery. While communication systems increasingly go digital, detection and estimation theory continues to be at the center of the response to demands from increasingly complex environments and varying signal qualities. One particularly important evolution in detection and estimation theory is how it extends into Integrated Sensing and Communications (ISAC). As evident through recent research work, such as in the IEEE Journal on Selected Areas in Communications, ISAC is an innovative integration of sensing and communication technologies (Liu et al., 2022). The technology not only enhances the performance of traditional wireless networks but also facilitates the leverage of advanced algorithms to gain improved detection and estimation outcomes.

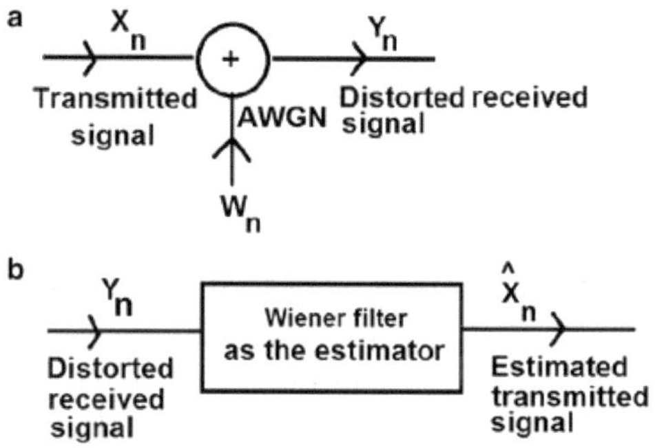

Figure 9.1. *Detection and estimation theory of digital communication.*

Source: https://media.springernature.com/lw685/springer-static/image/chp%3A10.1007%2F978-3-319-20651-6_2/MediaObjects/324663_1_En_2_Fig12_HTML.gif

By developing a perceptive network with sensing and communication synergistically, ISAC is the next generation of communication in digital systems, serving the requirement of multimodal capability in present systems. ISAC eventually exposes the dynamic relationship between detection and estimation and offers room for powerful and more capable communication infrastructure. Aside from theoretical advances, real-world applications of detection and estimation theories require rigorous calibration and verification processes. For instance, while producing robust calibration curves in radiocarbon dating, the data will have to be corrected to account for variations in atmospheric carbon levels as this affects the data in a way explained in academic literature related to C calibration (Reimer et al., 2020). This train of argument parallels that used in detection and estimation theory where various system and environmental parameters must be accounted for in order to enhance precision. Continuing innovation in statistical methods has a direct impact on the reliability of communication systems, making them capable of properly understanding and responding to a great range of signal conditions. As digital communications intersect with next-generation technologies, the application of detection and estimation theory will be ever more accountable for defining future success within the field (Figure 9.1).

Importance in Digital Communication

The modern communication has revolutionized the process of transferring information via various mediums to a large extent, and its importance is further increasing in personal and professional life. The key to this revolution lies in applying robust detection and estimation theories that enhance signal processing capability. Through effective parameter estimation and signal detection in noisy environments, these theories impart stability and credibility to information transfer. Advances in communications technology have made possible wider bandwidths and higher rates of data, so unprecedented flow of information is now achievable. Moreover, the use of adaptive importance sampling techniques offers a novel approach to performance evaluation of communication systems, showing its feasibility in approximating bit-error rate (BER) and system parameter optimization. These developments are of utmost importance in learning how powerful digital communication can be achieved, thereby impacting other industries like telecommunications and data management. Application of advanced methods in detection theory and estimation theory shows the importance of continuous improvement in digital communication systems. For instance, new empirical models that account for realistic operating constraints are crucial since they provide valuable insights on potential performance limits. In this case, measurement of crosstalk in wavelength-division multiplexed (WDM) systems is a sample of analytical breakthroughs in determining optimal detector thresholds, whose influence directly impacts on error rate performances. The findings suggest that performance limits for acceptable levels of crosstalk are higher than previously projected, leading to a reassessment of WDM network design and performance. Thus, the marriage of advanced statistical methods and practical applications serves to emphasize the significant role detection and estimation theories play in creating and furthering digital communications systems. Furthermore, the exploration of supplementary information using soft biometrics demonstrates a fresh horizon of digital communication that exceeds ordinary measurement. When scientists move into the application of general user characteristics to enhancing user authentication, the importance of soft biometrics is revealed. The discipline involves studying identifiable characteristics such as age or mood states in a bid to enhance predictive capability of biometric systems. The effects of communication systems within the digital environment are profound as the use of such information can enhance security protocols, offer enhanced user experience, and foster trust in communication systems. Addressing challenges of this predictive biometrics field will be high on the agenda of future innovation to make the detection and estimation theory progress not only to meet current requirements but also to predict future trends in digital communication technologies (Figure 9.2).

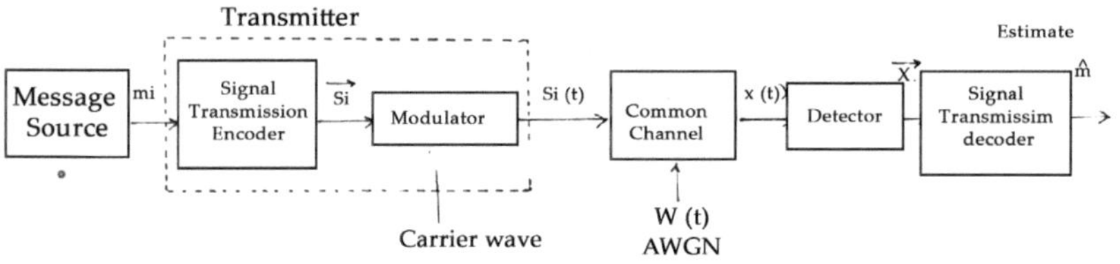

Figure 9.2. *Detection and estimation theory importance in digital communication.*

Source: https://i.ytimg.com/vi/bmtew4kzHXU/maxresdefault.jpg.

Overview of Key Concepts

The foundation of detection and estimation theory in digital communications lies in its ability to demodulate signals under uncertainty and noise. These tasks borrow two primary concepts: hypothesis testing and parameter estimation. Hypothesis testing allows for the ability to choose the most likely hypothesis regarding received signals based on observed data, which is essential in ensuring effective communication. Parameter estimation enhances this by refining estimates of critical model parameters that characterize transmitted information, further enhancing the accuracy of interpretation. Interaction between these elements not only facilitates more effective decision-making in noisy environments but also serves to emphasize the mathematical foundations that make modern communication systems possible. Understand the working of these hypothetical models provides insight into the continuos innovation of digital communications technologies and setting up grounds for future work and applications of high level. Also, as technology continues to advance, the methods based on detection and estimation have evolved further. Detection performance under challenging conditions is optimized using methods such as maximum likelihood estimation and Bayesian inference. Artificial intelligence and machine learning play a role of paramount importance in the development of these estimation methods, leading to breakthroughs like predictive algorithms and adaptive filtering. Particularly, the greater application of these advanced techniques within the theoretical framework of Integrated Sensing and Communications (ISAC) indicates the crossroads of detection theory with new digital paradigms for the purpose of facilitating enhanced communication ability. Continuous progress in these areas has the potential to integrate traditional detection methods with new technological needs, thereby enhancing operational performance and efficiency across different communication applications. Conversely, continuous progress under the 5G umbrella and conceptualized 6G technologies also necessitate in-depth understanding of detection and estimation principles. The integration of sensing capabilities within communication networks brings to the fore an evolutionary turn towards perceptive networks that leverage spatial awareness to improve the efficiency and reliability of transmission. In this context, there has been increased research on cross-layer design trade-offs, linking detection and estimation to physical layer performance as well as theoretical communication system boundaries. The emphasis on developing robust frameworks for Integrated Sensing and Communications establishes this connection and tends towards the future direction of digital communication theory. An in-depth knowledge of detection and estimation is thus not only relevant for research purposes but also for practical use in advancing network technologies, particularly in addressing the particular challenges of dense wireless environments (Ahmed, 2022; Liu et al., 2022).

9.1. FUNDAMENTALS OF DIGITAL COMMUNICATION

The digital communication principle is fundamentally rooted in the dependable inter-exchange of data between systems based on digital signals. The nature of digital communication is based on the capability of effective signal detection and estimation techniques, which form the core of minimizing transmission errors. Based on the growing demand for higher data rates, techniques such as Massive MIMO (multiple-input multiple-output) systems have remained a significant component of today's communication networks. These systems offer improved spectral efficiency through the utilization of numerous antennas in the base station, thereby enhancing the signal's clarity and immunity to noise, a critical consideration for both mobile broadband and other uses. The ongoing research work on the use of gigantic antenna arrays introduces infinite dimensions of spatial processing capability, opening the door to a new era of technology-driven digital communications based on innovative technology that takes transmission reliability and capacity problems head-on (Björnson et al., 2019). In order to continue building up an understanding of the theoretical foundation of digital communication, there is a need to establish the contribution of accurate detection mechanisms. Detection theory is interested in statistical detection of signals amidst noise and interference and requires effective mathematical models to function optimally. By employing signal-enhancing detection algorithms, error probability can be significantly reduced. The advancements in stable regression algorithms, such as the 2-D iteratively reweighted least squares lattice algorithm, illustrate the way in which the integration of advanced mathematical frameworks can contribute to detection processes in a variety of applications, such as defect detection in textured images. The process has a strong similarity to the problems in communication systems with outlier management and noise removal in order to facilitate proper communication. With improved false alarm reduction and better classification performance, these advanced detection models contribute effectively to the field of digital communication (Ercil et al., 2006). The combination of detection and estimation theory with future technologies marks a new millennium in digital communication. Signal processing methodologies keep advancing, not merely increasing data transmission rates but also allowing an entire range of varied applications, from mobile communications to ultra-reliable low-power communications. This exchange demands ongoing research questions into new methodologies, such as exploring very large aperture arrays and intelligent Massive MIMO, that can potentially transform communication standards. Ongoing developing of such technologies is crucial toward addressing the expanding complexity of modern communication scapes. In essence, as we continue in the development of robust detection and estimation techniques, we provide avenues for more efficient, reliable, and innovative digital communication systems that can meet the ever-growing demands of users all over the world (Björnson et al., 2019).

9.1.1. Basic Principles of Digital Communication Systems

The fundamental principles of digital communication systems are the foundation on which detection and estimation theory concepts in general are understood. The core of such systems is transmission, processing, and reception of data in digital form, usually by methods of modulation converting digital signals into waveforms of analog form for transmission over various media. Techniques such as pulse-code modulation (PCM) and quadrature amplitude modulation (QAM) demonstrate how information is represented and transmitted in an orderly manner to provide accuracy through noisy channels. When thinking about the nature of such techniques, particularly under adverse conditions such as in multiuser environments, complexity arises in the application of detection techniques. It is of utmost significance to understand how these principles work, as they directly impact performance metrics like bit error rate and signal-to-noise ratio, which feedback into detection algorithms and estimation methods that should optimize system efficiency and reliability. As digital communication systems continue to evolve, the inclusion of advanced techniques like machine learning in detection and estimation theory presents new avenues for research and application. Having algorithms that can work with large amounts of data and adapt to changed signal conditions provides an additional level of complexity in the control of the detection process for varying scenarios. Utilizing adaptive equalization and multiuser detection, for instance, can significantly help in enhancing the effectiveness of current systems, particularly in context-dependent environments when traditional methods would be inadequate. Moreover, the research into genetic algorithms, as evidenced in recent studies, indicates a shift towards optimization techniques that enhance detection power while reducing computational requirements, whereby they become more suitable for real-time application.

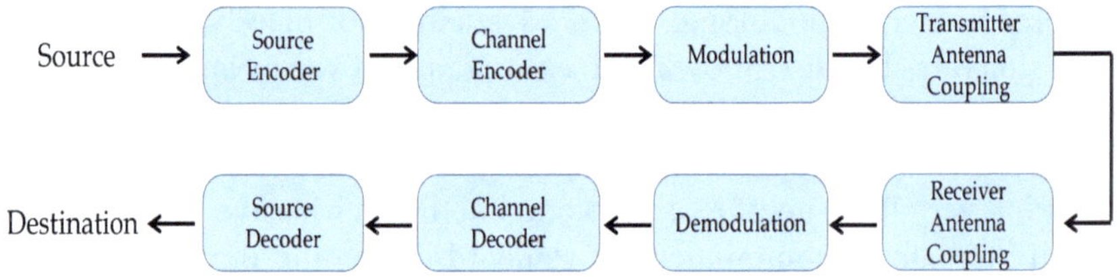

Figure 9.3. *Basic principles of digital communication systems.*

Source: https://external-content.duckduckgo.com/iu/?u=https%3A%2F%2Fwww.researchgate.net%2Fprofile%2FGeoffrey_Li4%2Fpublication%2F326144915-%2Ffigure%2Ffig1%2FAS%3A644145286766592%401530587612641%2FA-typical-block-diagram-of-digital-communication-systems.png&f=1&nofb=1&ipt=9bc15400e761fcaecef27c5dd3f9922896c324f166410ff565b46ef234425cbe.

This interdisciplinary research is critical to the development of next-generation communication technology that can tolerate higher data rates without compromising on performance strength. Notably, testing and measurement of these principles in the channel of digital communication rely on continuous development of channel estimation methods, which facilitate effective transmission and reception of the signal. Research has highlighted the importance of sound channel modeling and development of sophisticated detection algorithms that can mitigate multipath propagation and fading. For example, soft decision decoding and iterative detection research assist in optimizing the system's interference and distortion robustness. The distance and localization measurement techniques of RFID systems and their correlative degenerate channel modeling (Povalač et al., 2013) also provide insight into the complexity of modern communication systems. The understanding of these elements forms the basis for the practitioner to innovate within detection and estimation theory's framework, to eventually make wireless communications more reliable and efficient (Hanzo et al., 2007) (Figure 9.3).

9.1.2. Role of Signals in Communication

The role of signals in communication systems is foundational, as they serve as the primary means by which information is transmitted across various media. In the realm of digital communication, signals encapsulate data into a format suitable for transmission, which encompasses modulation techniques that convert digital data into analog waveforms. This transformation ensures that information can traverse distances effectively, be it through wired or wireless channels. The emergence of sophisticated modulation strategies, such as those incorporated into orthogonal frequency division multiplexing (OFDM), has significantly enhanced spectral efficiency while mitigating inter-symbol interference. However, despite these advancements, the limitations inherent in traditional detection and channel estimation techniques pose challenges in contexts like multiuser environments, where the complexity of the system may lead to rank-deficient scenarios that complicate signal interpretation and data recovery (Hanzo et al., 2007).

As communication systems evolved, the integration of multiple-input multiple-output (MIMO) technology further exemplified the critical role of signals in enhancing communication quality and capacity. MIMO systems utilize multiple antennas at both the transmitter and receiver to create multiple signal paths, thus improving reliability and throughput., The ability to transmit multiple signals simultaneously opens avenues for higher data rates and improved estimates of transmitted information, even under challenging conditions. Significant advancements in processing algorithms have empowered practitioners to develop robust detectors that leverage these capabilities effectively, promoting the transition towards more complex architectures like Massive MIMO. In this context, the resolution of issues such as pilot contamination illustrates how signal processing advancements continuously redefine communication efficiencies, aligning with the ongoing needs of modern wireless networks (Björnson et al., 2019). Moreover, the evolving landscape of communication technologies underscores the importance of signal processing in facilitating robust and reliable interactions. The advent of Massive MIMO has catalyzed transformative changes, merging large-scale antenna arrays with advanced techniques that enable unprecedented spectral efficiencies. These systems not only improve mobile broadband services but also expand applications beyond traditional communication, such as radar and sensing. As researchers explore new frontiers in signal processing—including holographic MIMO and intelligent systems—there is a growing recognition of how signals can be optimized for diverse applications. This continual innovation signifies a shift in understanding the broader implications of signals in communication, heralding new research directions that will shape the future landscape of digital communication systems, challenging existing paradigms while offering solutions to emerging technological demands (Björnson et al., 2019) (Figure 9.4).

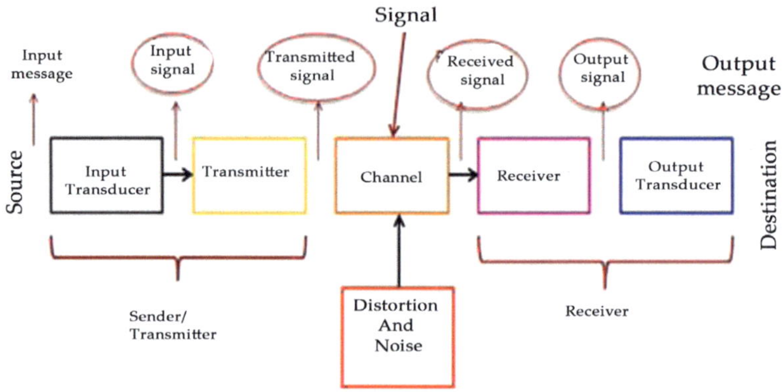

Figure 9.4. *Role of signals in communication.*

Source: https://external-content.duckduckgo.com/iu/?u=https%3A%2F%2Fthfvnext.bing.com%2Fth%2Fid%2FOIP.GZBlJFQgduPYVWFOrC7KUAHaFj%3Fcb%3Dthfvnext%26pid%3DApi&f=1&ipt=2fc289beeedc341c4e4259d797185008ad0f706a89829f496ba19438459f7846.

9.1.3. Types of Digital Modulation Techniques

In the fast-changing sector of digital communications, different modulation methods are essential for facilitating effective data transfer. Among them, amplitude shift keying (ASK), frequency shift keying (FSK), and phase shift keying (PSK) are basic methods that have led the way to more complex methods. ASK works by changing the carrier wave's amplitude and is easy and efficient for low-bandwidth applications. FSK, by contrast, uses alternative frequencies to represent binary data and is therefore more resistant to noise. PSK, extremely fashionable in today's communications, modulates the phase of the carrier signal to represent information and hence exploits the bandwidith more efficiently. Each of these approaches comes at a trade-off of complexity, power efficiency, and sensitivity to channel distortion, and the choice of modulation is thus extremely application-dependent. With the rapid advancement in technology, newer and more superior modulation schemes have been proposed to surpass the limitations of traditional techniques. Quadrature amplitude modulation (QAM) raises the signaling state count by combining the alteration of amplitude and phase, creating a highly adaptable technique capable of transmitting greater data rates at lower bandwidth. High-order QAM can support numerous bits per symbol for efficient data transmission. Also, methods such as orthogonal frequency-division multiplexing (OFDM) are increasingly used in broadband communications because they can combat frequency-selective fading. The technique divides the frequency spectrum into numerous closely spaced subcarriers, which can independently be modulated in such a way that the transmission becomes more reliable. Their development demands an understanding of their detection and estimation mechanisms, particularly when optimizing performance in complex communication environments, such as with MIMO OFDM systems (Hanzo et al., 2007). New digital modulation techniques also encompass advanced techniques that leverage the power of digital signal processing (DSP) tools to manage impairments like noise and distortion in channels. Coherent detection systems, for instance, have revolutionized optical communications to enable the use of multi-level modulation formats that enhance spectral efficiency. Through the use of sophisticated algorithms, these systems are able to effectively counteract linear fiber impairments and nonlinear effects due to mechanisms such as the Kerr effect, which is of great significance in high-speed WDM systems (Amari et al., 2017). The advent of nonlinearity compensation techniques enhances the potential performance of such systems, allowing higher rates of data and greater transmission distances. With progress in digital communication, the integration of complex modulation techniques and DSP techniques becomes the central focus for achieving maximum detection and estimation, revolutionizing the future of digital communication at its very foundation.

9.2. DETECTION THEORY

Detection Theory is a foundational notion in digital communication studies, providing the tools for detection of signal patterns in the midst of noise. Detection Theory is, at the core, concerned with methods applied to make decisions regarding signal presence or absence, referred to as hypothesis testing. In digital communication systems, signal detection transmitted can be greatly affected by several factors like noise characteristics, channel impairments, and the modulation scheme employed. Hence, developing effective detection algorithms that will perform under any given conditions is very important. With evolving detection methods, they need to catch up with the increasingly sophisticated communication systems of the modern age, necessitating studying both the theoretical foundations and the real-world applications in detail in order to enhance reliability and efficiency in real-world scenarios.

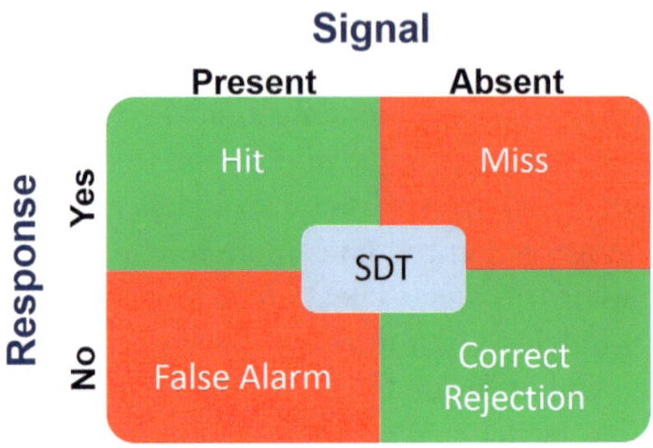

Figure 9.5. *Signal detection theory.*

Source: https://media.licdn.com/dms/image/v2/C5112AQH57bk2YqrI_w/article-cover_image-shrink_720_1280/article-cover_image-shrink_720_1280/0/1520046454355?e=2147483647&v=beta&t=hkcdD-fX95CYlZwvBzzGjZUvz-wRXa80DrOdtNyrcxo8.

The advent of advanced technologies such as Massive MIMO has profoundly influenced the uses of Detection Theory in modern communication systems. Massive MIMO, which involves the use of numerous antennas in the base station, has transformed signal detection and processing. Having the ability to support high spatial multiplexing, this technology supports improved spectral efficiency as well as interference resilience. Detection Theory incorporation into Massive MIMO systems plays a significant role in dealing with problems such as pilot contamination and signal separation. The performance enhancement caused by sophisticated detection algorithms is essential in reaching the potential of such technologies to the maximum, picturing a future where robust detection methods are the signature of effective communication methods (Björnson et al., 2019). With advancing research in Detection Theory, the evolution in multiuser detection methods has grown in significance, particularly in multiuser environments. The challenges of rank deficiency in multiuser MIMO OFDM systems are a testament to the necessity for new detection methods. Conventional detection algorithms, while effective, prove to be insufficient for application under higher user density and lower receiver antenna arrays. Novel strategies such as genetic algorithm (GA)-assisted detectors exhibit significant potential in circumventing these issues by rendering near-optimum performance with reduced complexity. This marks a new direction in detection system design, adjusting to the dynamic nature of digital communication. With embracing advanced detection approaches, investigators can deepen investigations of the interweavings of machine learning, optimization, and signal processing, which will reveal the gateway to more complex communication attributes of future wireless networks (Hanzo et al., 2007) (Figure 9.5).

9.2.1. Definition and Purpose of Detection Theory

Detection theory is a subliminal analytical framework in the field of digital communication, which has to do with the extraction of information signals from noisy environments. Detection theory basically attempts to determine if a signal is present or not against the environment of uncertainty or interference. This theoretical structure encompasses some algorithms and procedures that facilitate the decision-making process involved in the detection of signals, building a systematic approach to the measurement of the performance of many detection methods. Scientists can optimize communication systems, therefore, making more reliable transmission of data in a range of applications from telecommunications to remote sensing technologies through systematic assessment of the probabilities of detecting true signals and labelling noise as false. One of the primary functions of detection theory is to provide a structured description of trade-offs between signal detection. For instance, in raising detection sensitivity to register extremely weak signals, one's false-alarm rate may also increase, necessitating a balance in system design to be employed with care. This juggling has a deep impact on actual-world applications, such as ISAC systems, where signal processing must change based on environmental fluctuations while obtaining optimum efficiency (Liu et al., 2022). In the long term, the theoretical underpinnings of detection theory foster innovation in the digital communication realm, and it steers innovations that enhance signal purity, data reliability, and the seamless coordination between sensing and communication processes. Moreover, detection theory has profound effects on future technologies, particularly to prepare future wireless networks better. Since various industries are greatly reliant on data-sustained environments, the need for sophisticated detection mechanisms is imperative. The convergence of emerging paradigms, including the Metaverse, requires a redefinition of detection frameworks for the inclusion of new social and technological dynamics (Park et al., 2022). Through the adoption of detection theory principles, system designers are able to design adaptive models that are capable of supporting diverse user interactions while minimizing latency and maximizing network performance. In practice, detection theory not only enables contemporary digital communications practices but also leads the way towards future developments that will transform information transfer, reception, and consumption across platforms.

9.2.2. Types of Detection: Coherent and Non-Coherent

Amongst the principles of digital communication, signal detection efficiency is the most important in achieving optimal information reception and transmission. A few of the most common techniques are coherent and non-coherent detection techniques. Coherent detection requires precise knowledge regarding the phase of the signal so that it can use signals amplitude and phase information to the fullest. It is particularly helpful in systems with high frequency communication, where signal detail is concealed by noise. With the employment of synchronized receivers, interference can be effectively fought by coherent detection, resulting in better error performance.

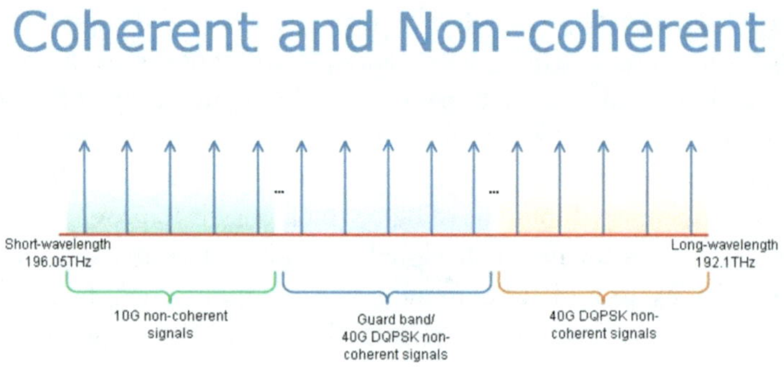

Figure 9.6. *Coherent and non-coherent.*

Source: https://www.thunder-link.com/tl-content/uploads/2022/12/3999wdm-basics-coherent-and-non-coherent-introduction.webp.

The method is pivotal in advanced applications such as Integrated Sensing and Communications, where accurate processing of signals allows for robust system operation in increasingly complex communication environments (Liu et al., 2022). Non-coherent detection, however, does not need phase synchronization, thus simplifying receiver design while allowing greater variations in signal amplitudes. This technique relies predominantly on the envelope of the entering signal and is therefore less susceptible to certain kinds of noise but is less than ideal for high-fidelity systems. Non-coherent detection is particularly valuable when signal environment changes are high speed, thereby allowing more flexible approaches. It can be used in many fields, including mobile communication systems and radar science, where the ability to operate without strict phase alignment can play a major role in system resilience (Tataria et al., 2021). Understanding the variations between types of detection can guide engineers to select the appropriate methods based on the specific needs of the communication task at hand. Development of detection techniques is necessary as we progress toward the next-generation wireless system, for example, 6G, with increased capacity and efficiency through novel paradigms of communications. Integration of coherent and non-coherent approaches within the detection framework is capable of resulting in useful performance improvement, particularly in the application of the rich environments to be found in future networks. Emerging concepts such as Integrated Sensing and Communications (ISAC) highlight the significance of addressing signal detection issues well. Confronting coherent and non-coherent detection as complementary techniques, scholars aim to enhance system robustness and flexibility to the utmost with significant frequency extension and the varied needs of present and future communications applications (Liu et al., 2022; Tataria et al., 2021). Here, continually improving detection theory is not abstract; it is absolutely essential to realizing the promise of next-generation digital communications networks (Figure 9.6).

9.2.3. Performance Metrics in Detection

In the digital communication scenario, detection performance measures are crucial in ascertaining the effectiveness of different detection techniques. Symbol error probability (SEP) and detection rate are essentially the parameters used as measures to compare the performance of communication systems under different circumstances. An explicit comprehension of these measures allows designers and builders of detectors to design and build detectors to effectively accommodate noise and interference, ubiquitous in practical applications. For instance, the maximum-likelihood (ML) detection architecture, under the Gaussian assumption of phase error distribution, constitutes a rudimentary framework for computation of SEP. Analytical error probability derivations are useful in performance quantification for a range of constellation designs and signal-to-noise ratio (SNR) configurations to enable informed decision-making in employing detection techniques (Eriksson et al., 2013). Additional research on the performance metrics indicates that complex channel conditions such as sparse intersymbol interference (ISI) channels must be addressed. In systems with high data rate, channel memory is important and requires advanced detection algorithms to maintain the performance efficient. Trellis-based methods of equalization, including multi-trellis and parallel-trellis Viterbi algorithms, illustrate how performance requirements can determine algorithm choice by balancing complexity versus optimality. Notably, while the former is not yielding the desired computational advantage, the latter provides substantial complexity reduction without sacrificing detection accuracy in some types of sparse channels. The above discussion establishes the importance of selecting an appropriate detection metric that is commensurate with the channel statistics and system requirements (Badri-Hoeher et al., 2005). Lastly, taking into consideration multiple performance metrics in detection constitutes the topology of digital communication systems. By critical analysis of theoretical models and practical implementations, researchers are able to further advance detection mechanisms that interventionally adapt across different environments. Development of hybrid solutions, such as reduced-state trellis-based equalizers, points towards the potential to improve detection performance while reducing computational load. With progress in the area, continual improvement of these measures will be a major contributor towards the progress of communication systems' reliability and capability, so that they may meet the growing requirements of today's digital worlds. The examination of sophisticated detection methods on a performance measure basis offers a vital foundation for future generations of breakthroughs in the theory of detection and estimation (Badri-Hoeher et al., 2005) (Figure 9.7).

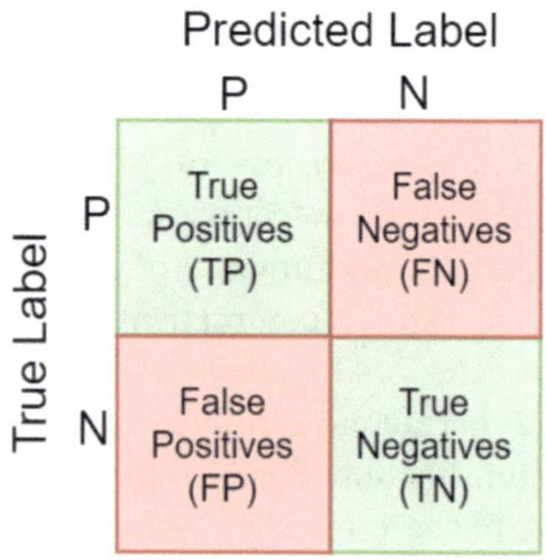

Figure 9.7. *Performance metrics in detection.*

Source: https://maucher.home.hdm-stuttgart.de/Pics/confusionMat.png.

9.3. ESTIMATION THEORY

Estimation Theory is also highly significant in enhancing the performance of digital communication systems by providing paradigms for extracting information from observed data. In this, special focus is placed on optimizing parameter estimation to characterize signal distributions and average channel conditions. This theory forms the basis for a lot of advanced detection and processing techniques required in communicating over noisy and interference-prone channels. With advancements in wireless communication systems, strong estimation methods are increasingly important. Traditional methods, however necessary, can be limited in the event of multi-input multi-output (MIMO) systems, necessitating new models that can handle such complexities, as outlined in the extensive multiuser detection strategies documented in current literature (Hanzo et al., 2007).

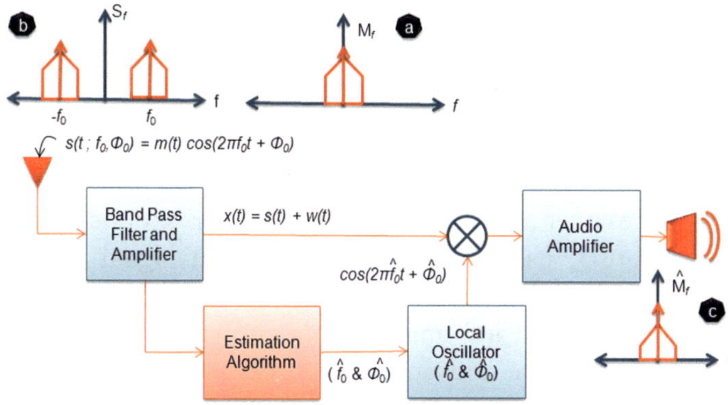

Figure 9.8. *Estimation theory.*

Source: https://i0.wp.com/www.gaussianwaves.com/gaussianwaves/wp-content/uploads/2020/07/DSB-AM-receiver.png?w=751&ssl=1.

Apart from that, innovations such as Massive MIMO technology demonstrate the ongoing advancement of Estimation Theory in wireless communications. The introduction of massive arrays of antennas has transformed wireless communication by enabling simultaneous transmission and reception of signals by numerous users. This calls for extremely precise estimation techniques in order to fight pilot contamination and achieve maximum spectral efficiency. In pursuit of this, the successful deployment of gigantic MIMO systems with full-digital transceiver chains has shown that theoretical principles of estimation could have practical implementations with tangible effects (Björnson et al., 2019). Therefore, continued research into improving estimation methods needs to be in accordance with these innovations, investigating new detection methods that are both computationally effective and able to take advantage of the inherent multi-dimensionality of contemporary communication networks. In the light of these advances, it can be seen that Estimation Theory is not a purely abstract mathematical field but an essential part of useful wireless communication applications. Sophisticated estimation techniques must be incorporated in the system so that the quality of channel state information is improved and for reliable detection of signals even in bad conditions. Employing approaches such as genetic algorithms and other optimization methods, researchers are discovering newer ways to accomplish greater system performance at the same time as reducing the complexity of calculations

(Hanzo et al., 2007). The complementarity between theory developments and applications promises a rich field for ongoing research, eventually leading to additional innovations in Estimation Theory in digital communications systems. Consistent with the scenario, the interactions among detection, estimation, and emerging technologies like massive MIMO will, likely, reveal additional paths for efficient communication solutions (Figure 9.8).

9.3.1. Overview of Estimation Theory

Estimation theory serves as a foundational pillar for various applications in digital communication, particularly in the realm of signal processing and detection systems. Its core objective is to devise algorithms and techniques that enable the extraction of information from observed data, especially when there is a presence of noise and uncertainty. This discipline emphasizes the development of statistical methods, such as maximum likelihood estimation and Bayesian estimation, which are pivotal in determining the parameters of interest from noisy observations. The effectiveness of these techniques relies heavily on the underlying models of the communication systems, which dictate the relationships between transmitted signals and the received data. In the context of digital communications, the accurate estimation of signals directly influences the performance of various detection algorithms, making it imperative to have a robust grasp of estimation theory for enhancing communication reliability and efficiency. In modern digital communication systems, especially those employing multiple-input multiple-output (MIMO) architectures, the importance of accurate channel estimation becomes increasingly pronounced. MIMO systems, characterized by the utilization of multiple antennas at both the transmitter and receiver ends, inherently introduce complexities in the detection process due to the potential for interference and multipath fading. Estimation theory provides a framework for understanding and mitigating these challenges, particularly in scenarios where conventional detection techniques falter, such as rank-deficient conditions. This situation arises when the number of users or transmit antennas exceeds the number of available receiver antennas, complicating the detection process further (Hanzo et al., 2007). Therefore, advanced methods developed through estimation theory can significantly enhance the detection capabilities of MIMO systems, thereby facilitating more effective data transmission and improving overall system performance. Moreover, the advancements in digital signal processing (DSP) have synergistically complemented estimation theory, particularly in scenarios influenced by fiber optics and nonlinear impairments. In recent years, coherent optical communication systems have emerged as a standard for high-speed transmissions, utilizing estimation algorithms to counteract both linear and nonlinear distortions present in optical fibers. As the demand for increased spectral efficiency escalates, so does the complexity of estimating and compensating for fiber nonlinearity, often exacerbated by factors such as the Kerr effect (Amari et al., 2017). Techniques explored within estimation theory play a vital role in developing strategies for nonlinear compensation, ensuring that the integrity of the transmitted signal is preserved even in high-capacity systems. This convergence of estimation theory with advancements in both DSP and optical communications exemplifies the pervasive impact of estimation methodologies across various facets of digital communication.

9.3.2. Types of Estimators: Point and Interval Estimators

The point and interval estimators are distinct in a fundamental manner within the theoretical framework of detection and estimation theory for digital communications. The point estimators provide a single number as an estimate of some unknown parameter to render data easily accessible in a concise form. For instance, in a power transmission system where it is not known, a point estimator can provide a system with an expected power level to use when operating at optimal. Point estimates, however, can be misleading since they do not capture the data's natural variability. Hence, while point estimators are useful since they are easy, they present a limited view, particularly in cases where uncertainty plays a significant role in the result. This is a vital factor in advanced communication systems where reliability and accuracy are crucial for effective operations. Interval estimators, however, provide an interval of probable values for an unknown parameter, encompassing uncertainty in the estimation. By giving a more sophisticated estimate, these estimators not only give an estimate at the center but also estimate the size of uncertainty with which it is charged and thus give a better idea of the values of the parameters. This is particularly crucial in the field of detection and estimation theory, where communication systems must deal with different conditions and noise. For example, in integrated sensing and communication systems, interval estimators play a significant role in the sensitivity-robustness trade-off as they allow one to estimate potential variation in signal power or levels of noise. Finally, the use of interval estimators offers better insight into the reliability and performance of systems so that communications protocols are effectively determined and enforced. Point and interval estimator unification is particularly crucial in technologies such as Integrated Sensing and Communications (ISAC). As has been expounded in recent studies, the existing advancements in dual-function radar-communications technologies necessitate joint designs of high quality that optimize functionality and performance. For instance, in highly populated frequency bands, which are addressed by Liu et al. (2020), employing an interval estimator may improve radar and communication signals' detectability by enabling the uncertainties of transmission on their part to be resolved. This serves to prevent interference and improve the quality of the signal in areas where there are several signals. Therefore, the strategic application of point and interval estimators enhances not just the analytical technique of detection and estimation but also the evolution of future communication systems as significant progresses toward more efficient and coexisting technologies in the digital communications field (Figure 9.9).

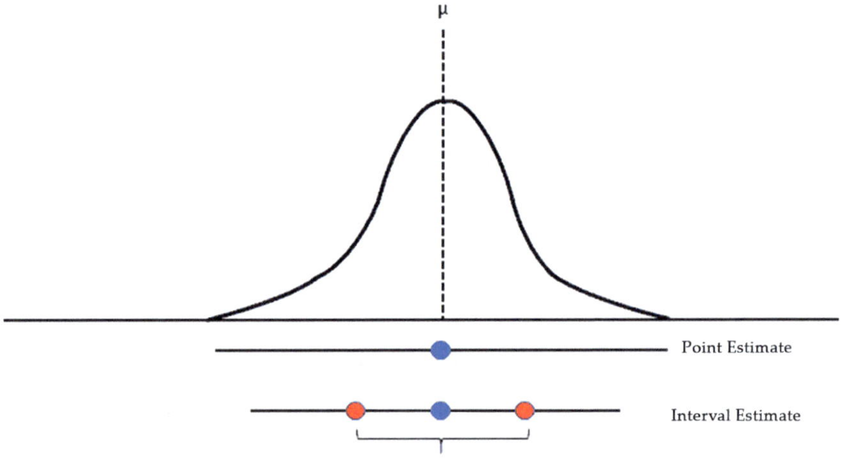

Figure 9.9. *Point and interval estimation.*

Source: https://sixsigmastudyguide.com/wp-content/uploads/2021/11/pi1.png.

9.3.3. Criteria for Evaluating Estimators

Evaluation of estimators in detection and estimation theory is a critical process that is dependent on several well-proven conditions. These include unbiasedness, efficiency, consistency, and sufficiency, all of which play a role in the evaluation of the quality and reliability of estimators. An unbiased estimator ensures that the expected value is the same as the true parameter with repeated samples, thereby avoiding systematic errors. Efficiency, normally measured by the Cramér-Rao bound, is an indication of how much information the estimator is utilizing relative to the estimator's variance. Consistency is the property that guarantees estimates converge to the actual parameter as the sample size tends to infinity, while sufficiency is the property that claims the estimator is obtaining all the data information there is. Collectively, these requirements give a firm basis for the performance analysis of various estimators used in digital communication systems to maintain transparent communication and integrity in data transfer. The inter-relationship between these evaluating parameters is most striking in advanced communication technologies such as Integrated Sensing and Communications (ISAC). In 6G system development, where sensing and communications have to coexist harmoniously, the ability to achieve the balance between efficiency and fairness is very much in a critical condition. According to Liu et al. (2022), ISAC presents a paradigm where the dual ability of sensing and communications can enhance network overall performance. The challenge is to ensure that estimators used in such systems are maximized in their performance without compromising one criterion for the sake of another, paving the way for the necessity of sophisticated design approaches. Recent research has identified that an understanding of the trade-offs in terms of performance between different criteria can produce breakthrough solutions in the field of wireless communications, injecting a higher degree of robustness into network operation. A crucial aspect of invoking such criteria successfully involves the application of mathematical models and signal processing algorithms that are the backbone of estimation methods today. With digital communications systems evolving—expanding into fields like quantum computing featured in Bharti et al. (2022)—paradigms of evaluation for estimators will have to enhance as well. In this noisy intermediate scale quantum (NISQ) age, innovations in machine learning and optimization algorithms are changing the way estimators are conceived and deployed. As such, the applicability of the defined evaluation metrics needs to be routinely reviewed to ascertain that they are consistent with evolving technologies. Future work in this field will undoubtedly be the integration of novel computational methods with traditional statistical norms, eventually generating more precise and accurate estimators of increasingly complex digital communication networks.

9.4. SIGNAL PROCESSING METHODS

The field of signal processing occupies a pivotal position in the development of detection and estimation theory in digital communication systems. Signal processing techniques support the enhancement of the efficiency of communication systems through better data transmission accuracy and reliability. OFDM is one of the key techniques in this field that employs multiple frequency channels to convey data in parallel. This technique is highly effective in minimizing interference and obtaining maximum throughput in multipath propagation conditions. The interaction between MIMO systems and OFDM is intriguing since it can carry out multiuser transmission at the same time, which results in better overall system performance under rank-deficient situations. As noted by Hanzo et al. (2007), the evolution of OFDM research has unveiled both advances and shortfalls of conventional detection techniques, stressing the need for novel solutions to mitigate such issues in actual implementation. The introduction of advanced algorithms has revolutionized signal processing, particularly digital communication. Techniques such as genetic algorithms (GA) have gained prominence due to their effectiveness in optimizing multiuser detection techniques.

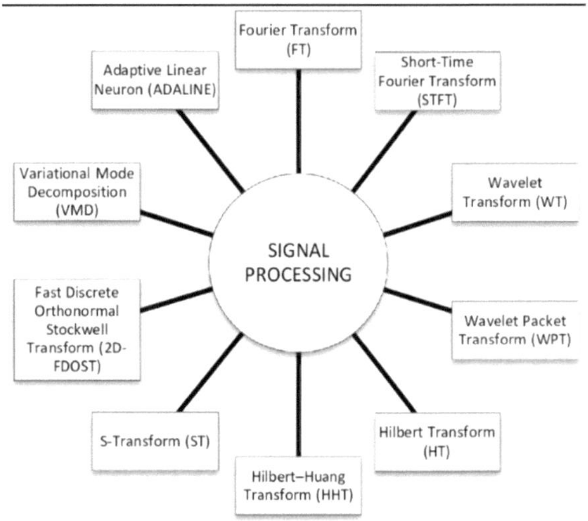

Figure 9.10. *Signal processing techniques.*

Source: https://www.researchgate.net/profile/Mohd-Jusoh-14/publication/375154103/figure/fig1/AS:11431281202466018@1698850638340/Signal-Processing-Methods.png.

This is particularly significant in MIMO-OFDM, where transmission antennas may be more than reception antennas, thus increasing the complexity of channel estimation and detection. As discussed in Hanzo et al. (2007), GA-aided detectors accommodate a near-optimum performance that makes computational complexity easier compared to traditional maximum-likelihood detection procedures. This optimization is essential for use in real-world applications such as cellular networks, where accuracy and efficiency are essential. Through the use of efficient computational algorithms that are components of genetic algorithms, signal processing is able to deal with issues arising from expanding data demands and channel diversity. Additionally, new architectures of signal sampling, such as Xampling, are essential in modern signal processing methods used in digital communication. The Xampling framework allows for effective analog compression before signal sampling, making spectrally-sparse signal processing with lower complexities possible. The approach utilizes sub-Nyquist sampling techniques, as discussed in, which enhance signal processing resilience against model mismatches with lower hardware complexities. The universal framework allows researchers and engineers to develop algorithms that facilitate low-rate signal processing, which is critical in real-time wireless communications systems. The versatility of Xampling across various categories of signals emphasizes its strength in handling the diversity of challenges found in digital signal processing, and calls for the need of continuous innovation in techniques in order to maintain pace with technological advancements (Figure 9.10).

9.4.1. Filtering Techniques in Detection and Estimation

In digital communication, filtering techniques are essential in enhancing detection and estimation algorithm performance. Desired signals can be successfully isolated from undesired noises using filtering techniques, which is critical across applications such as radar and communication systems. Filtering techniques have the ability to facilitate signal processing capability enhancements by engineers in order to provide for better targets or information to be detected. A variety of filtering techniques, including matched filtering, Kalman filtering, and adaptive filtering, enables practitioners to tailor the processing in light of the signal properties and environmental conditions. The increased complexity of communication systems necessitates these advanced techniques as they enable signal fidelity preservation within noisy and interference environments. One of the most promising filters in methodologies is derived based on the concept of under spread linear systems (ULSs) as discussed in recent publications. The identification of ULSs aims at systems that have clearly defined delay and Doppler properties, hence making detection simple. Using known input signals to drive these systems, it has been demonstrated that single-measurement valid parameter identification is achievable, provided the input signals time bandwidth product meets some specifications. This technology enhances detection capability in scenarios with targets close together, having more resolution than matched-filtering techniques used. Thus, these techniques are of utmost importance in radar systems, where the detection of numerous targets in challenging conditions is required for effectiveness (Gedalyahu et al., 2010). The combination of filtering algorithms with novel paradigms of signal acquisition, including Xampling, reflects the evolution of signal processing. Xampling presents a combined approach to signal acquisition where input bandwidth is compressed by analog compression before sampling such that commercial hardware is used to maximum capacity. The method enables subspace detection to be effective prior to conventional processing such that there is a platform for compressive sampling as well as low-rate signal processing. Compressive signal processing at rates below the Nyquist rate can provide a dramatic reduction in hardware requirements with commensurate improvement in system performance overall. By capitalizing on performance measures such as model mismatch robustness and hardware accuracy, researchers can form logical deductions about the most suitable compression methods, serving the interests of detection and estimation in digital communication (Eldar et al., 2011).

9.4.2. Role of Sampling in Digital Communication

Sampling is the key operation in digital communication, and the primary method for converting continuous-time signals to a form that can be transmitted and processed in a digital manner. The Shannon-Nyquist theorem is the foundation of this procedure, specifying the conditions under which a signal can be sampled without losing information. The theorem not only dictates the minimum sampling rate but also gives the importance of bandwidth in digital signal processing. With increasingly sophisticated digital communication networks and more advanced algorithms and computational processes, the role of sampling becomes increasingly important to understand. Sub-Nyquist sampling techniques are a case in point for the advancements in the subject. The aim of such techniques is to reduce the necessary analog-to-digital converter (ADC) sampling rates while maintaining signal quality, enhancing overall system efficiency and performance, a factor detailed in Eldar et al. (2011). Besides, current research has extended beyond conventional sampling theory limits and introduced new models to meet the requirements of contemporary digital communications systems. The Xampling theory offers a unified view with analog compression and non-linear signal processing as central aspects to optimize the way signals get recorded and processed in practice. By analyzing different methods of sub-Nyquist acquisition, such as the random demodulator and modulated wideband converter, researchers are developing techniques to minimize hardware cost and complexity and maximize performance. This multi-faceted sampling approach allows greater flexibility in numerous applications, especially with communication technologies that need higher data rates and more flexible processing. The implications of these sample methodologies emphasize the importance of rigorous mathematical modeling combined with realistic applications, as explicated in Eldar et al. (2011). Overall, the role of sampling in digital communication is pivotal to signal detection and estimation efficiency and accuracy. As digital systems increasingly rely on sophisticated software algorithms, the necessity for effective sampling procedures becomes all the more indispensable. The tendency to apply sub-Nyquist sampling frameworks is an indication of a sharp emphasis on better resource utilization and signal processing optimality. Such approaches not only address real-world implementation issues but also connect theoretical achievements to actual technology, making innovation in the technology domain possible. As more studies shed light on the possibilities of such strategies, it becomes obvious that sampling will be a key element of digital communication, supporting the ongoing development of high-capacity communication systems to respond to the increasing challenges of the digital era, as summarized through the combination of results in Eldar et al. (2011) and Eldar et al. (2011) (Figure 9.11).

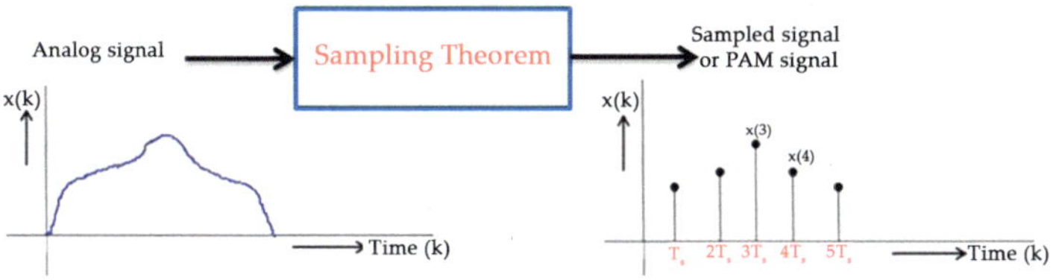

Figure 9.11. *Role of sampling in digital communication.*

Source: https://engineerstutor.com/wp-content/uploads/2018/08/9-2.jpg.

9.4.3. Impact of Noise on Signal Processing

Noise occurrence in signal processing definitely alters the quality and trustworthiness of transmitted information. Noise in digital communication comes in the form of interference that seeds uncertainty in recovering the original signal, which makes detection and estimation difficult. The impact of noise gains strength in intricate systems relying on precise signal transmission where even minor variations result in enormous data-interpretation errors. In wavelength-division multiplexing (WDM) networks, for instance, the cumulative impact of noise becomes an essential consideration in quantifying bit-error rates (BER). The ability to accurately estimate BER in noisy situations is not only important for the direct measurement of communication system performance but also for designing robust systems with the ability to cope with the inherent uncertainties associated with noise (Remondo et al., 2000). Beyond the basic concerns introduced by noise, its effects cut deeper into the effectiveness of communication systems. With technologies such as massive multiple-input multiple-output (MIMO), the power to combat the impact of noise largely relies on sophisticated signal processing techniques. Massive MIMO, which includes the usage of an overexcess number of antennas with respect to the active terminals, is a breakthrough concept that foresees the common noise problems. The technology necessitates strong computation power in order to process numerous signals at a time. The application of advanced algorithms enhances the ability of the system to rid itself of the effects of noise, thus its ability and energy efficiency. Importantly, algorithm development that is effective under noisy conditions remains a prominent research area, seeking to optimize performance with simplicity (Larsson et al., 2018). Detection and estimation system design and optimization must also incorporate preventive measures to combat noise. Both factors, computational complexity and power consumption, remain fundamental characteristics that impact the performance of the system. With development of communication technology and evolution of communication nature, it is necessary to understand how statistics lie behind noise. Stochastic techniques and adaptive filter methods are essential in assessing the robustness of signal processing systems against noise. Techniques such as importance sampling enable us to precisely estimate BER, which in turn determines appropriate settings for system detectors under noisy environments. These methods not only aid in measuring performance in real operations but also allow threshold levels to be realized which exceed normal expectations, thus redefining design specifications for communication networks (Remondo et al., 2000). While industry and academia continue to investigate the dynamics further, noise effects on signal processing will certainly outline future innovations in digital communications.

9.5. DETECTION AND ESTIMATION THEORY APPLICATIONS

Detection and estimation theory is a paradigm cornerstone in the area of digital communication that governs how communication systems are designed and deployed with high levels of sophistication. An example of one of many applications includes the design of signal detection techniques crucial for optimum performance under noisy and interference-prone conditions. The advancements of multiuser multiple-input multiple-output (MIMO) systems and orthogonal frequency division multiplexing (OFDM) systems demonstrate the most important role played by detection theory. It has been established that the normal detection methods do not work in rank-deficient scenarios, that is, scenarios where receiver-site antennas are less than user capacity. The research on conventional multiuser detectors (MUDs) in combination with cutting-edge algorithms promotes greater efficiency and performance, leading to better digital communication through ever-increasing network complexities (Hanzo et al., 2007). The emergence of Massive MIMO as a paradigm-shifting component of fifth-generation cellular networks further propels the significance of estimation theory and its applications in the real world. Previously a theoretical idea, Massive MIMO has evolved into a broad technology application, which allows base stations to service many antennas efficiently while providing high spectral efficiency. The intricacies of pilot contamination and other limitations have been tackled systematically by sophisticated signal processing techniques. This development marks the maturity of estimation theory, which now comes into play to overcome real-time communications' practical limitations. With the upsurge of Massive MIMO technologies, overlapping between detection and estimation theories presents abundant potential for research and innovation, pushing possible application areas beyond mobile broadband into radar, sensing, and positioning (Björnson et al., 2019). Finally, application of detection and estimation theory is far-reaching, touching many areas of digital communication technology. With the advent of genetic algorithm-based multiuser detection optimization techniques, researchers have now begun to venture out of traditional design paradigms. These new techniques leapfrog computational efficiency while simultaneously reaching near-optimal performance levels. By sparking a multidisciplinary cross-fertilization of ideas among areas of study such as machine learning, signal processing, and wireless communications, the future of detection and estimation theory promises much excitement and promise. As researchers continue to research the interaction between these fields, they are going to find new findings that will drive innovations in wireless networks and enhance the performance of communication systems. The applications of detection and estimation theory not only lie at the foundational level but are also critical to identifying the future trend of digital communication (Hanzo et al., 2007).

9.5.1. Use in Wireless Communication Systems

In wireless communication, detection of the signal and estimation form the basis of the success of today's digital communication systems. OFDM technology used in conjunction with MIMO technologies has formed the backbone of improving signal transmission efficiency.

The joint approach of these technologies facilitates the transmission of multiple data streams simultaneously, significantly improving wireless network capacity and reliability. However, as the demand for higher data rates continues to increase, conventional channel detection and estimation techniques are greatly strained, particularly in multiuser applications where the number of transmit antennas exceeds the capacity of the receiver. This is a prerequisite for advanced detection theory methods that can handle rank-deficient situations common in high-density communication scenarios, as the literature on conventional multiuser receivers for MIMO-OFDM systems (Hanzo et al., 2007) establishes. For further enhancing wireless communication efficiency, adaptive learning techniques are emerging as leading tools. The rapidly evolving wireless network environment with heterogeneous infrastructures and diverse applications demands learning and intelligent adaptability along with performance optimization.

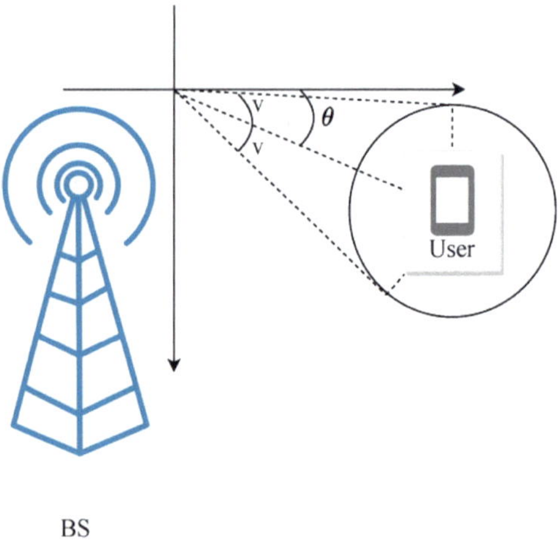

Figure 9.12. *Detection and estimation theory use in wireless communication systems.*

Source: https://www.mdpi.com/electronics/electronics-14-00750/article_deploy/html/images/electronics-14-00750-g002-550.jpg.

Machine Learning (ML) is the impetus behind this change, with state-of-the-art methodology in big data analytics, parameter estimation, and interactive decision-making. Over the past 30 years, innovation in ML techniques has led to enormous advancements in enabling advanced applications in a variety of domains, military, and civilian. Wireless systems can provide high-rate and low-latency communication with greater reliability by leveraging supervised, unsupervised, and reinforcement learning frameworks. This evolutionary direction is an expression of future wireless service needs, where intelligent networks can dynamically change to suit customer demands and environmental factors (Chen et al., 2019). Incorporation of the latest methods such as Genetic Algorithms (GAs) into detection algorithms has further revolutionized wireless communication systems. The algorithms provide huge gains in computational complexity, and near-optimum performance is possible without the behemoth complexity of the conventional maximum-likelihood detection methods. Applications of GAs in multiuser detection algorithms for MIMO-OFDM systems are only one instance of signal processing and machine learning innovation cross-pollinating. As communication environments become increasingly complex, the contribution of GAs to facilitating adaptive learning and optimization is at the forefront of priority. These advances not only result in better detection performance but also lay the ground for the development of robust frameworks that can be tailored to support various applications, including the Internet of Things (IoT) and cognitive radio networks, which further underscore the necessity for further research in this regard (Hanzo et al., 2007; Chen et al., 2019) (Figure 9.12).

9.5.2. Applications in Radar and Sonar Systems

In the detection and estimation theory field, sonar and radar systems are perfect examples to illustrate the practical application of such principles. Both systems rely on the ability to accurately detect and estimate targets' position and velocity in many conditions. Radar, for instance, employs electromagnetic waves to reflect signals off objects, while sonar employs sound waves under water conditions to achieve the same objective. The detection theory-based intricate algorithms enable such systems to separate the target signal and noise and thus are more accurate. Furthermore, they influence the design of complex subspace interference control, as in Besson et al. (2005). Radar and sonar technology hence both play to illustrate the central role detection and estimation play in today's navigation, surveillance, and communication technology. Even further expanding radar and sonar systems' capabilities, the introduction of Massive MIMO technology represents a humongous leap. Based on the use of a vast number of antennas at base stations, Massive MIMO represents an unprecedented source of spectral efficiency and spatial resolution benefits. Besides transforming cellular networks, the technology boosts radar and sonar applications with improved signal-processing capacity. The deployment of Massive MIMO has already begun to change these systems by allowing for greater detection accuracy and less interference in complex environments, according to Björnson et al. (2019). The increasing deployment at sub-6 GHz and millimeter-wave frequencies allows for the achievement of high-end applications like low-power machine-type communications and ultra-reliable sensing operations, and it is a game-changer for this industry. Besides, the ever-evolving revolution in detection and estimation techniques continues to redefine the radar and sonar applications realm. Techniques born out of estimation theory remain a key to understanding the uncertainties that come with real-world environments, like environmental noise and multipath propagation. For instance, Maximum-Likelihood Estimation (MLE) and Generalized Likelihood Ratio Tests (GLRT) present robust methodologies for signal detection under varying interference parameters, as highlighted in Besson et al. (2005). Theoretical advances like these not only aid in enhancing the estimating potential of radar and sonar platforms but also open up new avenues of research such as large-scale MIMO radar, which will most likely form the cornerstone of next-generation navigation and surveillance strategies. Consequently, this collaboration between detection theory and field applications supports the effectiveness and reliability of radar and sonar operations in a range of scenarios (Figure 9.13).

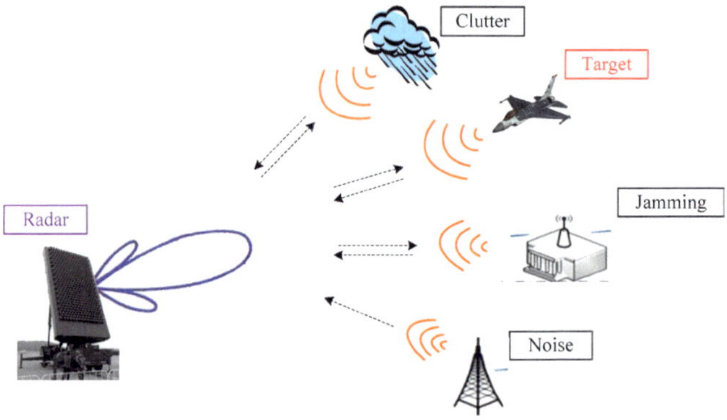

Figure 9.13. *Detection and estimation theory applications in radar and sonar systems.*

Source: https://www.mdpi.com/electronics/electronics-11-00156/article_deploy/html/images/electronics-11-00156-g003-550.jpg.

9.5.3. Role in Image and Video Processing

The application of detection and estimation theory to image and video processing is pivotal, especially since advances in digital communication technologies continue to evolve. At the core of such application is the accuracy to compress and analyze visual data, where algorithms are designed to detect central patterns and features in images. One such example is where motion detection techniques employ such concepts to pick out significant differences between frames, thus making the compression effective. By focusing on moving conspicuous locations rather than static backgrounds, systems can achieve very substantial savings in data size while maintaining important visual quality. These methods emphasize the practicality of detection theory in optimizing the power of image processing tools, particularly in bandwidth- and storage-constrained environments. Yet another illustration of this concept is the proposed low-complexity video compression algorithm that deals with motion edge blocks using solely Discrete Cosine Transform (DCT) coefficients. This new approach enhances motion estimation by restricting the search to varying macro-blocks, thereby reducing the encoding process computational demands.

Even though this method has a lower Peak Signal-to-Noise Ratio (PSNR) than conventional codes like H.264/AVC, it takes significantly shorter to execute and is thus particularly valuable for real-time applications like surveillance. The ability to recognize and process only beneficial blocks and skip static elements demonstrates the manner in which detection and estimation theories can transform the efficiency of video processing to the needs of today's digital communication environments (Kim et al., 2009). Within the broader imaging and video system applications field, the use of soft biometrics enhances the predictive power of the discipline, enhancing user authentication problems. Through the use of non-unique, generalizable characteristics such as age or emotional state, these systems embody the versatility of detection theory in use outside identification. As the environmental context of biometrics shifts, the emergence of soft biometrics is a fundamental step toward understanding contextual user information. Not only does this shift improve authentication but also supports the development of easy-to-use systems that can dynamically react to user behavior. These developments underscore the continued relevance of detection and estimation theory in the processing of visual data, evidencing its role in the innovation of more sophisticated, adaptive image and video processing technology (MCC et al., 2017).

9.6. CHALLENGES IN DETECTION AND ESTIMATION

Detection and estimation issues in digital communication are inherent, particularly as we approach technologies of the high-end kind like Massive MIMO. Pilot contamination, which is perhaps one of the biggest challenges here due to its notoriety, stems from the misuse of pilot signals in multi-user environments and leads to lower accuracy in estimation. As the deployment of extremely sophisticated signal processing techniques is gradually becoming the standard in 5G-class systems and other next-generation communication networks, effective estimation techniques are increasingly needed. Pilot contamination is a problem that needs innovative solutions capable of accommodating a large number of users without sacrificing high spectral efficiency. Such adaptations are crucial, as Massive MIMO is able to accommodate a more dense array of antennas, and thus detection and estimation complexities are unresolved (Björnson et al., 2019). Despite the revolution in signal processing technologies, realization of mmWave Massive MIMO is impaired due to practical challenges in detection and estimation tasks. Facilitating massive MIMO architectures translates to an immense number of radio frequency (RF) chains in proportion to the antennas, involving stupendous cost and design challenges, particularly at the mmWave frequencies. Hybrid beamforming relieves some of these limitations by splitting

MIMO processing between analog and digital. However, this introduces new constraints, for example, the uni-modulus property of beamforming coefficients, that limit the application of conventional MIMO strategies. Due to this, there is considerable necessity for systematic design paradigms that are capable of meeting these new challenges while preserving low interference levels and efficient channel estimation (Huang et al., 2017). Also, with the increased application of large MIMO systems in practical applications, detection and estimation problems evolve with technology development. The integration of sophisticated spatial processing models adds complexity to inter-cell interference problems, particularly with the use of large-scale arrays of antennas. Addressing such interference requires new methods of angle of arrival (AoA) estimation and controlling data-path gains efficiently with sophisticated beamforming algorithms. The complexity of the task is also increased by the sparsity of propagation paths typical in mmWave environments, which necessitates unique estimation approaches that can realistically model the dynamics of multiple signal paths. With emerging research, study of these challenges will be central to developing detection and estimation methods in digital communication systems (Huang et al., 2017).

9.6.1. Effects of Channel Impairments

The impact of channel impairments in digital communication systems is profound, with far-reaching implications on the detection and estimation processes which are integral to such technologies. Noise, for example, is an impairment which can result in the distortion of the signals as they move through the medium. This distorts the receiver's capacity to appropriately identify the signal emanating, resulting in increased error rates and overall performance loss. The development of statistical models to model these impairments is essential as they provide the groundwork for developing effective detection algorithms. Without a critical assessment, as seen in Arslan et al. (2014), adaptive receiver algorithms may become less effective, as they are developed based on the mathematical comprehension of signal properties under the impact of such impairments. Channel impairments in optical communications emerge via various aspects, including linear distortions and nonlinearities. Specifically, dispersion in optical fibers can cause pulse broadening, affecting the cleanliness of the received signals. Fiber nonlinearity due to the Kerr effect is another critical factor to be taken into account; at high power levels, the distortion introduced can severely impair signal integrity. More advanced coherent detection methods have made the application of digital signal processing (DSP) to mitigate these effects more possible, enhancing the transmission performance and integrity. Techniques particularly for handling channel impairments, highlighted in Amari et al. (2017), are an integral component of existing detection and estimation systems, enabling the systems to deal with variability in the transmission environment. Channel impairment analysis and compensation is not only vital from the perspective of performance improvement but also to improve the functionality of communication networks. Performance measures must consider signals-to-noise ratios, modulation types, and channel conditions. For instance, in wireless communications, channel dynamics necessitate robust estimation mechanisms to maintain signal fidelity. Algorithms like the chi-square constant false alarm rate Gaussianity test of Arslan et al. (2014) are robust innovations in signal detection under these impairments. Thus, continuous research in this field keeps enhancing approaches addressing channel impairment challenges, with the ultimate perspective of possessing more effective and resilient digital communication systems (Figure 9.14).

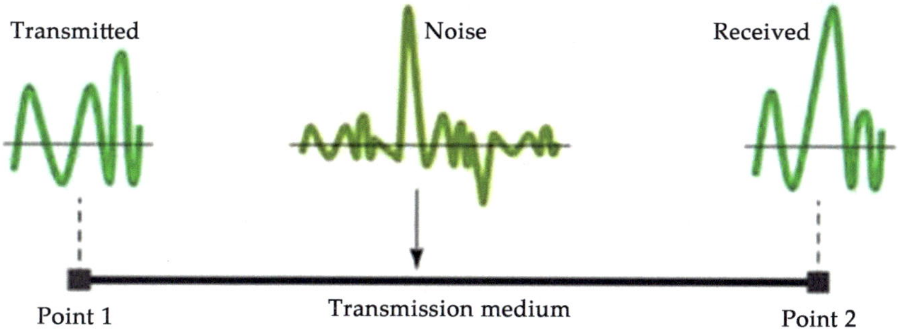

Figure 9.14. *Transmission impairment in data communication.*

Source: https://media.geeksforgeeks.org/wp-content/uploads/20201004223417/nise.jpg.

9.6.2. Computational Complexity in Real-Time Systems

Research in computational complexity for real-time systems is crucial to enhancing the efficiency of detection and estimation theory in digital communication systems. With increasing demands for speedy processing and accurate transmission of data, it is essential to understand the complexity of computational constraints. Algorithms in real-time systems, such as those used in phase retrieval and signal processing, must take into account the scalability of implementation complexities. For instance, convex optimization methods have been shown to be a critical solution in handling phase retrieval problems in high-dimensional space. Though ensuring high accuracy, the methods require a computation of complexity to determine feasibility in real-time systems

(Arik et al., 2018). The efficiency of these algorithms relies a lot on their performance, such that a scrutiny of computational complexity is not just an academic issue but a critical factor for applicability. Real-time systems often have to make compromises between computation load and signal estimation quality. With the development of advanced techniques such as the projected gradient method and the alternating direction method of multipliers (ADMM), the balance can be moved in favor of high accuracy and minimal computation cost. These techniques involve minimal estimation penalty with still quadratic complexity scaling relative to the number of parameters estimated (Arik et al., 2018). This is a significant aspect in environments where resources are limited and there is a great demand for speed of processing. By effectively minimizing computational complexity, such methods enable advanced detection and estimation techniques to be used in real-time systems for enhancing the overall reliability and effectiveness of communication. Massive MIMO technology becomes more significant, bringing forth the merger of real-time system performance and computational complexity. With advancements in Massive MIMO technology and deployment in the 5G ecosystem, it offers unmatched capabilities in spectral efficiency and signal processing. Implementation of all-digital transceiver chains in base stations represents a new frontier where spatial processing elements can improve not only mobile broadband services but also many applications including radar and sensing. As data streams and user connections increase, computational requirements emanating from them require understanding the inherent complexity of such systems. It ensures that future progress in the detection and estimation theory is founded on efficient computational methods, thus laying the grounds for the feasibility of safe and high-performance communications networks.

9.6.3. Trade-Offs Between Accuracy and Efficiency

It is extremely difficult to balance accuracy and efficiency in digital communication systems. Data detection and estimation accuracy is highly important since high accuracy ensures correct communication with low rates of errors. Though high accuracy tends to involve sophisticated algorithms and high computational efforts, these will most likely detract from operational efficiency. For instance, sophisticated techniques such as adaptive filtering and multiple signal classification improve accuracy by successful management of interference and noise but have the ability to introduce latency and increased processing times. The pursuit of greater accuracy hence compromises the ability of the system to transmit information in real-time and operate within real-time environments, calling for precise balancing of trade-offs when designing communication systems. Efficiency is critical in the computer case, especially because there is also an emerging need for higher data rate and more dependable networks. Efficient techniques that utilize less complicated modulation schemes and less complex signal processing are typically used to achieve maximal throughput without raising response time of the system. These methods have the potential to enable fast passing of data such that increased capacity and increased users can be supported in a network. With trade-offs, a more efficient system may compromise accuracy such that increased error probabilities can occur. This trade-off is especially essential in emerging technologies like 5G networks, where the ratio between accuracy and efficiency has a direct implication on the user experience and system performance overall, as observed in the evolution of 6G technology (Cheng-Wang et al., 2023). Moreover, as we navigate advancements in digital communication, acquiring the trade-off between accuracy and efficiency becomes essential to maximizing the boundaries of technology. Today's developments, such as Swarm Intelligence and Particle Swarm Optimization algorithms, are proof of this balancing act, where the optimization of performance indicators must be rigorously tested to address both sides of the equation. The balancing act between efficiency and accuracy dictates methodology development, where researchers constantly try to improve algorithms to strike the best balance between the two generally conflicting goals. By looking at existing usage and observing where points of improvement can be made, including healthcare and smart cities use, we are capable of discovering a path towards achieving more robust communication systems that address the fine-grained challenges wrought by modern demands (Ahmed, 2022).

SUMMARY

The detection theory and estimation of digital communication highlight the significance of advanced detection systems and their use. With digital communication systems being developed with higher sophistication, efficient detection methods become a more pressing need. Employing techniques like the Neyman-Pearson test and overcoming the limitation of binary data representation, scientists have been able to contribute immensely towards making communication more reliable. Notably, low-complexity systems, like single-bit output systems, are reflective of an interesting line of future potential. These techniques not only increase the energy efficiency of detection but also enable reliable monitoring systems to be instituted in dynamic environments, paving the way for enhancement in real-time applications (Stein et al., 2018). With the advancements in detection structures, channel estimation techniques emerge as important to ensure optimal performance in digital communication systems. The groundwork provided by multicarrier systems such as orthogonal frequency division multiplexing (OFDM) with multiple-input multiple-output (MIMO) capabilities has revolutionized the landscape of wireless communication. However, there are problems remaining, particularly rank-deficient cases that arise when the number of users exceeds the receiver antenna capacity. Bypassing such limitations by using cutting-edge multiuser detectors (MUDs) is critical to providing reliable communication in practice. Furthermore, the advent of genetic algorithm (GA)-aided detection methods underscores the promising unification of computational intelligence and communication engineering, enhancing performance at the cost of complexity (Hanzo et al., 2007). The future potential of detection and estimation theory in the digital communication area is bright, with considerable potential for future research and advances. New techniques and technologies for the enhancement of detection accuracy and efficiency will undoubtedly shape the future wireless communication systems. Research into new detection algorithms, particularly in combination with machine learning methods, can result in revolutionary leaps in performance. In coping with the complexities of the rapidly evolving user environments and high-rate data communication, we will need to adapt current theoretical frameworks to accommodate new situations. Ultimately, the marriage of underpinning detection theory with innovative applications will play a key role in defining the future direction of digital communication systems, rendering them robust and reliable in increasingly challenging environments.

The study of detection and estimation theory in digital communications is a foundation for studying ways information can be successfully transmitted and received across various communications environments. A key component of the theory is the detection of signal characteristics in the presence of interference and noise, an area that is pivotal in the dependability of communication networks. Through statistics, signal detection is optimized by engineers, and overall system reliability is improved. This theory not only corrects the signal detection principles but also provides tools to estimate received signal parameters so that the receivers can interpret transmitted information sufficiently. As wireless technology continues to improve, these detection and estimation theories play a progressively vital role in optimizing communication in the more advanced settings of fifth-generation (5G) systems and the projected union of sensing and communications in sixth-generation (6G) networks (Liu et al., 2022). An essential component of detection and estimation theory is the theory of performance trade-offs associated with designing communication systems. Engineers generally have to be torn between the detection accuracy factors and processing complexity as well as resource limitations. An example is that higher signal-to-noise ratios can deliver enhanced detection performance but at the cost of more advanced algorithms, hence increased computational requirements. A consideration of these trade-offs is especially important when dealing with integrated systems, which combine sensing and communication functionalities. By exploring information theoretic limits, physical layer optimization, and cross-layer design needs, academics can identify the optimal methods to achieve peak efficiency while consuming minimal resources. This is particularly applicable since the market is planning on deploying Integrated Sensing and Communications (ISAC), which will attempt to harmonize data sharing with environmental sensing (Liu et al., 2022). Lastly, the aspects of signal detection and estimation in environments with clutter hold maximum significance for the future development of communication systems. With advancements in digital communication due to the advent of immersive technologies, such as the Metaverse, these theories gain more prominence. Advances in deep learning and artificial intelligence can facilitate signal processing ability more than ever before, allowing for greater and more dynamic communication methods. For instance, the Metaverse integrates several technologically diverse elements needing easy communication experience, presenting the need for proper detection and estimation techniques appropriate for extensive applications. Analyzing the social values and environments for which these technologies are being created, one can realize the evolving scenario of digital communication and the immediacy of integral schemes addressing new problems in this dynamic field.

REFERENCES

1. Abreu, M. C. C., Abreu, M. C. C., Abreu, M. C. D. C., Chang, T. Y., Chen, Y. L., Dobry, G., Gao, Y., et al. (2017). Predictive biometrics: A review and analysis of predicting personal characteristics from biometric data. https://core.ac.uk/download/286269435.pdf (accessed on 18th September 2025).
2. Ahmed, G. G. (2022). Particle swarm optimization algorithm and its applications: A systematic review. *Archives of Computational Methods in Engineering, 29*, 2531–2561. https://doi.org/10.1007/s11831-021-09694-4 (accessed on 18th September 2025).
3. Amari, A., Ciblat, P., Dobre, O. A., et al. (2017). A survey on fiber nonlinearity compensation for 400 Gbps and beyond optical communication systems. http://arxiv.org/abs/1708.06313 (accessed on 18th September 2025).
4. Arik, S. O., & Kahn, J. M. (2018). Low-complexity implementation of convex optimization-based phase retrieval. http://arxiv.org/abs/1707.05797 (accessed on 18th September 2025).
5. Arslan, H., & Gorcin, A. (2014). An OFDM signal identification method for wireless communications systems. http://arxiv.org/abs/1501.00001 (accessed on 18th September 2025).
6. Badri-Hoeher, S., Hoeher, P. A., Land, I., et al. (2005). Trellis-based equalization for sparse ISI channels revisited. http://arxiv.org/abs/cs/0508051 (accessed on 18th September 2025).
7. Besson, O., Scharf, L., & Vincent, F. (2005). Matched direction detectors and estimators for array processing with subspace steering vector uncertainties. https://core.ac.uk/download/12038840.pdf (accessed on 18th September 2025).
8. Bharti, K., Cervera-Lierta, A., Kyaw, T. H., Haug, T., Alperin-Lea, S., Anand, A., Degroote, M., et al. (2022). Noisy intermediate-scale quantum algorithms. *Reviews of Modern Physics, 94*. https://doi.org/10.1103/revmodphys.94.015004 (accessed on 18th September 2025).
9. Björnson, E., Hoydis, J., Marzetta, T. L., et al. (2019). Massive MIMO is a reality–What is next? Five promising research directions for antenna arrays. http://arxiv.org/abs/1902.07678 (accessed on 18th September 2025).
10. Chen, K. C., Hanzo, L., Jiang, C., et al. (2019). Thirty years of machine learning: The road to Pareto-optimal wireless networks. http://arxiv.org/abs/1902.01946 (accessed on 18th September 2025).

11. Eldar, Y. C., & Mishali, M. (2011). Sub-Nyquist sampling: Bridging theory and practice. http://arxiv.org/abs/1106.4514 (accessed on 18th September 2025).
12. Eldar, Y. C., Elron, A., & Mishali, M. (2011). Xampling: Signal acquisition and processing in union of subspaces. http://arxiv.org/abs/0911.0519 (accessed on 18th September 2025).
13. Ercil, A., Ertuzun, A., Ertüzün, A., et al. (2006). 2-D iteratively reweighted least squares lattice algorithm and its application to defect detection in textured images. https://core.ac.uk/download/pdf/11738461.pdf (accessed on 18th September 2025).
14. Eriksson, T., Khanzadi, M. R., Krishnan, R., et al. (2013). Soft metrics and their performance analysis for optimal data detection in the presence of strong oscillator phase noise. https://core.ac.uk/download/pdf/70599800.pdf (accessed on 18th September 2025).
15. Gedalyahu, K., Bajwa, W. U., & Eldar, Y. C. (2010). Identification of parametric under spread linear systems and super-resolution radar. http://arxiv.org/abs/1008.0851 (accessed on 18th September 2025).
16. Hanzo, L., & Jiang, M. (2007). *Multiuser MIMO-OFDM for Next-Generation Wireless Systems.*
17. Huang, K., Lau, V. K. N., Li, X., et al. (2017). Hybrid beamforming via the Kronecker decomposition for the millimeter-wave massive MIMO systems. https://core.ac.uk/download/95560735.pdf (accessed on 18th September 2025).
18. Kim, C., & O'Connor, N. E. (2009). Low complexity video compression using moving edge detection based on DCT coefficients. https://core.ac.uk/download/11308803.pdf (accessed on 18th September 2025).
19. Larsson, E. G., Liu, L., & Van der Perre, L. (2018). Efficient DSP and circuit architectures for massive MIMO: State-of-the-art and future directions. http://arxiv.org/abs/1807.05882 (accessed on 18th September 2025).
20. Liu, F., Cui, Y., Masouros, C., Xu, J., Han, T. X., Eldar, Y. C., & Buzzi, S. (2022). Integrated sensing and communications: Toward dual-functional wireless networks for 6G and beyond. *IEEE Journal on Selected Areas in Communications, 40*, 1728–1767. https://doi.org/10.1109/jsac.2022.3156632 (accessed on 18th September 2025).
21. Liu, F., Masouros, C., Petropulu, A. P., Griffiths, H., & Hanzo, L. (2020). Joint radar and communication design: Applications, state-of-the-art, and the road ahead. *IEEE Transactions on Communications, 68*, 3834–3862. https://doi.org/10.1109/tcomm.2020.2973976 (accessed on 18th September 2025).

22. Park, S., & Kim, Y. G. (2022). A metaverse: Taxonomy, components, applications, and open challenges. *IEEE Access, 10*, 4209–4251. https://doi.org/10.1109/access.2021.3140175 (accessed on 18th September 2025).
23. Povalač, A. (2013). Spatial identification methods and systems for RFID tags. https://core.ac.uk/download/30311076.pdf (accessed on 18th September 2025).
24. Reimer, P., Austin, W. E. N., Bard, É., Bayliss, A., Blackwell, P. G., Bronk Ramsey, C., Butzin, M., et al. (2020). The IntCal20 Northern Hemisphere radiocarbon age calibration curve (0–55 cal kBP). *Radiocarbon, 62*, 725–757. https://doi.org/10.1017/rdc.2020.41 (accessed on 18th September 2025).
25. Remondo, D., Tattje, H. E. P., Srinivasan, R., Nicola, V. F., & Van Etten, W. C. (2000). Adaptive importance sampling for performance evaluation and parameter optimization of communication systems. https://core.ac.uk/download/pdf/11453536.pdf (accessed on 18th September 2025).
26. Stein, M. S. (2018). Asymptotic signal detection rates with 1-bit array measurements. http://arxiv.org/abs/1711.00739 (accessed on 18th September 2025).
27. Tataria, H., Shafi, M., Molisch, A. F., Döhler, M., Sjöland, H., & Tufvesson, F. (2021). 6G wireless systems: Vision, requirements, challenges, insights, and opportunities. *Proceedings of the IEEE, 109*, 1166–1199. https://doi.org/10.1109/jproc.2021.3061701 (accessed on 18th September 2025).
28. Wang, C. X., You, X., Gao, X., Zhu, X., Li, Z., Zhang, C., Wang, H., et al. (2023). On the road to 6G: Visions, requirements, key technologies, and testbeds. *IEEE Communications Surveys & Tutorials, 25*, 905–974. https://doi.org/10.1109/comst.2023.3249835 (accessed on 18th September 2025).
29. Wu, Q., Zhang, S., Zheng, B., You, C., & Zhang, R. (2021). Intelligent reflecting surface-aided wireless communications: A tutorial. *IEEE Transactions on Communications, 69*, 3313–3351. https://doi.org/10.1109/tcomm.2021.3051897 (accessed on 18th September 2025).

CHAPTER 10

APPLICATIONS AND FUTURE TRENDS IN DIGITAL COMMUNICATION

Contents

INTRODUCTION

With the advent of digital communication, the way human beings and organizations interact has changed, paving the way for innovative uses and upcoming trends that significantly boost connectivity and productivity. The change is thanks to the rapid development of technologies such as 5G, artificial intelligence, and the Internet of Things (IoT). Not only do these technologies enable real-time conversation but also intelligent applications supporting varied user needs across sectors. As increasing numbers of companies go online, it is essential to monitor such evolving trends in order to stay ahead. Thus, the research into digital communication applications opens up a broad spectrum of opportunities that hold the potential to revolutionize how we communicate and interact with one another in forthcoming years. The implications of such advances extend well past mere refinements of communication, and profoundly affect underlying market dynamics and operational tactics. The union of low latency communication, for instance, has been advanced into a foundation for numerous applications dealing with real-time data handling and interaction. As identified in current studies, many new industrial and consumer uses will remake the old operators business models and present enormous new markets to tap (Cuevas et al., 2017).

Figure 10.1. *Future of digital communication.*

Source: https://sendbird.imgix.net/cms/20230112_The-Future-of-Digital-Communications_-3-Predictions-for-2023-1.png?auto=format,compress&crop=faces.

This calls for significant investments in infrastructure to meet the technical challenges posed by new technologies. But the opportunities for market-facing operators are large, with clear potential to generate value in new services that leverage these developments. Moreover, digital communications growth is concomitant with the developments in geospatial technology, which has seen spectacular developments in geographic information systems and data analysis. Increased geoinformation availability, complemented by technological developments that enable its analysis, is an important trend in policy-making and spatial planning. The trends, as discussed in emerging literature, intensify the demand for an increased role for geotechnology in planning and decision-making (Geertman et al., 2000). As digital communication technologies advance, their integration with spatial analytics will probably set a new standard for applications that enhances an even more responsive and informed approach to planning in many settings. Where the two fields meet, future digital communication trends will not merely change the way people interact, but also enhance the strategic use of data in fields beyond (Figure 10.1).

Definition of Digital Communication

Digital communication is a paradigm that fundamentally changes the manner in which individuals and organizations exchange information with one another. Essentially, digital communication entails encoding, transmission, and decoding messages via electronic media, thereby simplifying communication across multiple platforms and devices. This mode of communication is motivated by technologies such as the internet, wireless networks, and digital hardware, which have revolutionized the culture of communication paradigms. The ability to exchange text, pictures, sound, and film instantly across large geographical distances depicts the immediacy and effectiveness of digital communication. In addition, the innovation indicates the shift from analog to digital methods of presentation, guaranteeing higher levels of sophistication in information processing capabilities. As such, digital communication helps facilitate personal communication, business processes, and data sharing, setting the way for future applications and innovations in this area. The implications of electronic communication extend far beyond mere information exchange; it reconfigures relationships and business practices across sectors. For instance, businesses leverage the potential of electronic communication to communicate with customers and stakeholders through social networking sites, email marketing, and live chats. The interactive web pages enable businesses to establish customer relationships and react quickly to customer feedback, thereby enhancing overall user experience and satisfaction. In addition, digital communication promotes remote collaboration through means that support real-time sharing of ideas and projects between geographically spread teams. This trend has further gained increased visibility in light of present global shifts toward remote working. The mutualism between organizational dynamics and digital communication thus promotes a more responsive and agile environment that enables firms to thrive in more competitive markets. Future digital communication will be based on the platform established by existing technology with developments that will enhance efficiency and accessibility. The advent of machine learning and artificial intelligence will be a dominant force in making communication experiences more personalized, enhancing user experience, and automating mundane work. In addition, connectivity benefits, such as the advent of 5G technology, will make communication faster and more reliable, creating the way for applications depending on real-time exchanges of data. As users become more adept at navigating digital realms, information literacy, encompassing literacy competencies such as assessing the credibility of a source, becomes more vital (Macgregor et al., 2005). Moreover, the use of monitoring and controlling systems, seen in most sectors, will increasingly increase-tune digital communication applications (Obeed et al., 2018). This trend indicates that digital communication may also redefine interpersonal communication and business operations boundaries as well as set the futures technological landscape.

Importance in Contemporary Society

In the rapidly evolving context of computer-mediated communication, the importance of analyzing its applications in contemporary society cannot be overemphasized. With the advent of emerging media technologies, the patterns of people communicating, learning, and sharing information have undergone a drastic shift. With increasing reliance being put on screen media, particularly among the younger generation, an imperative arises to explore the implications of these transformations on the learning settings. For instance, integration of digital literacy in higher education contexts has made instructors revisit their teaching pedagogy. As researchers have established the case for first-year university students, there exists incredible heterogeneity in employing various types of screen media, and hence, there is a need for pedagogical frameworks that support these different practices and needs in the digital era (Brown et al., 2015). Consequently, today's discourse on digital communication captures a pivotal moment of juncture among technology, education, and social growth. Additionally, with an era where digital media is omnipresent, issues of credibility and trust in information have been at the center of importance. The complexity is further heightened among teenagers, who must negotiate navigating through a digital world replete with disinformation and contradictory sources of information. The challenge of separating reliable content from defective information is an important pedagogical problem that has arisen during educational debates. It is important to understand the intricacies of credibility in online media in order to enable young audiences to make educated choices. Scholars aver that the critical examination not only enhances students' digital literacy but also enhances their ability to think positively within contemporary discourse. The focus on credibility in school curricula therefore emerges as the central component in preparing subsequent generations with the ability to succeed in a digital world (Andrew et al., 2008). As a result, as society becomes more accepting of digital communication, the ramifications of these technological innovations extend beyond simple convenience; they restructure fundamentally our social interactions and pedagogical practices.

Cyber communication is a complex space that enhances human encounter while concurrently raising questions about authenticity and participation. Emphasizing the role of digital literacy and dexterity in navigating this space becomes an imperative for teaching and learning. By infusing critical understanding of cyber media encounters into teaching platforms, we prepare learners not merely to catch up with current trends but to chart the course of communication into the future. A comprehensive exposure to these matters illustrates the potency of digital communication as a pillar of development in contemporary society, capable of both change and issue (Figure 10.2).

Figure 10.2. *Digital communication importance in contemporary society.*

Source: https://lh7-us.googleusercontent.com/6yOmwrR-xoxTlrPJXS6z0eiy8zsAC_2PilIPUmha04InqrL5DAXt2CNahVYJDeaQnYTYyTqz5IsNCYFjXvOwXT8LToU435GK_t-DPjRSVzKzqWOmIrFAgp2daFhSOHvgQ_edpG_yhkJltAcJJHYYhw.

10.1. HISTORICAL CONTEXT OF DIGITAL COMMUNICATION

The history of communication has been significantly influenced by innovations over the course of history that have impacted technology and interactions between people. Telecommunication innovation early on established the precedent for how people communicated long distances apart. These innovations were critical in making real-time communication possible and surmounting the barriers geography put in its way. When digital technology came into being, especially with the invention of the internet in the late 20th century, the world of communication changed dramatically. With the facilitation of sending and receiving information within seconds, interpersonal relationships were revolutionized to the extent that people were able to reach out to one another from all around the globe. Therefore, this background of history explains how digital communication has been a part of modern society and will have more sophisticated applications in the times to come. In addition to the technological advancements, the context of digital communication is replete with cultural shifts in attitudes and practices. This transition from print to digital media was a landmark development in information sharing. The arrival of the World Wide Web witnessed users transcending their roles as passive recipients of information but instead active participants in content shaping. This democratization of communication made individuals able to upload their voices, accommodating various expressions and global interactions. Second, as studies state, the focus on e-Heritage Systems enhances the incorporation of next-generation information and communication technology in cultural heritage, which reflects a fundamental change in information management and dissemination to the public. By recognizing these cultural shifts, the broader impacts of digital communication beyond technology exploitation can be recognized (Klein et al., 2025). The development of the digital communication landscape is continuous, reflecting the advancement of technological innovations and social trends. Integration of artificial intelligence and machine learning algorithms into communication apps has introduced novel interaction paradigms that optimize user experience and personalization. As brought into perspective through collaborative research work by organizations like Glasgow Caledonian University and the University of Alicante, interdisciplinarity in research on digital communication reflects the growing importance of research on its effects across various disciplines such as marketing, economics, and the social sciences. These trends reflect that digital communication is not only a tool for everyday communication but increasingly emerging as a part of strategic planning within various sectors. Thus, when we look towards the future, the past provides us with essential insights about the potential paths and applications of digital communication technologies (Claver-Cortés et al., 2017).

10.1.1. Analog-to-Digital Evolution

The transition to digital technology has dramatically reshaped communication systems to attain previously unprecedented levels of efficiency and accessibility. Analog communication, characterized by continuous signals, was constrained both in terms of clarity and capacity for data. With the beginning of information digitization during the late 20th century, information was digitized into binary form so that it would be easier to manipulate and transmit across various media.

This change enhanced the quality of communication along with providing new scope for innovation, particularly in terms of wireless technologies. As mobile networks went digital, so did user expectation significantly with higher speed and reliability being called for. Thus, the industry players focused on improving the underlying infrastructure, which has set off a hungry spate of theoretical and applied innovation in telecommunications, fully leveraging digitization and meeting the needs of a growing digital population (Adesuyi et al., 2014). The migration to digital communication also dramatically altered the information dissemination environment.

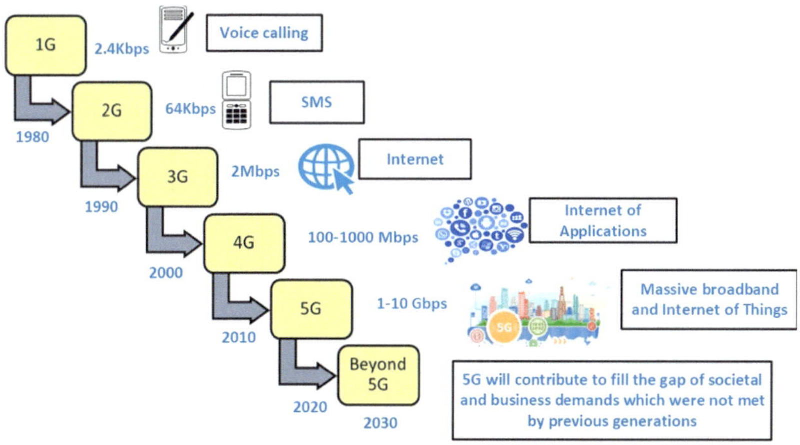

Figure 10.3. *Evolution from analog to digital technology.*

Source: https://external-content.duckduckgo.com/iu/?u=https%3A%2F%2Fwww.researchgate.net%2Fpublication%2F349740908%2Ffigure%2Ffig2%2FAS%3A-997362986131466%401614801274765%2FTimeline-of-Mobile-communication-generations-27.ppm&f=1&nofb=1&ipt=15c4de602c2c6353f0add8403d33100e25c5daafa7af84b97137cadb6df731da.

Computational systems are characterized by their ability to compress data, in which large amounts of data are contained in a much smaller space, hence minimizing storage requirements and optimizing bandwidth utilization. For the case of the Deep Space Network (DSN), advances in computer processing and network technologies have facilitated effective processing of advanced streams of data. The architecture recommended for the DSN involves giving precedence to open-systems standards to meet interoperability and to allow for cooperative efforts between NASA and its non-NASA stakeholders (Atkinson et al., 1992). Such innovations not only allow for seamless flows of information between the spacecraft teams but also ensure that data integrity and security are maintained in the process of communication. This transition to one further integrated digital platform demonstrates the overall importance of digital transformation in increasing both reliability and efficiency across communication systems. As the move from analog to digital continues, the augur for communications into the future is significant. With the increasing reliance on digital infrastructures, industries will have to deal with concerns of cybersecurity, data privacy, and the digital divide. This context calls for an aggressive approach to the development of innovative solutions that do not only uphold standards of security but also ensure access to digital technology on a balanced basis. The rate of innovation in wireless technology, as evidenced by the ongoing convergence of Wireless Intelligence Network (WIN) technology, presents a future with enhanced connectivity and communication ability (Adesuyi et al., 2014). With these technologies advancing, the focus will increasingly be on sustainable approaches and enhancing communication accessibility in developing countries. Generally, the evolution of digital communication promises an age that looks not only to technological progress but also to the inclusive embedding of societal requirements into the very crux of communication thinking in the future (Figure 10.3).

10.1.2. Key Milestones in Digital Communication Technology

The evolution of electronic communication technology may be traced through a series of milestones that have redefined the face of human interaction. The advent of the internet during the latter half of the 20th century was a revolutionary shift in information communication, unifying the people of the world in ways unknown before. Particularly noteworthy were the developments in email and early instant messaging systems, which paved the ground for more flexible and real-time forms of interaction. In addition, the merging of multimedia information into digital communication, with video conferencing software such as Skype and Zoom, further eased interactions, revolutionizing business and personal communications. These technological advancements not only enhanced communication effectiveness but also created innovative applications in various industries, enabling real-time collaboration and decision-making previously unattainable. The emergence of mobile communication technologies in the 21st century was another turning point in digital communication. The proliferation of smartphones with internet access boosted the shift towards mobile-first communication, allowing people to communicate anywhere and anytime. With social media websites gaining popularity, people started to use such sites more for socialization and information sharing. The convergence of mobile technologies and digital communication also saw the emergence of applications that support multiple functions such as messaging, file sharing, and shared workspaces, providing more interactivity and user involvement. These innovations are all examples of a paradigm shift in which communication is no longer restricted to traditional media, but instead becomes a multimodal platform that is geared towards meeting the heterogeneity needs of contemporary society. As digital communication technology keeps making advances, future trends keep lingering in our minds with innovation and social demands. The introduction of machine learning and artificial intelligence is going to revolutionize the communication process further by creating even more customization, response automation, and the presentation of customized content to users. Entirely unrelated industries like retail, education, and health are now beginning to use these technologies to streamline the communications strategy and enhance the user experience. Besides, as identified in current studies, the implementation of high-technology integration in digital communication can increase its energy efficiency and smart design procedures, thus contributing to sustainability goals in various sectors (Ang et al., 2017). Likewise, this focus on experiential learning and entrepreneurship in the school and the classroom highlights a developing sense of urgency in the development of digital communication for dynamic learning settings, indicating that the advancement of technology is getting well aligned with the progress within schooling (Arroteia et al., 2018). These trends clarify that the development of digital communication technologies will evolve with a transformative edge in the coming years ahead as indicated by Figure 10.4.

Figure 10.4. *Key milestones in digital communication technology.*

Source: https://media.licdn.com/dms/image/v2/C5612AQGkHQIy7zhKiA/article-cover_image-shrink_720_1280/article-cover_image-shrink_720_1280/0/1643331052792?e=2147483647&v=beta&t=fbrHllmOP3hT9XwPPc6W77oWUFm9_g95nfWKA_ySt0I.

10.1.3. Impact of the Internet on Communication Methods

The advent of the internet has transformed modes of communication fundamentally, redefining not only interpersonal communications but also institutional communication. Traditional communication modes such as postal mail and landline telephone have been replaced by instant messaging, video conferencing, and social media platforms. This rapid change has enabled users to interact in real time over vast distances, connecting others around the globe in seconds. There are important business and social effects, allowing for a more active and responsive range of communications, allowing team collaboration that can be physically dispersed. As outlined in previous research, the penetration of internet-based communication tools has compelled it to redesign communication plans in both professional and personal arenas with an emphasis on the flexibility to fit into a fast-changing digital arena. Further, the power of the internet also lies in its ability to democratize information dissemination and its concomitant effects on communication equality and accessibility. In contrast to the past that required much resource to disseminate, online spaces make it possible for individuals and institutions to share information, distribute news, or host discourse with low barrier to entry. This has made it possible for different voices in public discourses, thus making the narrative landscape more diverse. Greater accessibility is accompanied by disadvantages, in the way that misinformation can spread quickly along with actual communication. Safeguarding the integrity and credibility of online content has been a concern of paramount importance for platforms and users alike. Here, the internet not just changes the means of information dissemination but also raises important questions on the issue of responsibility and communication quality in the age of the internet, as discussed in new frontiers in geotechnology and virtual participation (Geertman et al., 2000). Future-looking, the ongoing progress of communication processes pushed by the internet suggests continuous innovation in online communication technologies. Technological developments such as artificial intelligence, machine learning, and augmented reality have the ability to create new channels of interaction that will radically alter user experience and expectation. The emergence of geographically enabled technologies constitutes a paradigm shift in the processing and utilization of geographical information, with policy-making and planning implications at all levels. Such innovations not only make communication more effective but also integrate advanced data analysis into everyday communication, making communication more relevant and informed. Integration of these technologies opens up promising opportunities but also calls for a critical assessment of ethical concerns and possible digital divides as seen in recent arguments on the future of digital communication (Garkavenko et al., 2015).

10.2. CURRENT APPLICATIONS OF DIGITAL COMMUNICATION

The technological evolution in communication has made available an expansive array of applications today that have an enormous influence in various sectors of the industries. One of the most significant developments is the integration of 5G technology, offering ultra-low latency and high-speed connections. This makes new business models impossible to fathom before, particularly in the healthcare sector, where real-time data transmission can enhance patient monitoring and telemedicine services.

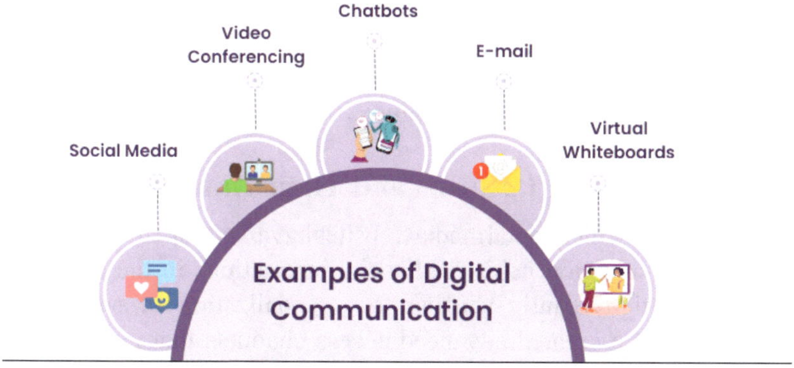

Figure 10.5. *Current applications of digital communication.*

Source: https://www.theknowledgeacademy.com/_files/images/Ways_of_Digital_Communication.png.

According to Cuevas et al. (2017), as the networks develop to accommodate the stringent requirements of low latency services, operators must grapple with justifying their investments. Yet the prospective market returns are significant as the new use cases are revolutionizing the communications world, compelling operators to position themselves strategically in the evolving landscape. The use of internet and mobile phone has also enhanced the efficiency of business operations, making it possible for organizations to engage in digital communication regardless of time and space. The flexibility provides many advantages with regard to optimizing workflows and improving collaborative work in distributed teams located in various geographical regions. As phrased by Herdon et al. (2006), mobile technology integration into business processes reflects a multi-disciplinary impact—encompassing economic, social, and technological dimensions. Companies can assess their operational expenses more strategically, knowing the costs and benefits of adopting mobile applications in industries like agriculture and logistics, whose enhanced access to information spurs innovation and productivity. Moreover, advancements in digital communication devices have revolutionized industries like education and marketing. In educational settings, technologies like Learning Management Systems facilitate remote learning, allowing educators to connect with more individuals while providing personalized learning experiences to students. The shift compelled institutions to redefine their pedagogical practices in an attempt to realize the full potential of electronic communication technologies. In marketing, the diffusion of social media and internet-based advertisement platforms has enabled businesses to reach specific groups with unprecedented precision, transforming business-customer interaction. With such technologies becoming more developed, their implications for both sectors run deep, signifying a change of revolution in how information is disseminated and consumed in contemporary society (Figure 10.5).

10.2.1. Social Media Platforms and Their Influence

The rapid speed with which social media websites have developed has significantly influenced communication habits and social interactions in the online world. With customers significantly shifting to personalization of content, social media websites have created advanced access channels that are personalized based on tastes. The result is profound, as customers now receive media in manners that verge on moving beyond traditional means of communication. The Australian Communications and Media Authority (ACMA) also indicates that this shift brings out both new opportunities and challenges to users and service providers alike, as they arise in an OTT content world and constantly shifting technology. Users are not receivers but constantly cause impact on media stories through their engagement, and this creates complex networks of influence that disrupt existing media and communications regulatory models. This trend reflects a deep change in media access and usage, toward a more participatory culture. Moreover, the social media sites' participatory culture enables a new fusion of content, communication, commerce, and community, as defined by new user-controlled business models. These models capitalize on interactive features embedded in digital technologies to facilitate users co-authoring content and participating intensively in media discourse. These sites empower individuals to be heard, therefore undermining the mainstream media hierarchies and opening up content creation to everyone. Public media history, informed by such participatory cultures, reveals how social media is a major social activism and participation channel. By encouraging cooperation and communication between users, social media both reinforces personal expression and builds a rich community culture, which is critical for the digital communication future. In examining the implications of these social media phenomena, it is critical to consider the potential stress they impose on regulation models built for another era of media use. The ACMAs research proposes that as consumers are engaging with media across different devices and platforms, the efficiency of the conventional regulatory models can be compromised. It can lead to inefficiencies in delivering the best user experience and protecting consumers from disinformation as well as harming content. With the dynamics of digital communication, policymakers urgently need to develop adaptive regulating strategies that respond to the complexities that come with having a multi-perspective media environment. Such endeavors must achieve a balance between innovation and responsibility, keeping in mind the transformative nature of social media platforms in shaping public discourse and community action and ultimately providing the platform for a more informed and engaged citizenry.

10.2.2. Digital Communication in Business and Marketing

The convergence of business and marketing with digital communication has changed the manner in which organizations communicate with their customers. In today's fast-paced world, organizations utilize digital platforms to develop more interactive and customized experiences. For instance, social media channels offer real-time communications, through which businesses can directly respond to questions and comments from their customers in real time. Such urgency not only establishes stronger customer relationships but also enhances brand loyalty. Besides, companies use data-driven advertisement techniques to target specified customers using analytics to personalize messages that meet individual customer needs for optimized engagement and conversion. Effective implementation of digital communication is crucial in managing dynamics in modern markets, and this puts a premium on companies to continually keep re-fashioning their approaches to stay relevant. With technological advancements continuing to shape the future of digital communication, companies should remain vigilant to access future trends. One of the innovations includes the utilization of 5G networks, which will provide ultra-low latency and more data-transmission rates. This innovation will support the innovation of new services and applications that rely on real-time connectivity, thereby transforming the functional strategies in marketing and customer engagement. For instance, companies are able to introduce virtual and augmented reality capabilities, which will support interactive experiences never imagined before. The challenge, though, is determining whether investments in infrastructure improvements will yield adequate returns on such new business models. Companies must strike a balance between the potential market returns and the costs in order to make sound decisions in their investment in technology (Cuevas et al., 2017). Moreover, digital communication accelerates internal organizational processes, which enhances collaboration and decision-making. Instant messaging apps, video conferencing, and project management apps have become part of maintaining the workflow. These apps not only improve team communication but also nurture an atmosphere of transparency and accountability. With advancements in digital communication through the introduction of AI-backed tools and chatbots, organizations are now better placed to automate repetitive tasks, allowing employees to focus on higher-level projects. This shift significantly enhances productivity as well as redefining roles and responsibilities. Typically, digital communication uses expand beyond customer-facing projects to unlock substantial opportunities to enhance operations in other business areas (Magwizi et al., 2020) (Figure 10.6).

Figure 10.6. *Digital communication in business and marketing.*

Source: https://marketcrosscompany.com/wp-content/uploads/2024/01/persona-que-usa-herramienta-ai-trabajo-2-1024x684.jpg.

10.2.3. Role of Digital Communication in Education

The role of digital communication in education has become increasingly paramount as technological advancements permeate every facet of learning environments. Educators harness digital tools to facilitate communication between students and instructors, allowing for more personalized and timely feedback. Such platforms encourage interactive learning where students can engage with course materials and peers through discussion forums, virtual meetings, and collaborative projects. Notably, this shift toward digital communication fosters a sense of community within online and hybrid learning setups, breaking the geographical barriers that traditionally hindered access to education. Furthermore, the seamless integration of various communication technologies enhances the overall learning experience, as students can adapt to diverse learning styles and methodologies that suit their individual needs. As digital communication continues to evolve, its applications within the educational landscape are broadening, paving the way for innovative pedagogical approaches. In addition to enhancing communication, digital platforms offer robust resources for educational enrichment that were previously unavailable. Online databases, educational apps, and e-learning modules provide students with the ability to explore subjects at greater depths. With the rise of multimedia content and interactive learning tools, learners can engage with information dynamically, reinforcing their understanding and retention of complex concepts. Such advancements align with the assertions made in Geertman et al. (2000), which highlights the necessity for advancements in technology that facilitate improved data handling and geoinformation in educational contexts. As students increasingly become digital natives, their familiarity with online communication tools translates into higher levels of comfort and interactivity in educational pursuits.

Figure 10.7. *Role of digital communication in education.*

Source: https://digitalpromise.org/wp-content/uploads/2016/02/blog_pearson_3.jpg.

The implications of this trend are profound, suggesting that future educational frameworks will rely heavily on digital communication, fostering a culture of continuous learning. Moreover, digital communication has significant implications for lifelong learning and professional development within the educational sector. As the landscape of job requirements continues to evolve, continuous professional development is crucial for educators to stay current with pedagogical strategies and technological integrations. Online workshops, webinars, and professional forums serve as platforms where educators can share resources, insights, and experiences. This continuous exchange of ideas enables teachers to collectively adapt to changing educational standards and student needs. The engagement of educators in digital environments has the potential to create a dynamic community of practice, further exemplifying trends highlighted in Montgomery (2001). In this increasingly interconnected world, the role of digital communication extends beyond traditional education, fostering environments where ongoing professional growth and innovation become integral components of effective teaching and learning practices (Figure 10.7).

10.3. TECHNOLOGICAL INNOVATIONS IN DIGITAL COMMUNICATION

The digital communication field has been greatly revolutionized by technologies that further the manner in which individuals and organizations communicate with each other. One of these technologies is the spread of mobile communication technology, which allows users to talk at any moment and place. The transition from the use of fixed-line systems to mobile networks has not only enhanced accessibility but has also cultivated global connectivity. This shift allows real-time communication, allowing businesses to respond rapidly to evolving markets and consumer needs. Also, the innovation in mobile apps facilitates easy interaction with various platforms by the users, broadening social media and online marketing reach. With organizations employing such technology, they can leverage increased consumer engagement, thus driving expansion and innovation within their business. The impact of such innovations indicates the necessity for continuous adaptation to remain competitive in an ever-evolving digital landscape. With technology constantly altering the manner in which digital communication is accomplished, the use of Artificial Intelligence (AI) cannot be overlooked. AI has developed significantly to enhance the user experience through customized communication and efficient customer service solutions. Chatbots and virtual assistants exemplify how AI can streamline interactions by providing immediate responses to queries, effectively reducing the need for human intervention in routine customer service tasks. This shift has allowed organizations to allocate resources more efficiently, focusing on complex issues requiring human judgment while enhancing consumer satisfaction through quicker resolutions. Moreover, AI-based analytics are necessary in the understanding of consumer behavior to enable companies to tailor the marketing efforts accordingly. By utilizing the functionalities of AI, organizations can not only improve their communication strategies but also forecast consumer requirements, developing deeper relationships and interaction with target users. The issues of the digital age of communications also increasingly emphasize the need for innovative solutions. As noted in current writing, the wireless industry, particularly in Third World nations, is confronted with a range of issues, from infrastructure deficits to unstable market currents. In defending against these issues, leveraging aggressive innovation through means such as Wireless Intelligence Network (WIN) becomes an alternative for enhancing connectivity and offsetting deficiencies inherent in existing systems. These advancements not only improve communication in urban areas but also significantly improve rural access, which further catalyzes inclusiveness. In addition, as customers' needs evolve, firms must keep pace with these rising expectations by leveraging advanced technologies that facilitate communications. The ongoing innovation of digital communication is as much a question of technological advancement as it is one of guaranteeing that the innovations are consonant with the diverse needs of an interconnected global consumer market, enhancing the general level of connectivity and engagement within society (Claver-Cortés et al., 2017; Adesuyi et al., 2014).

10.3.1. Advancements in Mobile Communication Technology

Mobile communication technology advances have not only transformed personal relationships, but also the overall nature of urban administration. With the multifunctional mobile phone that allows people to access a variety of services like internet browsing and location-based services, cities are now starting to leverage these technologies to better manage and improve the quality of life. For instance, the coming together of telecommunication technologies facilitates city activity monitoring with wired, wireless, or optical networks. The information and communication technology convergence not only enhances service provision but also provides a basis for smart city projects, which enables effortless management of resources and infrastructure in an effective manner to better serve citizens (Yigitcanlar et al., 2009). Also, mobile communication technologies' application extends to fields such as Architecture, Engineering, and Construction (AEC) where careful application of information and communication technology (ICT) is increasingly being recognized as a competitive edge. Recent studies place special emphasis on the importance of interoperability and collaboration in the AEC sector, marking a growing awareness of ICT as a valuable resource. By opening up communication channels between stakeholders, from contractors to customers, mobile technologies enable robust project management structures that can keep up with changing requirements. Organizations with strategic ICT investments are better placed to innovate and improve the efficacy of operations in their business processes, a foundation for sustainable competitive edge (Acikalin et al., 2009). Looking ahead, the future of mobile communications is towards an even stronger presence of artificial intelligence (AI) and machine learning in digital communications environments. The technologies will be used to optimize the way information is processed, analyzed, and leveraged in real time, thus leading to more personalized user experiences. AI-driven predictive analytics can provide more insights into user behavior, which will drive service personalization and higher engagement. In addition, the potential of 5G and beyond will open new avenues in mobile communications, enabling faster speeds and connectivity. Stakeholders must stay in line with changes as and when they are released, revising their strategy templates to elicit the full potential brought about by these evolving mobile technologies, transforming the digital communication landscape within the next several years (Yigitcanlar et al., 2009) (Figure 10.8).

Figure 10.8. *Advancements in mobile communication technology.*

Source: https://researcherstore.com/wp-content/plugins/phastpress/phast.php/c2VydmljZT1pbWFnZXMmc3JjPWh0dHBzJTNBJTJGJTJGcmVzZWFyY2hlcnN-0b3JlLmNvbSUyRndwLWNvbnRlbnQlMkZ1cGxvYWRzJTJGMjAyMCUyR-jEwJTJGbW9iaWxlX3NtYXJ0cGhvbmUuanBnJmNhY2hlTWFya2VyPTE3MDk-3MDYxNTItMTM2MzY2JnRva2VuPTE2NTJhZjhiM2EyMDhmZmQ.q.jpg.

10.3.2. The Rise of Artificial Intelligence in Communication

The onset of artificial intelligence (AI) in communication has brought in a revolution in the manner in which information is being conveyed by organizations and individuals. With more incursions into the digital age, AI technologies are making communication planning more efficient through the use of advanced data analysis and content delivery that is finely calibrated. The deployment of smart algorithms and machine learning models has allowed developing messages that are calibrated to particular audiences, thereby optimizing interaction as well as efficiency. For example, machine learning is a critical component in wireless communication systems, especially when we are heading towards 6G networks, to enable the design of advanced communication systems that enable end-to-end seamless connectivity for humans and machines. These innovations accentuate the central transition away from traditional communication processes toward data processes, making the process adaptive and responsive to users' needs, testified to by the widespread application of AI in various communication tools, ranging from chatbots to autoresponders for email (Abbas et al., 2020). The impact of AI on communication is not merely efficiency and personalization; it also presents serious cognitive and social considerations.

Figure 10.9. *The rise of artificial intelligence in communication.*

Source: https://images.squarespace-cdn.com/content/v1/5748ef421d07c0e0855afd96/f14ae407-900e-4931-a9d3-fe15056c6f10/Screenshot+2023-08-16+9.51.05+AM.png.

The convergence of AI technologies changes the topography of information processing and user understanding. As emerging technologies revolutionize interactive interactions with the web, the cognitive facets of AI-human interaction gain more importance. Employing situated and ecological perspectives to cognition enables us to study how AI shapes thoughts and actions in digital contexts. Emphasizing the role of AI applications as shapers of our cognitive interactions transforms our understanding of communication dynamics and highlights the imperatives for cross-disciplinary research between cognitive science and mind philosophy. This observational method accentuates the evolving nature of communication as it responds to AI breakthroughs, with the need for further investigation into how these advance our knowledge acquisition as well as human-to-human interactions (Clowes et al., 2017). Besides, the rise of artificial intelligence in communication raises ethical concerns that must be addressed as technology continues to evolve. With communication increasingly being mediated by AI systems, concerns about transparency, bias, and privacy of data are paramount. The threat to credibility of communication channels comes with the potential for deception and machine-generated content. As businesses adopt AI-driven communication to streamline processes and enhance user experiences, they must also forge ethical guidelines so that the technology is responsibly used. The demand for accountability of AI systems needs to be met with proactive steps keeping ethical considerations above technological advancements. By examining the interaction between AI technologies and ethical issues, researchers and practitioners can design a communication environment that not only embraces innovation but also upholds elementary ethical standards, ultimately developing a more trustworthy digital world for all (Figure 10.9).

10.3.3. Development of Virtual and Augmented Reality Tools

The augmented and virtual reality technology has, in a significant manner, revolutionized the digital communication world through enhanced and more immersive interaction. With industries recognizing the potential these technologies possess, the process of development has proceeded in a fast speed, driving innovations tailored to applications. From enhancing learning experiences by allowing students to explore three-dimensional models to facilitating remote collaboration among groups, the confluence of virtual and augmented reality (VR/AR) is changing the manner in which information is presented and digested. Beyond enhancing communication to be more interactive, it is also contributing to improved understanding of abstract concepts through experiential learning. This is where the marriage of advanced hardware and software moves to the forefront, paving the way for applications that leverage the particular virtues of these technologies while making interactions with end-users simpler and more immersive. Another central aspect to the broad availability of VR/AR platforms is the increasing accessibility and affordability of enablement hardware components such as headsets and motion tracking modules. With such technologies becoming ever more mainstream, applications of these technologies are disseminating to the majority of industries, ranging from gaming, medicine, to real estate. Additionally, 5G connectivity is also central to enhancing the user experience by enabling the necessary bandwidth and ultra-low latency required by such applications. Ongoing innovation in network infrastructure makes seamless embedding of virtual experiences in day-to-day communications and redefining users' engagement with digital content possible. Based on recent studies, companies that evolve to address these changes have great potential to reap the benefits of new business models tied to 5G and VR/AR technologies. This fact underscores the imperative for organizations to incorporate proactive innovative communication strategies in order to remain competitive in a changing digital environment (Cuevas et al., 2017). Along with it, the new term of Tangible User Interfaces (TUIs) adds a new dimension to VR/AR tool design, integrating physical interaction with digital experience. TUIs leverage users native interaction capabilities, extending digital communication through creating a physical relationship with the information that is being communicated. This new method raises usability and experience, drawing a higher level of interaction with digital content. While promising, TUI technology is still relatively nascent, and much research would need to be conducted to fully understand their implications and optimize their design for various applications. With continued innovation in the convergence of VR/AR and TUIs, there is potential for it to reinvent user experience in unprecedented ways and pave the way for more innovations that bridge the divide between the physical and digital worlds in modes of communication (Shaer et al., 2009).

10.4. CHALLENGES OF DIGITAL COMMUNICATION

The increasing application of digital communication technologies has brought a vast array of challenges that impinge on users and organizations alike. One of the significant challenges that have risen is that of cyber safety, which has grown in significance as numerous trends in technology redefine online interactions. Users are primarily subjected to an array of threats like cyberbullying, data breaches, and harassment on the internet, particularly given that social media platforms dominate digital communication. The complexity of these threats is further aggravated by the accelerated rate at which technology innovations are taking place, thus making individuals and institutions respond through proactive measures. Enabling users to operate within this environment calls for encouragement of digital literacy because the ability to use the internet safely significantly impacts an individual's potential in virtual environments. Hence, there is a need to transition from the simple identification of risks to enhancing users resilience and trustworthiness in online interactions (Third et al., 2014). Moreover, uneven access to digital communications reinforces existing societal imbalances. For example, groups like the elderly, poor households, and small- and medium-sized businesses are typically handicapped in communication as a result of inadequate technology proficiency and limited internet coverage. This digital divide not only restricts their ability to fully engage in contemporary society but also makes them more susceptible to cyberattacks. Addressing these issues requires collective action towards developing inclusive programs that meet varied needs, so all users possess the skill sets necessary to use the technology safely and effectively to communicate digitally. By adopting a comprehensive approach that includes education, access, and awareness, stakeholders can assist in bridging the digital divide and facilitate greater inclusivity in digital spaces (Third et al., 2014). Lastly, the integration of digital communication into business processes introduces process challenges that can disrupt effectiveness. Firms such as Apple Inc. would often find it difficult to manage huge volumes of data and maintain cybersecurity while attempting to enhance their communications tactics. The intricacies of online communication platforms necessitate a transformation in operational management because firms have to align innovation with containing risk. As business enterprises establish their communication infrastructure, they must implement measures that not only prioritize effective communication but also consider the vulnerabilities of digital interactions. Proposed development should involve establishing broader distribution networks and encouraging innovation but prioritizing users' safety and trust to ensure a safe ecosystem for all stakeholders involved in the long term.

10.4.1. Issues of Privacy and Data Security

The landscape of online communication continues to evolve, with far-reaching changes that deepen interconnectivity and functionality. But as these changes occur, privacy concerns and data security have become an issue of utmost urgency. With increasingly more devices being interconnected and increasingly more social media sites used, information about individuals is exposed to intrusions, exploitation, and abuse. This is enhanced by the high adoption of mobile technology and cloud computing, through which users may be exposed to malicious attacks and data loss. According to studies that have been conducted in recent times, there exists an imperative urgency to promote prioritization of digital resilience development for users in order to effectively deal with these emerging threats of online participation (Third et al., 2014). Thus, having people with the capabilities to use the digital world safely is critical to creating a culture of sensitivity as far as cybersecurity is concerned. The other side to examine as far as privacy and protection of data is concerned is shifting attention from merely defining risks towards understanding the broader consequences of digital usage. Users are likely to grapple with the decision of whether or not to utilize digital technologies, weighed against potential privacy invasions. Here, it is important to emphasize the dual nature of benefits and risks stemming from heightened connectivity. As illustrated in recent debate around cyber safety, teaching diverse groups about digital literacy and online safety can give them the power to make informed decisions about their online presence and data sharing (Anneroth et al., 2002). Furthermore, stakeholders such as policymakers, educators, and technology firms should work together to preserve the users' rights while encouraging reasonable online behavior, thereby guaranteeing confidence and trust in online interactions.

Finally, the future of digital communication will certainly demand innovative solutions in addressing these pressing issues concerning privacy and data security. As structures deviate from traditional frameworks, the need for user-oriented models becomes increasingly evident. Such models must emphasize the interface of user interaction with data protection, encouraging a mode in which users remain safe but still leverage the benefits of digital innovation. The systems beyond 3G reference model stresses the necessity of a dual-plane model that integrates value and capability factors, putting the spotlight on user experiences (Anneroth et al., 2002). Finally, with communication via digital media continuously on the rise, a proactive strategy towards privacy and data protection is essential to having a secure and enriching online community for all (Figure 10.10).

Figure 10.10. *Digital communication issues of privacy and data security.*

Source: https://amtrustfinancial.com/getattachment/ANA-Blog/Blog/Small-Business/March-2019/Cybersecurity-vs-Data-Privacy/GettyImages-914788014-cyber-security-vs-data-privacy-1-min.jpg?width=600&height=319.

10.4.2. Digital Divide and Accessibility Concerns

The digital divide is a widespread issue in contemporary society, mirroring the disparities of access to digital resources and technologies among groups of individuals. With rapidly changing digital communication technologies, economically marginalized individuals are likely to be left behind. The digital divide not only hinders access to information but also prevents people from being active players in different critical aspects of modern life, including education, employment, and active citizenship. Furthermore, with increased use of digital platforms by more businesses and governments for service delivery and communication, the exclusion of the technology illiterate and people with poor access is also large. The impact of the divide is wide-reaching, and it creates a cycle of exclusion that is difficult to break. Solving these matters requires an integrated solution that includes investment in infrastructure and programs aimed at improving digital literacy through initiatives at the community level. In fixing problems of accessibility within the realm of digital communication, an astute awareness of user profiling emerges as an essential element. User profiling, with the process of collecting information regarding users and processing it in an effort to personalize digital interactions, holds the ability to develop gates toward inclusivity if properly handled. The methods of user profiling, however, also raise ethical concerns, especially concerning data privacy and the potential for discrimination. Organisations developing digital communication applications must balance these concerns in trying to respond to multiple user needs. Effective profiling has the potential not only to enhance user experience but, from both organisational and user perspectives, can identify access barriers so that interventions can be directed towards increasing inclusivity. This method highlights the need for ethical guidelines that regulate the way organizations collect and utilize user information so that disadvantaged groups are beneficiaries of digital technology progress and not yet further excluded from it (Dijk et al., 2005). New directions in digital communication require attention to accessibility's infrastructural and organizational dimensions. Public-private partnership programs pose the best chances for closing the digital divide since they can produce digital applications that are within reach of all. Ideas proposed by experts are increased community access to broadband, which is crucial in introducing more masses into the digital arena. In addition, the introduction of Web 2.0 and 3.0 technologies provides chances to create more interactive and user-driven platforms, but this development can only come to pass if the accompanying issues of accessibility are addressed definitively (Gant et al., 2011). The future of digital communication thus hinges on our collective willingness to address these issues so that advancements are equitably shared among different populations and individuals are able to participate fully in the digital age.

10.4.3. Misinformation and Its Impact on Communication

The dissemination of misinformation has emerged as a vital issue within the context of digital communication, profoundly reshaping the information dissemination and reception dynamics. With greater audiences turning to social media and online sources for information, rapid dissemination of incorrect data has the capability of intensifying public bewilderment and fomenting skepticism against credible sources. The implications of such disinformation extend beyond the topic; they destroy the integrity of well-informed public discourse. As in the case of discussing emotion analytics, an understanding of the emotional impact of disinformation illustrates how the emotional state of audiences distorts their perception of credibility and trustworthiness in the news, thus affecting how they use information sources. This shows the need for greater awareness and education about the significance of fact over fiction and that emotional responses towards news must be acknowledged while also requiring to be confronted in order to establish a more aware and educated group of individuals (Sivek et al., 2018). The complexity of information credibility in the online environment is particularly intimidating to the youth, whose engagement with technology comes to shape their understanding of accuracy and reliability. Digital natives as youths are, they possess some strengths in handling technology but also threats emanating from the sheer abundance of material available, such as vast amounts of disinformation.

Figure 10.11. *Misinformation and its impact on communication.*

Source: https://huridocs.org/wp-content/uploads/2023/12/Fake-news-1.png.

This generational setting calls for a respectful attitude towards teaching young people about credibility and media literacy. The solution to these problems calls for an understanding of the evolving nature of electronic communication, where misinformation can undermine their critical processing of information. The worth of integrating news literacy classes that not only encompass fact-checking but also emotional intelligence and critical thinking skills becomes evident, empowering youths to navigate and critique the information environments they encounter daily (Andrew et al., 2008). Furthermore, with the continued advancement of communication technology, fighting misinformation requires collaboration among sectors including journalism, education, and technology. Journalistic integrity must be preserved by an aggressive battle against disinformation and enabling transparency and accountability in reporting. Meanwhile, the education system must incorporate comprehensive curricula that engage students in arguments about disinformation, emotional responses, and psychological aspects of information consumption. The modern information age demands innovative methods that not only authenticate accuracy but also consider the psychological impact of disinformation on the audience. By focusing on critical engagement and emotional sensitivities, the stakeholders can optimize the publics ability to discern fact from fiction and eventually embrace trust within communication processes and create a better-educated society to cope with the intricacy of the information era (Sivek et al., 2018) (Figure 10.11).

10.5. FUTURE TRENDS IN DIGITAL COMMUNICATION

The advancements in digital communication technology have highlighted the potential of 5G networks, which promise to revolutionize traditional communication paradigms. With more industries relying on ultra-low latency services, operators will be compelled to re-examine traditional business models due to the accelerating demand for high-quality connections. Not only does 5G transition facilitate faster data transmission, but it also enhances the effectiveness of applications for real-time communications and hence provides new business opportunities for corporate business across a wide range of industries. However, as Cuevas et al. (2017) highlights, massive investments in changing the network infrastructures for enabling 5G capabilities warrant some reservation regarding the return on investment from operators. This challenge forces stakeholders to be innovative and experiment with other applications, showing how the future of digital communication relies on the ability to reconcile technological advancement with continued economic viability. Outside of telecommunications, the convergence of mobile communication and the Internet has revolutionized the business operating environment to increase flexibility and productivity. Creation of mobile applications for various industries is one of the best examples of the big economic advantages these platforms have to offer to businesses. According to Herdon et al. (2006), this proliferation of mobile technology not only enhances the efficiency of business but also provides more services, changing the pattern of doing business. Certain industries such as agriculture and logistics have been greatly assisted by this revolution brought about by digitalization, which has automated things and allowed improved decision-making with real-time information. As more organizations adopt mobile communications technologies, effective use of these tools will be key to maintaining competitive positions and driving product and service innovation. The future will rest on how emerging technology interacts with digital communication. The intersection of artificial intelligence, the Internet of Things (IoT), and augmented reality in digital communication spaces is as a challenge and an opportunity. As both customers and companies get accustomed to interactive, high-speed platforms, demand for simple connectivity will expand. Additionally, as Cuevas et al. (2017) suggests, not only do operators need to invest in technology but also have to modify their level of service according to evolving customer needs. This adaptability will be crucial to sustainability in a more interconnected world, as communication on the internet moves from entrenched habits to sophisticated, interconnected systems that redefine the user experience and facilitate levels of interaction unheard of heretofore across a broad spectrum of industries.

10.5.1. Predictions for the Evolution of Communication Technologies

The past history of communication technology has been one of rapid developments, and the future is to anticipate the pace continuing on a trajectory that will radically transform the way society interacts and interfaces. With the development of wireless technology, particularly mobile communications, radical transformations are envisioned. Contemporary trends indicate a shift away from legacy fixed wired networks towards more sophisticated wireless solutions addressing an increasingly global populace. As recorded in forthcoming studies of the telecommunications landscape, WIN technology integration is likely to be central to overcoming the challenges facing the industries, including scalability and consumer demands. In addition, there should be insight into how these technologies will shape modes of communication, particularly in the developing world like Nigeria, where mobile communication has previously rewritten social life and economic activity (Adesuyi et al., 2014). Even though predicting technological advances is fraught with inevitable uncertainties, it goes without saying that the dynamic between technology and society cannot be unidirectional. The snooker ball model whereby new technologies spark uncomplicated, predictable responses in social patterns is insufficient for describing the diversity of these interactions. Historical practice teaches us that uses of new communication technologies invariably vary from their initial purpose. Instead, it should be observed that human society co-evolves alongside technology, creating unique socio-economic spheres that reflect this co-evolutionary process. This offers a cry for a more holistic understanding of how communication technologies will advance in the next several years, with the need to emphasize flexibility in revenue streams and innovation processes (Anderson et al., 2008). Forward, the role of adaptive and participatory strategies for the evolution of communication technology becomes increasingly vital. As communication goes online, stakeholders must take responsibility to develop spaces to foster rooted innovation and openness. Future communication infrastructure needs to embrace open systems that facilitate collaboration between various sectors and actors to ensure that diverse voices and perspectives inform the technological agenda. This approach not only renders the technological reaction to the needs of society more nimble but also engages communities in an active sense of responsibility for their communication equipment. By focusing on flexibility and user engagement, the growth of communication technology can be an agent for beneficial social transformation, setting the foundation for an enduring future in a globalized world (Figure 10.12).

Figure 10.12. *Predictions for the evolution of communication technologies.*

Source: https://imageio.forbes.com/specials-images/imageserve/1064785368/0x0.jpg?format=jpg&crop=3558,2001,x0,y119,safe&height=600&width=1200&fit=bounds.

10.5.2. The Role of 5G and Beyond in Enhancing Connectivity

The advent of 5G technology is a milestone in increasing the height of connectivity in every industry. With its ultra-low latency and significantly increased bandwidth, 5G transcends the boundaries of existing technology. This groundbreaking capability results in innovation in use such as the Internet of Things (IoT), augmented reality (AR), and smart cities. The fusion of such technologies creates an environment where devices communicate with one another at minimal lag time, facilitating real-time decision-making. As businesses transform in order to adopt these enhancements, there is an emerging need for operators to invest in robust infrastructure that can support these new business models. The need for such investments is compounded by the need for high-performance connectivity, which, while perhaps costly, promises tremendous market gains and opportunities for service providers. Operators are thus faced with both challenge and opportunity in navigating this transitions space (Cuevas et al., 2017). Moreover, as communications networks evolve to the sixth generation (6G), the integration of machine learning (ML) principles is likely to be the foundation of future connectivity. These developments will enhance wireless communications reliability and efficiency, ultimately driving the digital transformation of societies. With characteristics such as zero-touch optimization, the network management capability can become increasingly autonomous, diminishing human touch and creating frictionless user experiences. Such self-management will allow networks to react rapidly to changing needs and eventually optimize resource allocation. The union of ML and 6G is poised to revolutionalize communications technologies, particularly for use in autonomous vehicles and real-time data processing applications that require strong, real-time connections (Abbas et al., 2020). Finally, the trajectory that has been set by 5G and the unprecedented leap of 6G reflects a final transformation in the landscape of digital communications. Increased connectivity is not simply increased speed; it is also a matter of flexibility, productivity, and the imbedding of smart systems that can self-controllingly manage and operate networks. Changes of such a radical nature are no less vital to consumer applications as they are to industrial uses, where automation and real-time monitoring would make possible previously unimaginable levels of operational efficiency. As the stakeholders in technology continue to explore these technologies, their impacts on everyday life will begin to show, reshaping the future of communication and connectivity in all aspects (Cuevas et al., 2017; Abbas et al., 2020).

10.5.3. Potential Impact of Quantum Computing on Communication

As quantum computing advances, its potential to influence communications systems grows in significance. Quantum computing has the potential to introduce higher levels of processing as well as the capability to force speeds of data processing and transmission to record levels. This improvement targets directly the growing demands for more effective communication networks, which are called for in a world where data from digital applications and devices increase exponentially. Quantum computers' ability to deal with and analyze massive data sets can result in breakthroughs in communication technologies like the invention of new, stronger and faster protocols that cannot be done with classical computing. These will lead to a next-generation Internet service with seamless communications without sacrificing strong security measures. In addition to speed, quantum computing naturally transforms the security profile of digital communications. Quantum key distribution (QKD) is a very good example of how quantum technology can fundamentally alter secure communication in that data transmitted becomes more secure and tamper-proof. This is extremely relevant in the context of rising cyber attacks and intrusions into both individual and organizational information. Classical encryption methods, even if powerful, can at some point be cracked with the heavy computational capabilities of quantum systems. The development of quantum-driven communication systems can therefore serve to eradicate threats in such a manner through the use of quantum principles to develop unbreakable encryption methods. Governments and corporations are therefore investing in these technologies to assist in safeguarding confidential data, emphasizing the extreme importance of quantum computing in future communication systems. Furthermore, as suggested by recent research results, new communication protocols may emerge with the incorporation of quantum systems. Exploration of concepts such as decentralized processing of data and advanced features of information and communication technology (ICT) has the prospect to redefine the communication paradigms of the modern era. Merging of quantum computing with existing digital frameworks provides avenues for greater interoperability between diverse communication platforms. Through information flow optimization, quantum computing can potentially have a higher quality of service and higher user satisfaction. Companies already are beginning to realize quantum computing's potential to transform the way we engage with technology and identify paths that can reinvent communication for generations to come. With the quest for efficiency and security continuing, what quantum computing entails for communication will be a constant theme in technological discourse (Heng, 2025; N/A, 2017).

10.6. Ethical Considerations in Digital Communication

The ethical concern in electronic communication is of the most vital kind, since the speed at which technology is evolving is presenting chances for developers and users but also challenges. Among the primary ethical concerns is privacy and data protection. Electronic media have the tendency to collect vast personal information, hence raising concerns about the way in which data are stored, employed, and shared with third parties. The consequences of data breaches are disastrous, resulting in identity theft and extensive emotional trauma for the individuals whose data are breached. As explained, issues of security and privacy are paramount, particularly the potential for exploiting vulnerable groups in online settings. Therefore, the creation of stringent ethical frameworks focusing on the consent and responsibility of users in managing data is vital in creating a secure digital communication ecosystem. Another significant ethical field is equity and accessibility in Internet communication. With society increasingly reliant on technology to communicate, learn, and be employed, disparities in using technology can accentuate existing disparities. At-risk groups, like poor households and the disabled, may be drastically excluded from Internet communication, and it could seriously harm their social integration and personal growth. Further, the design of online platforms is usually not inclusive, and therefore is likely to leave out some groups of users. The need for equality in online communication highlights the imperative of human-centered technology solutions. By prioritizing inclusivity and equal access, developers can create platforms that enable all kinds of users to engage effectively in the digital space and consequently enhance their experience and prospects. Participatory research strategies are now a central strategy in addressing such ethical issues, namely in valuing how diverse user groups interact within digital spaces. In involving youth as co-researchers, research can reveal the ethics involved in young people's communication on digital platforms in networked societies. This approach not only empowers the participants but also improves the gathered data, where sweeping information is revealed about their experiences and impacts of digital communication on social relationships and identity. As rendered in recent debates within the scholarship, considering such perspectives is imperative when creating ethical standards capturing users' realities of interacting with technology. Lastly, engaging communities within research helps to bridge the gap between technological progress and ethical responsibilities, opening the way for future studies of the social implications of digital communication behaviors (Figure 10.13).

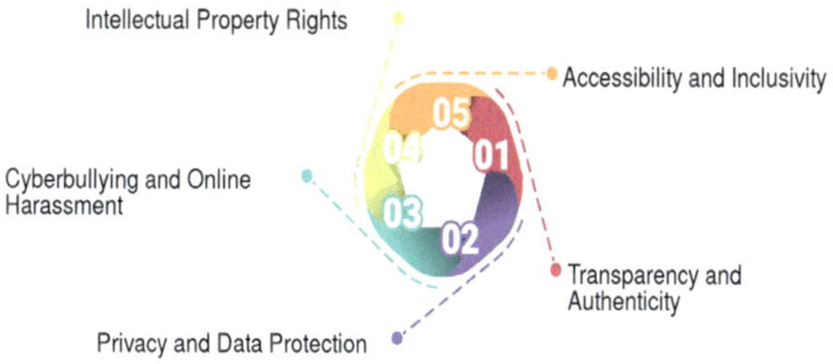

Figure 10.13. *The ethical considerations in digital communication.*

Source: https://fastercapital.co/i/Ethical-communication--How-to-communicate-ethically-and-respectfully-with-your-stakeholders--Ethical-Considerations-in-Digital-Communication.webp.

10.6.1. The Importance of Digital Literacy

Digital literacy is now a fundamental ability in this more networked world, where information flows continually in digital media. While individuals negotiate this complex system, the ability to critically assess and employ digital tools appropriately has become a requirement. Particularly in communication, digital literacy not only allows users to communicate safely online but also enhances the capacity to contribute significantly to digital discourse. Without such capabilities, users will not be able to protect themselves from online hazards and disinformation, which can destabilize both public debate and personal security. As remarked in the case of cyber safety, encouraging digital literacy is crucial for developing resilience among diversified user groups, including particularly at-risk groups who are not in general the target for current education systems (Third et al., 2014). In addition, the use of digital technologies in everyday living has made it more crucial than ever to emphasize the development of digital competencies across different groups. Through a practical learning strategy espoused by some researchers, individuals can comfortably interact with digital spaces. Such engagement is especially critical for older persons and small and medium businesses, who might also pose specific challenges and threats on the internet. As digital literacy is enhanced, not only are people empowered to understand and react to the risks involved in digital participation but also to realize the potential that comes from being connected. As digital technologies evolve, so should our learning activities, moving towards sustained behavior change rather than risk awareness so that all users can thrive in a digital economy (Third et al., 2014). Further, the role of digital literacy extends beyond individual and professional domains; it determines our general utilization of natural resources and knowledge management within society. Digitization of resources in academia and research requires a population ready to access sophisticated digital repositories and archives without any hitches. This is a challenge that comes with barriers as well as opportunities and needs an industry with a solid core in digital curation. As the natural resource management experts address complicated ecological and social issues, digital literacy equips them with effective reasoning and technical skills necessary to effectively solve problems. By developing a culture of digital competency, society can most effectively guarantee that new professionals not only oversee existing knowledge but contribute to its extension with innovative digital practices.

10.6.2. Ethical Implications of AI in Communication

The use of artificial intelligence (AI) in communication technologies has made unprecedented advances in terms of efficiency and accessibility. The improvements are, nevertheless, accompanied by resultant ethical issues which need close attention. The primary issue is the potential for AI to widen existing gaps in access to communication facilities and information. As emphasized by recent research, issues of equity, inclusion, and access remain top of mind while deploying technology-facilitated solutions, particularly among marginalized groups (Apesoa-Varano et al., 2020). The risk is that AI-facilitated platforms will choose to go with users who are already technologically savvy, thereby further marginalizing those who are not. In addition, ethical issues with regard to data manipulation can sustain disparities if the voice of marginalized groups is not taken into account, compelling stakeholders to take into consideration the need for inclusive AI-based communication. Privacy and security are another important ethical consideration in implementing AI-based communication. With algorithms still rummaging through individual data in order to tailor content and enhance users' experience, the possibility of abuse looms in the background. Commodification of the personal data raises significant ethical concerns regarding consent and ownership of the data. As AI systems become more complex, these systems are capable of making decisions on their own about what data to collect and how to apply it without explicit user consent (Apesoa-Varano et al., 2020). This is not open, and thus the user will not be able to decide anything about their online presence. Besides, such practices can potentially result in systemic discrimination as AI systems reproduce unconscious bias within training data they act upon. This calls for robust ethical foundations that value privacy and encourage transparency, ensuring user agency regarding communication practices that are AI-based. Moreover, the impact of AI overflows into its implications on social relationships and the very fabric of human interaction. The AIs ability to automate responses and generate human-like conversation has the potential to create difficulties in genuine communication. The empathetic subtleties and emotional intelligence that play such a crucial role in human communication may be lost in AI-mediated communication. This is an issue that gets prioritized while considering the intangibles of digital change, as determined by analysts who caution against side effects that might reshape societal norms. As AI is being integrated into communications more and more, there is a need to consider how the technologies have the potential to change user experiences and social norms. Most importantly, it is important to consider such ethical issues in shaping an environment that is supportive of truly human relationships as well as the optimum benefits of AI in digital communication.

10.6.3. Balancing Innovation with User Rights

Against the backdrop of the rapidly evolving environment of online communication, balancing innovation and the rights of users is a delicate matter that needs careful consideration. In the wake of emerging technologies, organizations and developers are typically motivated by the vision of improved user experiences and effective service provision. The innovations, however, should not come at the cost of users' privacy and security. It is important to note that although advanced multimedia search capabilities enable consumers to access information more effectively, they also challenge the gathering, storage, and use of personal information. Subsequently, systems that maintain ethical principles must be developed to support technological progress such that innovations respect consumers' rights while assisting in the development of an improved digital space. The socio-economic ramifications of these technological advancements also serve to highlight the need for a balanced perspective towards user rights. More than one billion online users today, the demand for innovative solutions—namely in multimedia content form—has picked up speed with a call for equal protection and access. Therefore, it is necessary to consider how new forms of media and distribution methods can foster inclusivity without diluting the rights of people. With researchers and technologists collaborating to improve multimedia search through techniques like indexing and retrieval technologies, legal implications also need to be addressed. By ensuring technical innovation is balanced with the safeguarding of user rights, stakeholders will be able to collaborate and produce an equitable and sustainable digital environment where innovation is a force for good that does not impinge on user agency (Boujemaa et al., 2007). As networked media's future is being crafted, it will be necessary to balance the tension between revolutionary advances and user rights. The future growth in digital content underscores the need for systems to safeguard against exploitation while enabling users to navigate the evolving digital landscape with confidence. This includes encouraging transparency in data utilization as well as encouraging users' rights in new uses and system design. Community-oriented strategies, such as peer-to-peer networks, should foster user management of personal information such that as multimedia is personalized, users retain their privacy and rights. Through such means, stakeholders are able to benefit from innovation in a manner respectful of users' rights and ultimately enhance the quality of life while promoting ethical practices (Aggoun et al., 2008).

SUMMARY

In short, the digital communication sphere is constantly in flux, as technology development progresses at a speedy pace and societies' actions also change. As companies incorporate more mobile communication and Internet technologies into their businesses, there is an increasingly dynamic and streamlined environment. The expansion of mobile applications makes it possible for organizations to communicate with customers and mechanize their business processes because organizations are able to operate with greater flexibility, which allows them to connect with stakeholders at any location and at any time. This has led to significant economic gains and also prompted research into new ways of communication that are appropriate to the evolving needs of the contemporary digital society (Herdon et al., 2006). Moreover, the proliferation of communications technologies has distant impacts across sectors, from agriculture to logistics. Such applications being used with growing popularity mean that they create a web of relationships with wide-ranging impacts on manufacturers, firms, customers, and workers. By facilitating collaborations between such diverse groups of players, firms can enhance their operational performance and responsiveness to customers' needs, which can ultimately result in growth and competitiveness in an over-competitive market. This dynamic underlines the necessity for organizations to strategically adapt in order to leverage digital resources, ensuring they remain competitive in an ever-changing economy (Herdon et al., 2006). The future of digital communication is very promising in the coming years, particularly because interdisciplinary collaborations, such as between institutions like Glasgow Caledonian University and the University of Alicante, demonstrate the effectiveness of collaborative research and the exchange of knowledge in driving innovation. Such collaborations emphasize the necessity of an interdisciplinary approach to address the intricate challenges and opportunities of digital communication. With researchers working on different topics—ranging from communication to marketing, and beyond—the results will be instrumental in shaping best digital strategies that engage global audiences. Hence, the convergence of technology, social interaction, and intellectual aspirations is central to creating developments that can navigate future trends in digital communication.

The changing digital communications environment has seen significant advancement, driven predominantly by advances in technology that affect education and workplaces. The integration of technology into these sectors is no added value but necessitates a re-examination of the manner in which staff work at its very essence. This transformation is analyzed through such projects as the Work-with-IT project of the University of Strathclyde, which focuses on the challenge and opportunities presented by these transformations. The study focuses on the importance of effective management of technological transformations because failure to do so could result in disengagement and ultimately stagnate the performance of institutions. This report not only underscores technology's central role in remaking communication but also sheds light on the necessary upgrading of skills among employees to surf these changes successfully (Comrie et al., 2010). Additionally, as communication platforms online continue to evolve, their impacts extend beyond mere operational efficiency, reaching into the psychological and pedagogical domains. The development of technology-based learning environments shows the intricate relationship between educational psychology and digital platforms, how it has evolved from basic theories to show how they are applied in contemporary environments. The model offered in books like Technology-based Learning Environments: Psychological and Educational Foundations shows how learners interact and convey through digital aids and the thinking activities that they employ in the processes. As educators and schools strive to enhance learning using technology, it becomes essential to understand such psychological considerations in order to be able to achieve effective communication and interaction (Barker et al., 1995). Such a mixture of instructional theory and computer use underscores the necessity for continuous research during the process of adapting pedagogy in terms of the advancements in computer communication. In the coming future, digital communication will naturally be influenced by ongoing technological advancements and their intersection within various industries. With evolving businesses in terms of new modes of communication, the need for innovative means to engage with stakeholders becomes increasingly critical. The upcoming future will naturally have greater emphasis on the application of data analytics, artificial intelligence, and user experience design to improve driving interaction and engage with diverse groups of audiences. The implications for digital communication practices are great, as they have the capacity to redefine how information is consumed and disseminated. By examining key points of current controversy and understanding short-term needs and long-term aspirations, stakeholders can enhance readiness for the ever-changing digital communication landscape in order to remain relevant and responsive to future needs (Comrie et al., 2010).

REFERENCES

1. Abbas, R., Ahmad, I., Ali, S., et al. (2020). 6G white paper on machine learning in wireless communication networks. http://arxiv.org/abs/2004.13875 (accessed on 18th September 2025).
2. Acikalin, U., Goulding, J. S., Isikdag, U., et al. (2009). Construction informatics in Turkey: Strategic role of ICT and future research directions. https://core.ac.uk/download/1666747.pdf (accessed on 18th September 2025).
3. Adesuyi, R. S. (2014). The future of mobile industry. https://core.ac.uk/download/19447812.pdf (accessed on 18th September 2025).
4. Aggoun, A., Amon, P., Arbel, I., et al. (2008). Multimedia delivery in the future internet. https://core.ac.uk/download/335423.pdf (accessed on 18th September 2025).
5. Anderson, B., & Stoneman, P. (2008). Predicting the socio-technical future (and other myths). https://core.ac.uk/download/9063918.pdf (accessed on 18th September 2025).
6. Ang, J. H., Saldivar, A. A. F., Goh, C., et al. (2017). Energy-efficient through-life smart design, manufacturing and operation of ships in an Industry 4.0 environment. https://core.ac.uk/download/82893542.pdf (accessed on 18th September 2025).
7. Anneroth, M., Crisler, K., Pulli, P., et al. (2002). A user-focused reference model for wireless systems beyond 3G. https://core.ac.uk/download/20483106.pdf (accessed on 18th September 2025).
8. Apesoa-Varano, E. C., Gladstone, C., Kim, K. K., et al. (2020). Technology and caregiving: Emerging interventions and directions for research. https://core.ac.uk/download/323070528.pdf (accessed on 18th September 2025).
9. Arroteia, N., Blesa, A., Curran, R., et al. (2018). Global board games project: A cross-border entrepreneurship experiential learning initiative. https://core.ac.uk/download/323303066.pdf (accessed on 18th September 2025).
10. Atkinson, D. J., Beswick, C. A., Cooper, L. P., et al. (1992). Deep Space Network information system architecture study. https://core.ac.uk/download/pdf/10438278.pdf (accessed on 18th September 2025).
11. Barker, P., Benest, I., Elliott, G., et al. (1995). Reviews. https://core.ac.uk/download/6937.pdf (accessed on 18th September 2025).
12. Boujemaa, N., Compañó, R., Dosch, C., et al. (2007). *CHORUS Deliverable 2.1: State of the Art on Multimedia Search Engines*. https://core.ac.uk/download/11434796.pdf (accessed on 18th September 2025).

13. Brown, A., & Cinque, T. (2015). Educating generation next: Screen media use, digital competencies and tertiary education. https://core.ac.uk/download/pdf/30672534.pdf (accessed on 18th September 2025).
14. Bullard, S. H., Kulhavy, D., McBroom, M. W., et al. (2017). Digital preservation and access of natural resources documents. https://core.ac.uk/download/84705487.pdf (accessed on 18th September 2025).
15. Claver-Cortés, E., García-Lillo, F., Marco-Lajara, B., et al. (2017). ICTs, disruptive forces and the production paradox in tourism: Present and future issues in the visitor attraction sector. https://core.ac.uk/download/293866911.pdf (accessed on 18th September 2025).
16. Clowes, R. W., Heersmink, R., & Smart, P. (2017). Minds online: The interface between web science, cognitive science, and the philosophy of mind. https://core.ac.uk/download/131214931.pdf (accessed on 18th September 2025).
17. Comrie, A., Cullen, D., & McDonald, D. (2010). Final report of work-with-IT: The JISC study into evolution of working practices. https://core.ac.uk/download/9033706.pdf (accessed on 18th September 2025).
18. Cuevas, M., Dohler, M., Laya, A., et al. (2017). Business case and technology analysis for 5G low latency applications. https://core.ac.uk/download/96762095.pdf (accessed on 18th September 2025).
19. Dijk, J. A. G. M. van, Ebbers, W. E., Fennis, B. M., et al. (2005). Alter ego, state of the art on user profiling: An overview of the most relevant organizational and behavioral aspects regarding user profiling. https://core.ac.uk/download/pdf/11463498.pdf (accessed on 18th September 2025).
20. Flanagin, A. J., & Metzger, M. J. (2008). Digital media and youth: Unparalleled opportunity and unprecedented responsibility. https://core.ac.uk/download/71339835.pdf (accessed on 18th September 2025).
21. Gant, J., & Turner-Lee, N. (2011). Government transparency: Six strategies for more open and participatory government. https://core.ac.uk/download/71357252.pdf (accessed on 18th September 2025).
22. Garcia, C. O., Cardona, D., Chesñevar, C., et al. (2020). South American expert roundtable: Increasing adaptive governance capacity for coping with unintended side effects of digital transformation. https://core.ac.uk/download/286359460.pdf (accessed on 18th September 2025).
23. Garkavenko, V., & Tiberghien, G. (2015). ICT diffusion and the digital divide in tourism: Kazakhstan perspective. https://core.ac.uk/download/42367170.pdf (accessed on 18th September 2025).

24. Geertman, S., & Stillwell, J. (2000). Geoinformation, geotechnology, and geoplanning in the 1990s. https://core.ac.uk/download/51004.pdf (accessed on 18th September 2025).
25. Giardina, N., Mallan, K., & Singh, P. (2010). The challenges of participatory research with 'tech-savvy' youth. https://core.ac.uk/download/10893570.pdf (accessed on 18th September 2025).
26. Heng, S. (2025). Rising stars in information and communication technology. https://core.ac.uk/download/pdf/9309277.pdf (accessed on 18th September 2025).
27. Herdon, M., & Szilágyi, R. (2006). Impacts for m-Internet applications and perspectives in agriculture. https://core.ac.uk/download/18405519.pdf (accessed on 18th September 2025).
28. Klein, H., & Monod, E. (2025). From e-heritage systems to interpretive archaeology systems. https://core.ac.uk/download/pdf/6464944.pdf (accessed on 18th September 2025).
29. Macgregor, G., & McGill, L. (2005). Digital libraries and information literacy issues within virtual learning environments: An e-learning impasse? https://core.ac.uk/download/9015448.pdf (accessed on 18th September 2025).
30. Magwizi, C. N., Магвізі, К. Н. (2020). Research on the operational management system of an organization: A case study of Apple Computer, Inc. https://core.ac.uk/download/286540915.pdf (accessed on 18th September 2025).
31. Mermigas, D. (2008). Mapping the money in public media. https://core.ac.uk/download/71348159.pdf (accessed on 18th September 2025).
32. Montgomery, K. (2001). *Teensites.com: A Field Guide to the New Digital Landscape*. https://core.ac.uk/download/71339873.pdf (accessed on 18th September 2025).
33. N/A. (2017). *SciTech News Volume 71, No. 1 (2017)*. https://core.ac.uk/download/83049835.pdf (accessed on 18th September 2025).
34. N/A. (2025). *Six Emerging Trends in Media and Communications–Occasional Paper*. https://core.ac.uk/download/pdf/30673242.pdf (accessed on 18th September 2025).
35. Obeed, K. M. (2018). Water level monitoring and controlling of water treatment plants using wireless sensors in LabVIEW. https://core.ac.uk/download/12005720.pdf (accessed on 18th September 2025).
36. Shaer, C. O., Hornecker, E., & Shaer, O. (2009). Tangible user interfaces: Past, present and future directions. https://core.ac.uk/download/9033674.pdf (accessed on 18th September 2025).

37. Sivek, S. C. (2018). Both facts and feelings: Emotion and news literacy. https://core.ac.uk/download/212894731.pdf (accessed on 18th September 2025).

38. Third, A., Collier, A., & Forrest-Lawrence, P. (2014). Addressing the cyber safety challenge: From risk to resilience. https://core.ac.uk/download/pdf/30674272.pdf (accessed on 18th September 2025).

39. Yigitcanlar, T. (2009). Managing ubiquitous eco cities: The role of urban telecommunication infrastructure networks and convergence technologies. https://core.ac.uk/download/10895362.pdf (accessed on 18th September 2025).

INDEX

E

F

G

H

I

L

M

N

O